INTRODUCTION TO
Logic Design

INTRODUCTION TO
Logic Design
Second Edition

Sajjan G. Shiva

University of Alabama in Huntsville
Huntsville, Alabama

MARCEL DEKKER, INC.   NEW YORK · BASEL · HONG KONG

Library of Congress Cataloging-in-Publication Data

Shiva, Sajjan G.
   Introduction to logic design / Sajjan G. Shiva — 2nd ed.
    p. cm.
   Includes index.
   ISBN 0-8247-0082-1
   1. Digital electronics. 2. Logic design. I. Title.
TK7868.D5S433 1998
621.39'5—dc21

            94-46945
            CIP

The publisher offers discounts on this book when ordered in bulk quantities. For more information, write to Special Sales/Professional Marketing at the address below.

This book is printed on acid-free paper.

Marcel Dekker, Inc.
270 Madison Avenue, New York, New York 10016
http://www.dekker.com

Current printing (last digit):
10 9 8 7 6 5 4 3 2 1

PRINTED IN THE UNITED STATES OF AMERICA

In Memory
of
My Mother

Preface

This book is an introduction to analysis and design of digital circuits. A course in digital systems design is now required in both electrical/computer engineering and computer science curricula. Computer science students in general are good at assimilating algorithmic solutions to problems, while electrical/computer engineering students absorb hardware concepts readily. I have therefore attempted to blend the theoretical and practical implementation aspects of digital systems in each chapter of this book. Thus the book is suitable for use in both curricula at the sophomore or junior level. It covers all the topics for the ACM curriculum 79 course CS4 and IEEE Computer Society's model program course SA6. It is also suitable for self-study, for professionals requiring a knowledge of digital hardware design, and as a reference for modern techniques in logic design for those familiar with older design methodologies.

This edition of the book covers the same set of topics covered in the first edition; however, the topics were rearranged to provide a smoother transition. The material on memory systems design from the first edition has been omitted, since that topic more aptly belongs to the first course on computer architecture. Details on integrated circuits (ICs) have been updated and several sections have been rewritten to clarify presentation. Also, the end-of-chapter problems have been updated.

This book covers topics ranging from number system theory to asynchronous logic design. Prior knowledge of electronics is not required to understand the topics discussed. Appendix B provides the fundamentals of electrical circuit analysis as a prerequisite to understanding the circuit-level hardware details discussed in Chapter 10.

The first three chapters are theoretical. Chapter 1 presents the three popular number systems: binary, octal, and hexadecimal, with an emphasis on various data representation schemes in binary. Chapter 2 is a formal introduction to Boolean algebra and its application to logic design, and provides the preliminary details of integrated circuits. Chapter 3 presents the two popular (graphic and tabular) logic minimization techniques. With the progress in hardware technology, hardware costs have come down drastically, thereby reducing the importance of logic minimization. Chapter 3 can therefore be skipped without loss of continuity.

Two chapters are devoted to combinational circuit analysis and design. Chapter 4 introduces the analysis and design procedures along with implementation schemes using ICs. Chapter 5 describes the popular ICs available for combinational circuit design and the modular, top-down design methodology.

Three chapters are devoted to sequential circuit analysis and design. Chapter 6 describes the classical analysis and design techniques for synchronous sequential circuits and Chapter 7 details the design techniques using the popular ICs. Design of control circuits is also discussed in Chapter 7. Chapter 8 describes the analysis and design techniques for asynchronous sequential circuits.

Chapter 9 is an introduction to designing with programmable logic devices. This mode of logic implementation has continued to be popular, with more powerful and flexible devices being introduced almost daily.

Chapter 10 describes the electrical characteristics of various IC technologies. Topics in this chapter do not directly influence the remaining chapters of the book, and hence this chapter can be skipped by readers not interested in such details.

Chapter 11 outlines 21 laboratory experiments using the standard TTL type of ICs. I have stayed away from providing step-by-step procedures for these experiments, in order to allow maximum flexibility for laboratory instruction.

Examples are included in the body of each chapter to reinforce the concepts presented. Each chapter ends with a set of references for further reading and a set of problems. A complete solutions manual is available from the publisher.

There are four appendixes. Appendix A introduces IEEE standard logic symbols. Appendix B is an introduction to electrical circuit analysis. Appendix C gives a logic simulation tool that can be utilized to reinforce the analysis and design techniques introduced in this book without actually building those circuits. Appendix D introduces computer-aided design (CAD) tools for programmable logic devices.

The book is a result of teaching logic design courses to science and engineering students over the last several years. The first edition was used by a number of universities around the world for a course in logic design. A typical one-semester course should include the material from Chapters 1 through 9. Chapter 1 can be omitted if the students have a background in number systems and codes. Chapter 3 can be de-emphasized to gain time to cover all the topics in Chapters 8 and 9. Chapter 10 is for engineering-oriented students. A combination of experiments selected from Chapter 11 and simulation exercises utilizing either the tool described in Appendix C or a similar simulation tool would enhance the understanding of the topics covered in this book.

I am indebted to many students and colleagues who have helped in developing and refining the material in this book. I would like to express my thanks for many useful comments and suggestions provided by users of the first edition and the reviewers during the development of this edition. Thanks are due to Russell Dekker for his patience and encouragement during the preparation of this manuscript and Eric Stannard for the superb production support. It is a pleasure to acknowledge the support of my home team: my wife, Kalpana, and daughters, Sruti and Sweta. Their assistance in the preparation of the manuscript and their love and understanding made my job easy.

Sajjan G. Shiva

A Note About the Integrated Circuit Examples

Details about various integrated circuits (ICs) taken from vendors' manuals are reprinted in this book. Most of these examples are from the TTL technology. An attempt was made to provide the most up-to-date data available. Because IC technology progresses so rapidly, the reader is referred to current manuals from the IC vendors for the latest information. Some of the ICs described in the book may no longer be available. Usually, alternative sources for these ICs can be found, or a later version of the same IC is now being produced by the vendor. Also, some IC manufacturers have merged with other companies (Monolithic Memories, Inc. is now a subsidiary of Advanced Micro Devices; Signetics Corporation is now part of Phillips Semiconductors, etc.). Nevertheless, the details given here are representative of the characteristics a designer would seek.

Contents

1

Number Systems and Codes

1.1 Digital System Organization

We are surrounded today by a myriad of digital devices. Digital watches, electronic calculators, digital meters, microprocessors, and digital computers are all examples of such systems. A *digital system* manipulates data that are composed of a finite number of discrete elements. The results that the digital system produces are also made up of a set of discrete elements. In contrast, an *analog system* manipulates data that are represented in a continuous form, producing results that also appear in continuous form. In electronic digital systems, the discrete elements of data correspond to *signals* which are either voltage levels or current magnitudes. Each specified voltage level can represent an element of data. The signal thus can be at only one of these specified levels. In an analog system, signals assume values in a continuous range of voltage.

For example, the signals in a digital system might be restricted to two levels (0 and +5 volts), corresponding to the two discrete elements of information, while analog signals may take any value in the range of +5 to −5 volts. Compared with analog systems, digital systems are more accurate and reliable. Hence, they are replacing analog systems wherever possible, although for certain applications analog systems are clearly superior. In order to introduce the terminology, we will now examine two popular digital devices.

An electronic calculator is a digital device in which input data are composed of discrete values entered through the keyboard, and the instructions to manipulate the data are also entered through the keyboard by means of the function keys. The output is a set of discrete values represented as digits on the display. In a programmable calculator, the sequence of instructions (i.e., the *program*) is stored in the calculator memory and used repeatedly on various sets of input data to produce results.

Figure 1.1 shows the components of a digital computer, the most general digital device. The program to manipulate the data is first brought into the *memory unit* through the *input device*. The data to be processed are then brought into the memory unit, also through the input device. The *control unit* fetches instructions from the program stored in the memory one at a time, analyzes each instruction, and instructs the *processing unit* to perform the operations called for by the instruction. The results produced by the processing unit are forwarded to the memory unit for storage and then transferred to the *output device*.

As mentioned earlier, the elements in the discrete data representation correspond to discrete voltage levels or current magnitudes in the digital system hardware. If the digital system is required to manipulate only numeric data, for instance, it will be best to use 10 voltage levels, with each level corresponding to a decimal digit. But the noise introduced by multiple levels make such representation impractical. Therefore, digital systems typ-

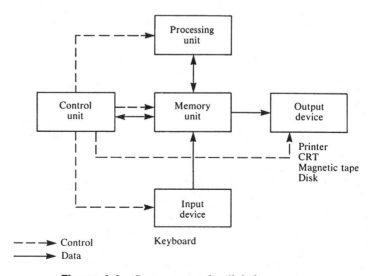

Figure 1.1 Components of a digital computer.

ically use a two-level representation, with one voltage level representing a 0 and the other representing a 1. To represent all 10 decimal digits using this *binary* (two-valued) alphabet of 0 and 1, a unique pattern of 0s and 1s is assigned to each digit. For example, in an electronic calculator, each keystroke should produce a pattern of 0s and 1s corresponding to the digit or the operation represented by that key.

Because the data elements and operations are all represented in binary form in all practical digital systems, a good understanding of the binary number system and data representation is basic to the analysis and design of digital system hardware. In this chapter, we will discuss the binary number system in detail. In addition we will discuss two other widely used systems: *octal* and *hexadecimal*. These two number systems are useful in representing binary information in a compact form. When the human user of the digital system works with data manipulated by the system, either to verify it or to communicate it to another user, the compactness provided by these systems is helpful. As we will see in this chapter, data conversion from one number system to the other can be performed in a straightforward manner.

The data to be processed by the digital system are made up of decimal digits, alphabetic characters, and special characters, such as +, −, *. The digital system uses a unique pattern of 0s and 1s to represent each of these digits and characters in the binary form. The collection of these binary patterns is called the *binary code*. Various binary codes have been devised by digital system designers over the years. Some popular codes will be discussed in this chapter.

1.2 Number Systems

Let us review the decimal number system, the system with which we are most familiar. There are 10 symbols (0 through 9), called digits, in the system—along with a set of relations defining the operations of addition (+), subtraction (−), multiplication (×), and division (/). The total number of digits in a number system is called the *radix* or *base* of the system. The digits in the system range in value from 0 through $r - 1$, where r is the radix. For the decimal system, $r = 10$ and the digits range in value from 0 through $(10 - 1) = 9$.

In the so-called *positional notation* of a number, the radix point separates the "integer" portion of the number from the "fraction" portion. If there is no fraction portion, the radix point is not explicitly shown in the positional notation. Furthermore, each position in the representation has a weight associated with it. The weight of each position is equivalent to the radix raised to a power. The power starts with a 0 at the position immediately to the left of the radix point and increases by 1 as we move each position toward the left, and decreases by 1 as we move each position toward the right. A typical number in the decimal system is shown in the following example.

Example 1.1

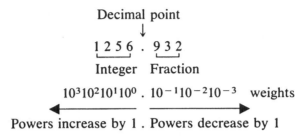

This number can also be represented as a polynominal:

$$1 \times 10^3 + 2 \times 10^2 + 5 \times 10^1 + 6 \times 10^0$$
$$+ 9 \times 10^{-1} + 3 \times 10^{-2} + 2 \times 10^{-3}$$

We can thus generalize these two representations to any number system. The general *positional notation* of a number N is

$$N = (a_n \cdots a_3 a_2 a_1 a_0 \cdot a_{-1} a_{-2} a_{-3} \cdots a_{-m})_r \qquad (1.1)$$

where r is the radix of the number systems; a_{-1}, a_0, a_1, a_2, and so on, are digits such that $0 \leq a_i \leq (r - 1)$ for all i; a_n is the most significant digit (MSD), and a_{-m} is the least significant digit (LSD). The *polynomial representation* of the above number is

$$N = \sum_{i=-m}^{n} a_i r^i \qquad (1.2)$$

There are $n + 1$ integer digits and m fraction digits in the number shown above.

Consider an integer with n digits. A finite range of values can be represented by this integer. The smallest value in this range is 0 and corresponds to each digit of the n-digit integer being equal to 0. When each digit corresponds in value to $r - 1$, the highest digit in the number system, the n-digit number attains the highest value in the range. This value is equal to $r^n - 1$. Table 1.1 lists the first few numbers in various systems. We will discuss binary, octal, and hexadecimal systems next.

TABLE 1.1 Number Systems

Decimal ($r = 10$)	Binary ($r = 2$)	Ternary ($r = 3$)	Quaternary ($r = 4$)	Octal ($r = 8$)	Hexadecimal ($r = 16$)
0	0	0	0	0	0
1	1	1	1	1	1
2	10	2	2	2	2
3	11	10	3	3	3
4	100	11	10	4	4
5	101	12	11	5	5
6	110	20	12	6	6
7	111	21	13	7	7
8	1000	22	20	10	8
9	1001	100	21	11	9
10	1010	101	22	12	A
11	1011	102	23	13	B
12	1100	110	30	14	C
13	1101	111	31	15	D
14	1110	112	32	16	E
15	1111	120	33	17	F
16	10000	121	100	20	10
17	10001	122	101	21	11
18	10010	200	102	22	12
19	10011	201	103	23	13
20	10100	202	110	24	14

1.2.1 Binary System

In this system, the radix is 2 and the two allowed digits are 0 and 1. *BI*nary digi*T* is abbreviated as BIT. A typical binary number is shown in the positional notation in the following example.

Example 1.2

$$N = (11010 . 1101)_2$$

$$2^4 2^3 2^2 2^1 2^0 . 2^{-1} 2^{-2} 2^{-3} 2^{-4} \quad \text{weights}$$

$$16\ 8\ 4\ 2\ 1 . \tfrac{1}{2} \tfrac{1}{4} \tfrac{1}{8} \tfrac{1}{16} \quad \text{weights in decimal}$$

⟵――――――――・――――――――⟶

Weights double for each move to the left from the binary point.　　Weights are halved for each move to the right from the binary point.

In polynomial form, this number is

$$N = 1 \times 2^4 + 1 \times 2^3 + 0 \times 2^2 + 1 \times 2^1 + 0 \times 2^0$$
$$+ 1 \times 2^{-1} + 1 \times 2^{-2} + 0 \times 2^{-3} + 1 \times 2^{-4}$$

$$= 16 + 8 + 0 + 2 + 0 + \frac{1}{2} + \frac{1}{4} + 0 + \frac{1}{16} \text{ (decimal)}$$

$$= 26 + \frac{1}{2} + \frac{1}{4} + \frac{1}{16} \text{ (decimal)}$$

$$= (26\tfrac{13}{16})_{10}$$

As we can see from the polynomial expansion and summation shown here, the positions containing a 0 do not contribute to the sum. To convert a binary number into decimal, we can simply accumulate the weights corresponding to each nonzero bit of the number.

Each bit can take either of the two values: 0 or 1. With two bits, we can derive 2^2, or 4, combinations: 00, 01, 10, and 11. The decimal values of these combinations (binary numbers) are 0, 1, 2, and 3, respectively. Similarly, with three bits we can derive 2^3, or 8, combinations ranging in value form 000 (0 in decimal) to 111 (7 in decimal). In general, with n bits it is possible to generate 2^n combinations of 0s and 1s, and these combinations when viewed as binary numbers range in value from 0 to $(2^n - 1)$. Table 1.2 shows some binary numbers for various values of n. The 2^n combinations possible for any n are obtained by starting with n 0s and counting in binary until the number with n 1s is reached. A more mechanical method of generating these combinations is described herein.

The first combination has n 0s and the last has n 1s. As we can see from Table 1.2, the value of the least significant bit (LSB)—i.e., bit position 0—alternates in value between 0 and 1 every row, as we move from row to row. Similarly, the value of the bit in position 1 alternates every two rows (i.e., two 0s followed by two 1s). In general, the value of the bit in position i alternates ever 2^i rows starting from 0s. This observation can be utilized in generating all the 2^n combinations.

1.2.2 Octal System

In this system, $r = 8$, and the allowed digits are 0, 1, 2, 3, 4, 5, 6, and 7. A typical number is shown in positional notation in the following example.

TABLE 1.2 Binary Numbers

$n = 2$	$n = 3$	$n = 4$

$\boxed{1\ \ 0}$	$\boxed{2\ \ 1\ \ 0}$	$\boxed{3\ \ 2\ \ 1\ \ 0}$ ← Bit position
00	000	0000
01	001	0001
10	010	0010
11	011	0011
	100	0100
	101	0101
	110	0110
	111	0111
		1000
		1001
		1010
		1011
		1100
		1101
		1110
		1111

Example 1.3

$$N = (4\ 5\ 2\ 6\ .2\ 3)_8$$
$$8^3 8^2 8^1 8^0 . 8^{-1} 8^{-2} \text{ weights}$$
$$= 4 \times 8^3 + 5 \times 8^2 + 2 \times 8^1 + 6 \times 8^0 + 2 \times 8^{-1}$$
$$+ 3 \times 8^{-2} \text{ polynomial form}$$
$$= 2048 + 320 + 16 + 6 + \tfrac{2}{8} + \tfrac{3}{64} \text{ (decimal)}$$
$$= (2390\tfrac{19}{64})_{10}$$

1.2.3 Hexadecimal System

In this system, $r = 16$, and the allowed digits are 0, 1, 2, 3, 4, 5, 6, 7, 8, 9, A, B, C, D, E, and F. Digits A through F correspond to decimal values 10 through 15, respectively. A typical number is shown in the following example.

Example 1.4

$N = (A\ 1\ F\ .\ 1\ C)_{16}$

$16^2 16^1 16^0 . 16^{-1} 16^{-2}$ weights

$= A \times 16^2 + 1 \times 16^1 + F \times 16^0 + 1 \times 16^{-1}$
$+ C \times 16^{-2}$ polynomial form

$= 10 \times 16^2 + 1 \times 16^1 + 15 \times 16^0 + 1 \times 16^{-1}$
$+ 12 \times 16^{-2}$ (decimal)

$= (2591\frac{28}{256})_{10}$

1.3 Conversion

To convert numbers from a nondecimal system to decimal, we simply expand the given number as a polynomial and evaluate the polynomial using decimal arithmetic, as shown in Examples 1.1 through 1.4. When a decimal number is converted to any other system, the integer and fraction portions of the number are handled separately. The *radix divide technique* is used to convert the integer portion, and the *radix multiply technique* is used for the fraction portion.

1.3.1 Radix Divide Technique

(i) Divide the given integer successively by the required radix, noting the remainder at each step. The quotient at each step becomes the new dividend for subsequent division. Stop the division process when the quotient becomes zero.

(ii) Collect the remainders from each step (last to first) and place them left to right to form the required number.

The following examples illustrate the procedure.

Example 1.5

$(245)_{10} = (?)_2$ (i.e., convert $(245)_{10}$ to binary)

Required ↗ base

Remainders

$= \underrightarrow{(11110101)}_2$

Here, 245 is first divided by 2, generating a quotient of 122 and a remainder of 1. Next 122 is divided, generating 61 as the quotient and 0 as the remainder. The division process is continued until the quotient is 0, with the remainders noted at each step. The remainder bits form each step (last to first) are then placed left to right to form the number in base 2.

To verify the validity of the radix divide technique, consider the polynomial representation of a four-bit integer $A = (a_4a_3a_2a_1)$:

$$A = \sum_{i=1}^{4} a_i \cdot r^i$$

This can be rewritten as

$$A = 2(2(2(a_3) + a_2) + a_1) + a_0$$

From this form, it can be seen that the bits of the binary number correspond to the remainder at each divide-by-two operation. Some examples follows.

Example 1.6

$$(245)_{10} = (?)_8$$

```
8 | 245
8 |  30   5 ↑
8 |   3   6 |
      0   3 |   = (365)_8
```

Example 1.7

$$(245)_{10} = (?)_{16}$$

```
16 | 245
16 |  15    5 = 5 ↑
       0   15 = F |   = (F5)_{16}
```

1.3.2 Radix Multiply Technique

As we move each position to the right of the radix point, the weight corresponding to each bit in the binary fraction is halved. The radix multiply technique uses this fact and multiplies the given decimal number by 2 (i.e., divides the given number by half) to obtain each fraction bit. The technique consists of the following steps:

(i)   Successively multiply the given fraction by the required base, noting the integer portion of the product at each step. Use the fractional part of the product as the multiplicand for subsequent steps. Stop when the fraction either reaches 0 or recurs.

(ii) Collect the integer digits at each step from first to last and arrange them left to right.

If the radix multiplication process does not converge to 0, it is not possible to represent a decimal fraction in binary exactly. Accuracy, then, depends on the number of bits used to represent the fraction. Some examples follow.

Example 1.8

$$(.250)_{10} = (?)_2$$

$$
\begin{array}{r}
.25 \\
\times 2 \\
\hline
0.50 \\
\times 2 \\
\hline
1.00 = (.01)_2 \\
\end{array}
$$

Example 1.9

$$(.345)_{10} = (?)_2$$

$$
\begin{array}{r}
.345 \\
\times\ 2 \\
\hline
0.690 \\
\times\ 2 \\
\hline
1.380 \\
\times\ 2 \\
\hline
0.760 \\
\times\ 2 \\
\hline
1.520 \\
\times\ 2 \\
\hline
1.040 \\
\times\ 2 \\
\hline
0.080 = (.010110)_2 \\
\end{array}
$$

Multiply fractions only

The fraction may never reach 0; stop when the required number of fraction digits is obtained; the fraction will not be accurate.

Example 1.10

$$(.345)_{10} = (?)_8$$

$$
\begin{array}{r}
.345 \\
\times\ 8 \\
\hline
2.760 \\
\times\ 8 \\
\hline
6.080 \\
\times\ 8 \\
\hline
0.640 \\
\times\ 8 \\
\hline
5.120 = (.2605)_8
\end{array}
$$

Example 1.11

$$(242.45)_{10} = (?)_2$$

2	242	
2	121	0
2	60	1
2	30	0
2	15	0
2	7	1
2	3	1
2	1	1
	0	1

$$
\begin{array}{r}
.45 \\
\times\ 2 \\
\hline
0.90 \\
\times\ 2 \\
\hline
1.80 \\
\times\ 2 \\
\hline
1.60 \quad * \\
\times\ 2 \\
\hline
1.20 \\
\times\ 2 \\
\hline
0.40 \\
\times\ 2 \\
\hline
0.80 \\
\times\ 2 \\
\hline
1.60 \quad *\text{repeats}
\end{array}
$$

$$= (1111\ 0010\ .\ 01\ \overline{11\ 00})_2$$

The radix divide and multiply algorithms are applicable to the conversion of numbers from any base to any other base. When a number is converted from base p to base q, the number in base p is divided (or multiplied) by q in *base p arithmetic*. Because of our familiarity with decimal arithmetic, these methods are convenient when p equals 10. In general, it is easier to convert a base p number to base q ($p \neq 10$, $q \neq 10$) by first converting the number to decimal from base p and then converting that decimal number to base q (i.e., $(N)_p \rightarrow (?)_{10} \rightarrow (?)_q$), as shown by the following example.

Example 1.12

$$(25.34)_8 = (?)_5$$

Convert to base 10:

$$(25.34)_8 = 2 \times 8^1 + 5 \times 8^0 + 3 \times 8^{-1} + 4 \times 8^{-2} \text{ decimal}$$

$$= 16 + 5 + \tfrac{3}{8} + \tfrac{4}{64} \text{ decimal}$$

$$= (21\tfrac{28}{64})_{10}$$

$$= (21.4375)_{10}$$

Convert to base 5:

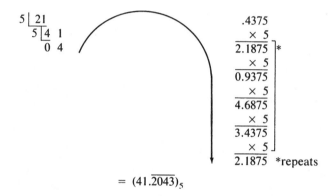

$$= (41.\overline{2043})_5$$

1.3.3   Base 2^k Conversion

Each of the eight octal digits can be represented by a three-bit binary number. Similarly, each of the 16 hexadecimal digits can be represented by a four-bit binary number. In general, each digit of the base p number system, where p is an integral power k of 2, can be represented by a k-bit binary number.

In converting a base p number to base q, if p and q are both integral powers of 2, the base p number can first be converted to binary, and this can in turn be converted to base q by inspection. This conversion procedure is called the *base 2^k conversion.*

Example 1.13

$$(4\ 2\ A\ 5\ 6\ .\ F\ 1)_{16} = (?)_8$$

$$p = 16 = 2^4, \qquad q = 8 = 2^3$$

Therefore,

$$k_1 = 4, \qquad k_2 = 3$$

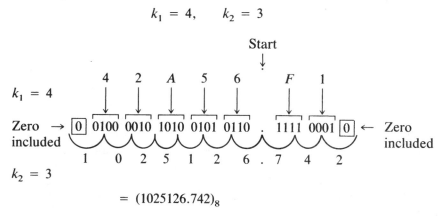

$$= (1025126.742)_8$$

Example 1.14

$$(AF5.2C)_{16} = (?)_4$$

	A	F	5	.	2	C		base 16 $= 2^4$ $k_1 = 4$

$$\underbrace{1010}\ \underbrace{1111}\ \underbrace{0101}.\underbrace{0010}\ \underbrace{1100} \qquad \text{base } 4 = 2^2 \quad \therefore k_2 = 2$$

$$2\ 2\ 3\ 3\ 1\ 1\ .\ 0\ 2\ 3\ 0$$

$$= (223311.0230)_4$$

Example 1.15

$$(567.23)_8 = (?)_{16}$$

base 8 $\quad \therefore k_1 = 3$

Zeros included $\boxed{0}\boxed{0}\boxed{0}\ 101\ 110\ 111.010\ 011\ \boxed{0}\boxed{0}$ base 2

$$1 \qquad 7 \qquad 7\ .\ 4 \qquad C \qquad \text{base 16} \quad \therefore k_2 = 4$$

$$= (177.4C)_{16}$$

It is thus possible to represent binary numbers in a very compact form by using octal and hexadecimal systems. The conversion between these systems is also straightforward. Because it is easier to work with fewer digits than with a large number of bits, digital system users prefer to work with octal or hexadecimal systems when understanding or verifying the results produced by the system or communicating data between users or between users and the machine.

1.4 Arithmetic

Arithmetic in all other number systems follows the same general rules as in decimal. Binary arithmetic is simpler than decimal arithmetic since only two digits (0 and 1) are involved. Arithmetic in octal and hexadecimal systems requires some practice because of the general unfamiliarity with those systems. In this section, we will describe binary arithmetic in detail, followed by a brief discussion of octal and hexadecimal arithmetic. For simplicity, integers will be used in all the examples in this section. Nontheless, the procedures are valid for fractions and numbers with both integer and fraction portions.

In the so-called *fixed-point representation* of binary numbers in digital systems, the radix point is assumed to be either at the right end or the left end of the field in which the number is represented. In the first case the number is an integer, and in the second it is a fraction. Fixed-point representation is the most common type of representation. In scientific computing applications, in which a large range of numbers must be represented, *floating-point representation* is used. Floating-point representation of numbers will be discussed in Section 1.6.

1.4.1 Binary Arithmetic

Table 1.3 illustrates the rules for binary addition, subtraction, and multiplication.

TABLE 1.3 Binary Arithmetic

(a) Addition			(b) Subtraction			(c) Multiplication		
		A			A			A
$A + B$			$A - B$			$A \times B$		
	0	1		0	1		0	1
	0	0	1		0	0	1		0	0	0
B			B			B		
	1	1	10		1	11	0		1	0	1

Carry Sum   Borrow Difference

Addition In Table 1.3(a), note that $0 + 0 = 0$, $0 + 1 = 1$, $1 + 0 = 1$, and $1 + 1 = 10$. Thus, the addition of two 1s results in a SUM of 0 and a CARRY of 1.

When two binary numbers are added, the carry from any position is included in the addition of bits in the next most significant position, as in decimal arithmetic. Example 1.16 illustrates this.

Example 1.16

```
  4 3 2 1 0    bit position

  0 1 1 0      carry
  1 0 1 1 0    augend       Note: The carry in the LSB
+ 0 0 1 1 1    addend       position is 0.
  1 1 1 0 1    sum
```

Here, bits in the LSB position (i.e., position 0) are first added, resulting in a sum bit of 1 and a carry of 0. The carry is included in the addition of bits at position 1. The three bits in position 1 are added using two steps $(0 + 1 = 1, 1 + 1 = 10)$, resulting in a sum bit of 0 and a carry bit of 1 to the next most significant position (position 2). This process is continued through the most significant bit (MSB).

In general, the addition of two n-bit numbers results in a number that is $n + 1$ bits long. If the number representation is to be confined to n bits, the operands of the addition should be kept small enough so that their sum does not exceed n bits.

Subtraction From Table 1.3(b), we can see that $0 - 0 = 0$, $1 - 0 = 1$, $1 - 1 = 0$, and $0 - 1 = 1$ with a BORROW of 1. That is, subtracting a 1 from a 0 results in a 1 with a borrow from the next most significant position, as in decimal arithmetic. Subtraction of two binary numbers is performed stage by stage as in decimal arithmetic, starting from the LSB to the MSB. Some examples follow.

Example 1.17

```
    5 4 3 2 1 0    bit position

    0     0
    ɫ 0 1 ɫ 0 1    minuend
  - 0 1 1 0 1 0    subtrahend
    0 1 0 0 1 1    difference
```

Bit position 1 requires a borrow from bit position 2. Because of this borrow, minuend bit 2 is a 0. The subtraction continues through the MSB.

Example 1.18

| 7 | 6 | 5 | 4 | 3 | 2 | 1 | 0 | bit position

```
  0 0       0
  + + θ θ + 0  1  1      minuend
      1 1
- 0 1 1 0 1 1 1 0      subtrahend
  ─────────────────
  0 1 0 1 1 1 0 1      difference
```

Bit 2 requires a borrow from bit 3; after this borrow, minuend bit 3 is 0. Then, bit 3 requires a borrow. Because bits 4 and 5 of the minuend are zeros, borrowing is from bit 6. In this process, the intermediate minuend bits 4 and 5 each attain a value of 1 (compare this with the decimal subtraction). The subtraction continues through the MSB.

Multiplication   Binary multiplication is similar to decimal multiplication. From Table 1.3(c), we can see that $0 \times 0 = 0$, $0 \times 1 = 0$, $1 \times 0 = 0$, and $1 \times 1 = 1$. An example follows.

Example 1.19

```
            1011   multiplicand
        ×   1100   multiplier
```

—multiplier bits

```
                 0000        (1011) × 0
Partial          0000        (1011) × 0
products         1011        (1011) × 1
                 1011        (1011) × 1
            ─────────
            10000100   product
```

In general, the product of two n-bit numbers is $2n$ bits long. In Example 1.19, there are two nonzero bits in the multiplier, one in position 2 corresponding to 2^2 and the other in position 3 corresponding to 2^3. These two bits yield partial products whose values are simply that of the multiplicand shifted left two and three bits, respectively. The 0 bits in the multiplier contribute partial products with 0 values. Thus, the following shift-and-add algorithm can be adopted to multiply two n-bit number A and B, where $B = (b_{n-1}\, b_{n-2} \cdots b_1 b_0)$.

1. Start with a $2n$-bit product with a value of 0.
2. For each b_i $(0 \leq i \leq n - 1) \neq 0$ shift A i positions to the left and add to the product.

This procedure reduces the multiplication to repeated shift and addition of the multiplicand.

Division The longhand (trial-and-error) procedure of decimal division can also be used in binary, as shown in Example 1.20.

Example 1.20

$$110101 \div 111 = ?$$

$$0111 \quad \text{Quotient}$$

$$
\begin{array}{r|l}
 & \quad\quad\quad X \quad\quad Y \\
111 & 110,101 \quad\quad 110 \; < 111 \quad\quad q_1 = 0 \quad \text{do not subtract} \\
 & -111 \\
 & \overline{1101} \quad\quad\quad 1101 > 111 \quad\quad q_2 = 1 \quad \text{subtract} \\
 & \;-111 \\
 & \overline{1100} \quad\quad\quad 1100 > 111 \quad\quad q_3 = 1 \quad \text{subtract} \\
 & \;-111 \\
 & \overline{1011} \quad\quad\quad 1011 > 111 \quad\quad q_4 = 1 \quad \text{subtract} \\
 & \;-111 \\
 & \overline{\;100} \quad\quad\quad \text{remainder}
\end{array}
$$

In this procedure, the divisor is compared with the dividend at each step. If the divisor is greater than the dividend, the corresponding quotient bit is 0; otherwise, the quotient bit is 1, and the divisor is subtracted from the dividend. The compare-and-subtract process is continued until the LSB of the dividend. The procedure is formalized in the following steps.

1. Align the divisor (Y) with the most significant end of the dividend. Let the portion of the dividend from its MSB to its bit aligned with the LSB of the divisor be denoted X. We will assume that there are n bits in the divisor and $2n$ bits in the dividend. Let $i = 0$.
2. Compare X and Y. If $X \geq Y$, the quotient bit is 1: perform $X - Y$. If $X < Y$, the quotient bit is 0.
3. Set $i = i + 1$. If $i \geq n$, stop. Otherwise, shift Y one bit to the right and go to step 2.

For the purposes of illustration, this procedure assumed the division of integers. If the divisor is greater than the dividend, the quotient is 0, and if the divisor is 0, the procedure should be stopped since dividing by 0 results in an error.

As we can see from these examples, multiplication and division operations can be reduced to repeated shift and addition (or subtraction). If the hardware can perform shift, add, and subtract operations, it can be programmed to perform multiplication and division as well. Older digital systems used such measures to reduce hardware costs. With the advances in digital hardware technology, it is now possible to implement these and more complex operations in an economical manner. (See the references at the end of this chapter for books that describe procedures for multiplication and division that are more elegant than the ones described here.)

Shifting   Generally, shifting a base r number left by one position (and inserting a 0 into the vacant LSD position) is equivalent to multiplying the number by r. Shifting the number right by one position (inserting a 0 into the vacant MSD position) generally is equivalent to dividing the number by r.

In binary system, each left shift multiplies the number by 2, and each right shift divides the number by 2, as shown in Example 1.21.

Example 1.21

	Binary	Decimal
N	01011.11 $\nearrow$ Insert	$11\frac{3}{4}$
$2 * N$	10111.1 $\boxed{0}$	$23\frac{1}{2}$
$N \div 2$	Insert $\searrow$ $\boxed{0}$0101.11$\boxed{1}$ $\searrow$ Discard	$5\frac{3}{4}$ (Inaccurate, since only two-bit accuracy is retained.)

If the MSB of an n-bit number is not 0, shifting it left would result in a number larger than the magnitude that can be accommodated in n bits— and the 1 shifted out of the MSB position cannot be discarded. If nonzero bits shifted out of the LSB position during a right shift are discarded, the accuracy is lost. Later in this chapter, we will discuss shifting in further detail.

1.4.2 Octal Arithmetic

Table 1.4 shows the octal addition and multiplication tables. The examples that follow illustrate the four arithmetic operations in octal and their similarity to decimal arithmetic. (Table 1.4 can be used to look up the result at each stage in the arithmetic.) An alternate method is used in the following examples. The operation is first performed in decimal and then the result is converted into octal, before proceeding to the next stage, as shown in the scratchpad.

TABLE 1.4 Octal Arithmetic

(a) Addition

					A			
A + *B*	0	1	2	3	4	5	6	7
0	0	1	2	3	4	5	6	7
1	1	2	3	4	5	6	7	10
2	2	3	4	5	6	7	10	11
B 3	3	4	5	6	7	10	11	12
4	4	5	6	7	10	11	12	13
5	5	6	7	10	11	12	13	14
6	6	7	10	11	12	13	14	15
7	7	10	11	12	13	14	15	16

(b) Multiplication

					A			
A × *B*	0	1	2	3	4	5	6	7
0	0	0	0	0	0	0	0	0
1	0	1	2	3	4	5	6	7
2	0	2	4	6	10	12	14	16
B 3	0	3	6	11	14	17	22	25
4	0	4	10	14	20	24	30	34
5	0	5	12	17	24	31	36	43
6	0	6	14	22	30	36	44	52
7	0	7	16	25	34	43	52	61

Example 1.22: Addition

```
    1 1 1  ←carries          Scratchpad
    1 4 7 6                       6
  + 3 5 5 4                     + 4
  ─────────                   ─────────
    5 2 5 2    sum            (10)₁₀ = (12)₈
                                  1
                                + 7
                                + 5
                              ─────────
                              (13)₁₀ = (15)₈
                                  1
                                + 4
                                + 5
                              ─────────
                              (10)₁₀ = (12)₈
                                  1
                                + 1
                                + 3
                              ─────────
                              (5)₁₀ = (5)₈
```

The addition uses $(10)_{10} = (12)_8$, $(13)_{10} = (15)_8$, $(10)_{10} = (12)_8$, and $(5)_{10} = (5)_8$.

Example 1.23: Subtraction

```
    4 14            Digit position 2 required a
    5̶ 4̶ 7 5         borrow from position 3
  − 3 7 6 4         ∴ Octal   Decimal
  ─────────            14₈      12
    1 5 1 1          − 7₈     − 7
                     ──────   ─────────
                      5₈       5₁₀ = 5₈
```

Example 1.24: Subtraction

```
        3 7 7
    5 4̶ 0̶ 0̶ 4 5
  − 3 2 5 6 5 4
  ─────────────
    2 1 2 1 7 1
```

The intermediate 0s become $r - 1$ or 7 when borrowed.

Example 1.25: Multiplication

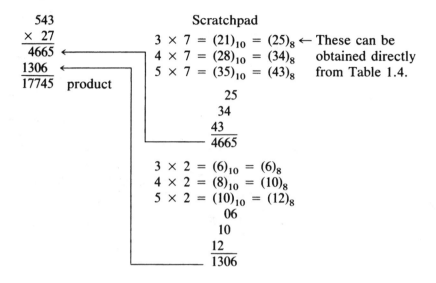

```
  543                    Scratchpad
× 27           3 × 7 = (21)₁₀ = (25)₈ ← These can be
 4665 ←        4 × 7 = (28)₁₀ = (34)₈    obtained directly
 1306 ←        5 × 7 = (35)₁₀ = (43)₈    from Table 1.4.
17745  product
                              25
                              34
                              43
                            4665

               3 × 2 = (6)₁₀ = (6)₈
               4 × 2 = (8)₁₀ = (10)₈
               5 × 2 = (10)₁₀ = (12)₈
                              06
                              10
                              12
                            1306
```

Example 1.26: Division

$$543 \div 7$$

```
        062
      ┌──────
    7 │ 543
      │ 0
      │ ───
      │ 543
      │ 52
      │ ──
      │ 23
      │ 16
      │ ──
      │  5
```

Use the multiplication table in Table 1.4 to derive the quotient digit (by trial and error).

1.4.3 Hexadecimal Arithmetic

Table 1.5 shows the addition and multiplication tables. The following examples illustrate hexadecimal arithmetic.

TABLE 1.5 Hexadecimal Arithmetic

(a) Addition

+	0	1	2	3	4	5	6	7	8	9	A	B	C	D	E	F
0	0	1	2	3	4	5	6	7	8	9	A	B	C	D	E	F
1	1	2	3	4	5	6	7	8	9	A	B	C	D	E	F	10
2	2	3	4	5	6	7	8	9	A	B	C	D	E	F	10	11
3	3	4	5	6	7	8	9	A	B	C	D	E	F	10	11	12
4	4	5	6	7	8	9	A	B	C	D	E	F	10	11	12	13
5	5	6	7	8	9	A	B	C	D	E	F	10	11	12	13	14
6	6	7	8	9	A	B	C	D	E	F	10	11	12	13	14	15
7	7	8	9	A	B	C	D	E	F	10	11	12	13	14	15	16
8	8	9	A	B	C	D	E	F	10	11	12	13	14	15	16	17
9	9	A	B	C	D	E	F	10	11	12	13	14	15	16	17	18
A	A	B	C	D	E	F	10	11	12	13	14	15	16	17	18	19
B	B	C	D	E	F	10	11	12	13	14	15	16	17	18	19	1A
C	C	D	E	F	10	11	12	13	14	15	16	17	18	19	1A	1B
D	D	E	F	10	11	12	13	14	15	16	17	18	19	1A	1B	1C
E	E	F	10	11	12	13	14	15	16	17	18	19	1A	1B	1C	1D
F	F	10	11	12	13	14	15	16	17	18	19	1A	1B	1C	1D	1E

(b) Multiplication

×	0	1	2	3	4	5	6	7	8	9	A	B	C	D	E	F
0	0	0	0	0	0	0	0	0	0	0	0	0	0	0	0	0
1	0	1	2	3	4	5	6	7	8	9	A	B	C	D	E	F
2	0	2	4	6	8	A	C	E	10	12	14	16	18	1A	1C	1E
3	0	3	6	9	C	F	12	15	18	1B	1E	21	24	27	2A	2D
4	0	4	8	C	10	14	18	1C	20	24	28	2C	30	34	38	3C
5	0	5	A	F	14	19	1E	23	28	2D	32	37	3C	41	46	4B
6	0	6	C	12	18	1E	24	2A	30	36	3C	42	48	4E	54	5A
7	0	7	E	15	1C	23	2A	31	38	3F	46	4D	54	5B	62	69
8	0	8	10	18	20	28	30	38	40	48	50	58	60	68	70	78
9	0	9	12	1B	24	2D	36	3F	48	51	5A	63	6C	75	7E	87
A	0	A	14	1E	28	32	3C	46	50	5A	64	6E	78	82	8C	96
B	0	B	16	21	2C	37	42	4D	58	63	6E	79	84	8F	9A	A5
C	0	C	18	24	30	3C	48	54	60	6C	78	84	90	9C	A8	B4
D	0	D	1A	27	34	41	4E	5B	68	75	82	8F	9C	A9	B6	C3
E	0	E	1C	2A	38	46	54	62	70	7E	8C	9A	A8	B6	C4	D2
F	0	F	1E	2D	3C	4B	5A	69	78	87	96	A5	B4	C3	D2	E1

Example 1.27: Addition

```
    1 1              Scratchpad
    1 5 F C          Decimal
  + 2 4 5 D          C = 12
    3 A 5 9          D = 13
                     16⌐25 = (19)₁₆  ← This can be obtained
                     16⌐1   9 ↑          directly from Table 1.5.
                         0   1
```

$$16 \overline{)25} = (19)_{16} \leftarrow$$

Decimal
$$1 = 1$$
$$F = 15$$
$$5 = \underline{5}$$
$$\overline{21} = (15)_{16}$$

Example 1.28: Subtraction

```
  1 13 15              Scratchpad
     3                 Decimal
   2 4 5 D   minuend   (15)₁₆ =   21
 −1 5 F C   subtrahend −(F)₁₆ = −15 = (6)₁₆
   0 E 6 1   difference           6

                      (13)₁₆ =   19
                      −(5)16 = − 5
                               14 = (E)₁₆
```

$$(15)_{16} = 21$$
$$-(F)_{16} = -15 = (6)_{16}$$
$$\phantom{-(F)_{16} = } \overline{6}$$

$$(13)_{16} = 19$$
$$-(5)16 = -5$$
$$\overline{14} = (E)_{16}$$

Example 1.29: Multiplication

```
      1 E 4 A                    Scratchpad
      × FA2          Decimal          Hexadecimal
        3            A × 2 = 20 =      14
    0 3 C 9 4   P₁   4 × 2 = 8  =      08
 +   1 2 E E 4  P₂   E × 2 = 28 =      1C
 +1 C 6 5 6     P₃   1 × 2 = 2  =     02
   1 D 9 8 0 D4                     03C94 = P₁

                     A × A = 100 =      64
                     4 × A = 40  =      28
                     E × A = 140 =     8 C
                     1 × A = 10  =    0A
                                    12EE4 = P₂

                     A × F = 150 =      96
                     4 × F = 60  =      3C
                     E × F = 210 =     D2
                     1 × F = 15  =    0F
                                    1C656 = P₃
```

Example 1.30: Division

$$1\,A\,F\,3 \div E$$

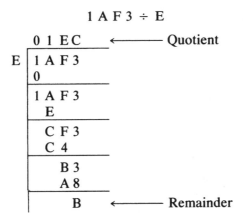

1.5 Representation of Negative Numbers

The examples shown so far have used only positive numbers. In practice, a digital system must represent both positive and negative numbers. To accommodate the sign of the number, an additional digit, called the sign digit, is included in the representation, along with the magnitude digits. Thus, to represent an n-digit number, we would need $n + 1$ digits. Typically, the sign digit is the MSD. Two popular representation schemes have been used: the *sign-magnitude system* and the *complement system*.

1.5.1 Sign-Magnitude System

In this representation, $n + 1$ digits are used to represent a number, where the MSD is the sign digit and the remaining n digits are magnitude digits. The value of the sign digit is 0 for positive numbers and $r - 1$ for negative numbers, where r is the radix of the number system. Some sample representations follow.

Example 1.31

Here, we assume that five digits are available to represent each number. The sign and magnitude portions of the number are separated by ",," for illusion purposes only. The ",," is not used in the actual representation.

Number	Representation
$(-2)_2$	1,0010
$(+56)_8$	0,0056
$(-56)_8$	7,0056
$(+1F)_{16}$	0,001F
$(-1F)_{16}$	F,001F

Sign Magnitude

All numbers are shown as five-digit numbers.

The sign and magnitude portions are handled separately in arithmetic using sign-magnitude numbers. The magnitude of the result is computed and then the appropriate sign is attached to the result, just as in decimal arithmetic. The sign-magnitude system has been used in such small digital systems as digital meters and typically when the decimal mode of arithmetic is used in digital computers. The decimal (or binary coded decimal) arithmetic mode will be described later in this chapter. Complement number representation is the most prevalent representation mode in modern-day computer systems.

1.5.2   Complement Number System

Consider the subtraction of a number A from a number B. This is equivalent to adding $(-A)$ to B. The complement number system provides a convenient way of representing negative numbers (i.e., complements of positive numbers), thus reducing the subtraction to an addition. Because multiplication and division correspond respectively to repeated addition and subtraction, it is possible to perform the four basic arithmetic operations using only the hardware for addition when the negative numbers are represented in complement form. The two popular complement number systems are radix complement and diminished radix complement.

The *radix complement* of a number $(N)_r$ is defined as

$$[N]_r = r^n - (N)_r \qquad \text{if } (N)_r \neq 0 \qquad (1.3)$$
$$= 0 \qquad \qquad \text{if } (N)_r = 0$$

where $[N]_r$ is the radix complement and n is the number of digits in the integer portion of the number $(N)_r$.

This system is commonly called either *twos complement* or *tens complement*, depending on which number system is used. This section will describe the twos complement system. Because the tens complement system displays the same characteristics as the twos complement system, it will not be discussed here.

Example 1.32

(a)   The twos complement of $(01010)_2$ is

$$2^5 - (01010) = 100000 - 01010 = 10110$$

Here $n = 5$ and $r = 2$.

(b)   The twos complement of $(0.0010)_2$ is

$$2^1 - (0.0010) = 10.0000 - 0.0010 = 1.1110$$

Here, $n = 1$ and $r = 2$.

(c)   The tens complement of $(4887)_{10}$ is

$$10^4 - 4887 = 5113$$

Here, $n = 4$ and $r = 10$.

(d)   The tens complement of $(48.87)_{10}$ is

$$10^2 - 48.87 = 51.13$$

Here, $n = 2$ and $r = 10$.

As can be verified by Example 1.32, there are two other methods for obtaining the radix complement of a number.

METHOD 1

$[01010]_2 = ?$

10101   a. Complement each bit (i.e., change each 0 to 1 and 1 to 0).

  + 1   b. Add 1 to the LSB to get the twos

¯¯¯¯¯¯

10110   complement.

METHOD 2

$[010{:}10]_2 = ?$

10   a. Copy the bits from the LSB until and including the first nonzero bit.

101   b. Complement the remaining bits through the MSB to get the twos complement.

101¦10

The *diminished radix complement* $[N]_{r-1}$ of a number $(N)_r$ is defined as:

$$[N]_{r-1} = r^n - r^{-m} - (N)_r \qquad (1.4)$$

where n and m are respectively the number of digits in integer and fraction portion of the number. Note that

$$[N]_r = [N]_{r-1} + r^{-m} \qquad (1.5)$$

That is, the radix complement of a number is obtained by adding a 1 to the LSB of the diminished radix complement form of the number.

The diminished radix complement is commonly called the *ones complement* or *nines complement*, depending on which number system is used.

Example 1.33

	$(N)_r$	r	n	m	$[N]_{r-1}$
(a)	1001	2	4	0	$2^4 - 2^0 - 1001$
					$= 1000 - 1 - 1001$
					$= 1111 - 1001 = 0110$
(b)	100.1	2	3	1	$= 2^3 - 2^{-1} - 100.1$
					$= 1000 - 0.1 = 100.1$
					$= 111.1 - 100.1 = 011.0$
(c)	486.7	10	3	1	$= 10^3 - 10^{-1} - 486.7$
					$= 1000 - 0.1 - 486.7$
					$= 999.9 - 486.7 = 513.2$

From Example 1.33 it can be seen that the ones complement of a number is obtained by subtracting each digit from the largest digit in the number system. In the binary system, this is equivalent to complementing (i.e., changing 1 to 0 and 0 to 1) each bit of the given number.

Example 1.34

$$N = \quad 10110.110$$
$$\text{Ones complement of } N = \quad 11111.111$$
$$-10110.110$$
$$\overline{\qquad\qquad\qquad}$$
$$01001.001$$

which can also be obtained by complementing each bit of N.

As in sign-magnitude representation, a sign bit is included in the representation of numbers in complement systems as well. Because the complement of a number corresponds to its negative, positive numbers that are represented in complement systems remain in the same form as in the sign-

magnitude system. Only negative numbers are represented in the complement form as shown by the following example.

Example 1.35

Here we assume that five bits are available for representation and that
the MSB is the sign bit.

Decimal	Sign-magnitude	Twos complement	Ones complement
+5	0,0101	0,0101	0,0101
−5	1,0101	1,1011	1,1010
+4	0,0100	0,0100	0,0100
−4	1,0100	1,1100	1,1011

To obtain the complement of a number, we can start with the sign-
magnitude form of the corresponding positive number and adopt the complementing procedures discussed here. In Example 1.35, the sign bit is separated from the magnitude bits by a ",'' for illustration purposes only. This
separation is not necessary in complement systems since the sign bit also
participates in the arithmetic as though it were a magnitude bit (as we will
see later in this section).

Table 1.6 shows the range of numbers that can be represented in five
bits, in the sign-magnitude, twos complement, and ones complement systems. Note that the sign-magnitude and ones complement systems have two
representations for 0 (+0 and −0), whereas the twos complement system
has a unique representation for 0. Note also the use of the combination
10000 to represent the largest negative number in the twos complement
system. In general, the ranges of integers that can be represented in an n-
bit field (using 1 sign bit and $n - 1$ magnitude bits) in the three systems
are

sign-magnitude: $-(2^{n-1} - 1)$ to $+(2^{n-1} - 1)$
ones complement: $(2^{n-1} - 1)$ to $(2^{n-1} - 1)$
twos complement: (2^{n-1}) to $(2^{n-1} - 1)$

We will now illustrate the arithmetic in these systems of number
representation.

Twos Complement Addition   Example 1.36 illustrates the addition of numbers represented in twos complement form.

TABLE 1.6 The Three Representation Schemes

Decimal	Sign-magnitude	Twos complement	Ones complement
+15	01111	01111	01111
+14	01110	01110	01110
+13	01101	01101	01101
+12	01100	01100	01100
+11	01011	01011	01011
+10	01010	01010	01010
+9	01001	01001	01001
+8	01000	01000	01000
+7	00111	00111	00111
+6	00110	00110	00110
+5	00101	00101	00101
+4	00100	00100	00100
+3	00011	00011	00011
+2	00010	00010	00010
+1	00001	00001	00001
+0	00000	00000	00000
−0	10000	00000	11111
−1	10001	11111	11110
−2	10010	11110	11101
−3	10011	11101	11100
−4	10100	11100	11011
−5	10101	11011	11010
−6	10110	11010	11001
−7	10111	11001	11000
−8	11000	11000	10111
−9	11001	10111	10110
−10	11010	10110	10101
−11	11011	10101	10100
−12	11100	10100	10011
−13	11101	10011	10010
−14	11110	10010	10001
−15	11111	10001	10000
−16		10000*	

*Twos complement uses 10000 to expand the range to (−16).

Example 1.36

(a) | Decimal | Sign-magnitude | Twos complement |
|---|---|---|
| 5 | 0,0101 | 0,0101 |
| 4 | 0,0100 | 0,0100 |
| | | 0,1001 = 9 |

Here the sign-magnitude and twos complement representations are the same, since both numbers are positive. In twos complement addition, the sign bit is also treated as a magnitude bit and participates in the addition process. *In this example, the sign and magnitude portions are separated for illustration purposes only.*

(b) | Decimal | Sign-magnitude | Twos complement |
|---|---|---|
| 5 | 0,0101 | 0,0101 |
| −4 | 1,0100 | 1,1100 |
| | | 10,0001 = + $(0001)_2$ |

Carry from the sign↗
position is ignored.

Here the negative number is represented in the complement form and the two numbers are added. The sign bits are also included in the addition process. There is a carry from the sign bit position, which is ignored. The sign bit is 0, indicating that the result is positive.

(c) | Decimal | Sign-magnitude | Twos complement |
|---|---|---|
| 4 | 0,0100 | 0,0100 |
| −5 | 1,0101 | 1,1011 |
| | | 1,1111 = $-(0001)_2$ |

The result is negative;
no carry.

Here, no carry is generated from the MSB during the addition. The result is negative since the sign bit is 1; the result is in the complement form and must be complemented to obtain the sign-magnitude representation.

(d) | Decimal | Sign-magnitude | Twos complement |
|---|---|---|
| −5 | 1,0101 | 1,1011 |
| −4 | 1,0100 | 1,1100 |
| | | 11,0111 = $-(1001)_2$ |

Ignore the carry; the
result is negative.

When subtraction is performed in decimal arithmetic (and in the sign-magnitude system), the number with the smaller magnitude is subtracted from the one with the larger magnitude, and the sign of the result is that of the larger number. Such comparison is not needed in the complement system as shown in Example 1.36.

In summary, in twos complement addition, the carry generated from the MSB is ignored. The sign bit of the result must be the same as that of the operands when both operands have the same sign. If it is not, the result is too large to fit into the magnitude field and hence an overflow occurs. When the signs of the operands are different, a carry from the sign bit indicates a positive result; if no carry is generated, the result is negative and must be complemented to obtain the sign-magnitude form. The sign bit participates in the arithmetic.

Ones Complement Addition   Example 1.37 illustrates ones complement addition. It is similar to that in twos complement representation, except that the carry generated from the sign bit is added to the LSB of the result to complete the addition.

Example 1.37

	(a) Decimal	Sign-magnitude	Ones complement	
	5	0, 0101	0, 0101	
	-4	1, 0100	1, 1011	
			1 0, 0000	sum

Add the end-around carry $\longrightarrow$ $\llcorner\!\longrightarrow 1$
$$0, 0001$$

	(b) Decimal	Sign-magnitude	Ones complement	
	4	0, 0100	0, 0100	
	-5	1, 0101	1, 1010	
			1, 1110	sum

No carry.
∴ Complement
and the result
is $-(0001)$.

1.5.3   Shifting Revisited

As we have seen before, shifting a binary number left one bit is equivalent to multiplying it by 2. Shifting a binary number right one bit is equivalent to dividing it by 2. Example 1.38 illustrates the effect of shifting an unsigned number.

Example 1.38

Consider the number N with six integer bits and four fraction bits:

$$N \quad \boxed{0\ 0\ 1\ 0\ 1\ 1\ .\ 1\ 0\ 1\ 0}$$

Shifting N one bit left, with a 0 inserted into the LSB:

$$2N \quad \boxed{0\ 1\ 0\ 1\ 1\ 1\ .\ 0\ 1\ 0\ 0} \longleftarrow \text{inserted}$$

Shifting left again:

$$4N \quad \boxed{1\ 0\ 1\ 1\ 1\ 0\ .\ 1\ 0\ 0\ 0} \longleftarrow \text{inserted}$$

If we shift this number left again, the 1 in the MSB would be lost, thereby resulting in an overflow.

Shifting N right one bit:

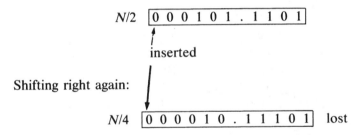

$$N/2 \quad \boxed{0\ 0\ 0\ 1\ 0\ 1\ .\ 1\ 1\ 0\ 1}$$

inserted

Shifting right again:

$$N/4 \quad \boxed{0\ 0\ 0\ 0\ 1\ 0\ .\ 1\ 1\ 1\ 0\ 1} \quad \text{lost}$$

The 1 in the LSB is lost because of this shift, thereby resulting in a less accurate fraction. If there are enough bits to retain all the nonzero bits in the fraction through the shift operation, no loss of accuracy will result. In practice, there will be a finite number of bits for number representation. Hence, care must be taken to see that shifting does not result in either overflow or inaccuracy.

Sign-magnitude Shifting When sign-magnitude numbers are shifted, the sign bit is not included in shift operations. Shifting follows the same procedure as in Example 1.38.

Twos Complement Shifting When twos complement numbers are shifted right, the sign bit value is copied into the vacant (MSB of the magnitude) bit position on the left, and a 0 is inserted in the vacant LSB position during the left shift. Example 1.39 illustrates this.

Example 1.39

| 1 | 0 0 1 0 1 0 0 0 0 |

1 1 0 0 1 0 1 0 0 0   right shift   (copy sign bit)

1 1 1 0 0 1 0 1 0 0   right shift   (copy sign bit)

1 1 0 0 1 0 1 0 0 |0| left shift   (insert 0)

1 0 0 1 0 1 0 0 0 |0| left shift   (insert 0)

A change in the value of the sign bit during a left shift indicates that there is an overflow (i.e., the result is too large).

Ones Complement Shifting   When ones complement numbers are shifted, a copy of the sign bit is inserted in the vacant LSB position during the left shift, or in the MSB position of the magnitude bits during the right shift. The sign bit receives the MSB of the magnitude during the left shift.

Example 1.40

	(a)	(b)	
N	1 1001	0 0001	
$2N$	1 0011	0 0010	insert sign bit
$4N$	0 0111	0 0100	
	overflow		
$N/2$	1 1100	0 0000	copy sign bit
$N/4$	1 1110		

1.5.4   Comparison of Complement Systems

Table 1.7 summarizes the operations in both the complement systems. The twos complement system is now used in almost all digital computer systems. The ones complement system has been used in older computer systems. The advantage of the ones complement system is that the ones complement can be obtained by inverting each bit of the original number from 1 to 0 and 0 to 1, which can be done very easily by the logic components of the digital system. The conversion of a number to twos complement system requires an addition operation after the ones complement of the number is obtained or a scheme to implement the copy/complement algorithm described earlier in this chapter. Twos complement is the most widely used system of representation.

One other popular representation—*biased* or *excess-radix representa-*

TABLE 1.7   Comparison of Complement Number Systems

		Ones complement	Twos complement	
Operation	If the carry from the MSB is:	Then perform:	Then perform:	Sign bit of the result
Add	0	(Result is in complement form); complement to convert to sign-magnitude form		1
	1	(Result is in sign-magnitude form); add 1 to the LSB of the result.	(Result is in sign-magnitude form); neglect the carry.	0
Left shift		Copy sign bit into the LSB.	Insert 0 into the LSB.	Sign bit = MSB of magnitude
Right shift		Copy sign bit into the MSB of the magnitude.		Sign bit unchanged

tion—is used to represent floating-point numbers. This representation is described in Section 1.6.

1.6   Floating-Point Numbers

Fixed-point representation is convenient for representing numbers with bounded orders of magnitude. For instance, in a digital computer that uses 32 bits to represent numbers, the range of integers that can be used is limited to $\pm(2^{31} - 1)$, which is approximately $\pm 10^{11}$. In scientific computing environments, a wider range of numbers may be needed, and floating-point representation may be used. The general form of a floating-point representation number N is

$$N = F \times r^{E} \qquad (1.6)$$

where F is the *fraction* (or *mantissa*), r is the radix, and E is the *exponent*.

Consider the number

$$N = 3560000$$
$$= (.356) \times 10^7$$
$$= (.0356) \times 10^8 = (3.56) \times 10^6$$

All of these forms are valid floating-point notation. The first two forms are preferred, however, since with these forms there is no need to represent the integer portion of the mantissa, which is 0. The first form requires the fewest digits to represent the mantissa, since all the significant zeros have been eliminated from the mantissa. This form is called the *normalized* form of floating-point representation. Note from the example above that the radix point *floats* within the mantissa incrementing the exponent by 1 for each move to the left and decrementing the exponent by 1 for each move to the right. This shifting of the mantissa and *scaling* of the exponent is frequently done in the manipulation of floating-point numbers.

Let us now concentrate on the representation of floating-point numbers. The radix is implied by the number system used and hence is not shown explicitly in the representation. The mantissa and the exponent can be either positive or negative. Hence, the floating-point representation consists of the four components E, F, S_E, and S_F, where S_E and S_F are signs of the exponent and mantissa, respectively.

The F is represented in the normalized form. True binary form is used in the representation of F (rather than any of the complement forms), and one bit is used to represent S_F (0 for positive, 1 for negative). Since the MSB of the normalized mantissa is always 1, the range of mantissa is

$$0.5 \leqslant F < 1$$

The floating-point representation of 0 is an exception and contains all 0s.

When two floating-point numbers are added, the exponents must be compared and equalized before the addition. To simplify the comparison operation without involving the sign of the exponent, the exponents are usually converted to positive numbers by adding a *bias constant* to the true exponent. The bias constant is usually the magnitude of the largest negative number that can be represented in the exponent field. Thus, in a floating-point representation with q bits for the exponent field, if a twos complement representation is used, the unbiased exponent E_n will be in the range

$$-2^{q-1} \leqslant E_n \leqslant 2^{q-1} - 1$$

If we add the bias constant of 2^{q-1}, the biased exponent E_b will be in the range.

$$0 \leqslant E_b \leqslant 2^{q-1}$$

The true (unbiased) exponent is obtained by subtracting the bias constant from the biased exponent. That is,

$$E_n = E_b - 2^{q-1}$$

For example, if $q = 9$, then the bias constant is 256. Three values of the exponent are shown below:

	-256	0	$+255$
E_n:	100000000	000000000	011111111
E_b:	000000000	100000000	111111111

So long as the mantissa is 0, theoretically the exponent can be anything, thereby making it possible to have several representations for floating-point representation numbers. In fixed-point representation, we represented a 0 by a sequence of 0s. To retain the uniqueness of 0 representation for both fixed- and floating-point representations, the mantissa is set to all 0s and the exponent is set to the most negative exponent in biased form (i.e., all 0s).

Figure 1.2 shows the Institute of Electrical and Electronics Engineers (IEEE) standard representation of floating-point numbers. Out of the 32 bits in the representation, the MSB is reserved for the sign (i.e., S_F), the next 8 bits represent the exponent (i.e., S_E and E), and the remaining bits are for the fraction. The sign bit is 0 for positive numbers and 1 for negative numbers. The exponent uses a bias component of 127.

The MSB of the fraction in the normalized floating-point form is always 1. This is not specifically represented in the IEEE representation but is assumed to exist. Thus, an extra bit of accuracy is obtained, since a 24-bit fraction can be represented in only 23 bits.

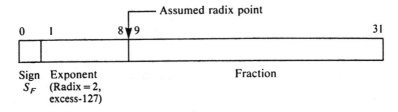

Figure 1.2   IEEE standard floating-point representation.

Example 1.41

Represent (10100.10011) in the floating-point form.

$$(10100.10011) = (.1010010011) \times 2^5$$

Hence,

sign bit = 0

exponent = 5 + 127 = 132 = (10000100)

fraction = 010010011 (The most significant

1 not shown)

Thus the presentation is

0 10000100 010010011000 . . . 00

This representation allows only 24 bits for the fraction. If the numbers must be represented with more accuracy, the representation extends to the second word containing the remaining fraction bits.

This example is an oversimplified description of the IEEE standard for the representation of floating-point numbers. Nevertheless, it represents a practical representation scheme. Most digital systems now use the IEEE standard. Earlier systems used various other formats. For example, the IBM format is shown in Figure 1.3.

1.7 Binary Codes

So far we have seen various ways of representing numeric data in binary form. A digital system requires that all information be in binary form. The external world, however, uses various other symbols such as alphabetic characters and special characters (e.g., +, =, −) to represent information. In order to represent these various symbols in binary form, a unique pattern of 0s and 1s is assigned to represent each symbol. This pattern is the *code word* corresponding to that symbol. As we have seen earlier, it is possible

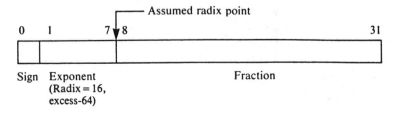

Figure 1.3 The IBM representation scheme.

to represent 2^n elements with a binary string containing n bits. That is, if q symbols are to be represented in binary form, the minimum number of bits n required in the code word is given by

$$2^{n-1} < q \leqslant 2^n \tag{1.7}$$

The code word might possibly contain more than n bits to accommodate error detection and correction. Once the number of bits in the code word is set, the assignment of the code words to the symbols of information to be represented could be arbitrary (in which case a table associating each element with its code word is needed) or might follow some general rule.

For example, if the code word is required to represent four symbols —say, dog, cat, tiger, and camel—we can use the four combinations of two bits (00, 01, 10, and 11). Assignments of these combinations to the four symbols can be arbitrary.

To represent the 26 letters of the alphabet, we would need a five-bit code. With five bits, it is possible to generate 32 combinations of 0s and 1s. Any of the 26 out of these 32 combinations can be used to represent the alphabet. Similarly, a four-bit code is needed to represent the 10 decimal digits. Any 10 of the 16 combinations possible can be used to represent the decimal digits. (We will examine some possibilities later in this section.)

The codes designed to represent only numeric data (i.e., decimal digits 0 through 9) can be classified into two categories: *weighted and non-weighted*. The *alphanumeric* codes can represent both alphabetic and numeric data. A third class of codes is designed for *error detection and correction* purposes.

1.7.1 Weighted Codes

As stated, we will need at least four bits to represent the 10 decimal digits. But with four bits it is possible to represent 16 elements. Since only 10 of the 16 possible combinations are used, numerous distinct codes are possible. Table 1.8 shows some of these possibilities. Note that all of these codes are weighted codes, since each bit position in the code has a weight associated with it. The sum of weights corresponding to each nonzero bit in the code is the decimal digit represented by it.

Note that each word in the (8 4 2 1) code in the table is a binary number whose decimal equivalent is the decimal digit it represents. This is a very commonly used code and is known as the Binary Coded Decimal (BCD) code. When a BCD-encoded number is used in arithmetic operations, each decimal digit is represented by four bits. For example, $(567)_{10}$ is represented in BCD as

TABLE **1.8**   Some Weighted Codes

Digit	Weights	8 4 2 1 code	2 4 2 1 code	6 4 2 −3 code
0		0 0 0 0	0 0 0 0	0 0 0 0
1		0 0 0 1	0 0 0 1	0 1 0 1
2		0 0 1 0	0 0 1 0	0 0 1 0
3		0 0 1 1	0 0 1 1	1 0 0 1
4		0 1 0 0	0 1 0 0	0 1 0 0
5		0 1 0 1	1 0 1 1	1 0 1 1
6		0 1 1 0	1 1 0 0	0 1 1 0
7		0 1 1 1	1 1 0 1	1 1 0 1
8		1 0 0 0	1 1 1 0	1 0 1 0
9		1 0 0 1	1 1 1 1	1 1 1 1

Note: The (8 4 2 1) code is the popular binary coded decimal (BCD) code.

$$5 \quad 6 \quad 7$$
$$(0101\ 0110\ 0111)_{BCD}$$

During the arithmetic, each four-bit unit is treated as a digit and the arithmetic is performed on a digit-by-digit basis, as shown in Example 1.42.

Example 1.42

Decimal	BCD
532	0101 0011 0010
+ 126	0001 0010 0110
658	0110 0101 1000 Binary arithmetic on each four-bit unit.

When the sum of two digits is greater than 9, the resulting four-bit pattern is not a valid code word in BCD. In such cases, a correlation factor of 6 is added to that digit, to derive the valid code word.

Example 1.43

Decimal	BCD
532	0101 0011 0010
+ 268	0010 0110 1000
	0111 1001 1010
	+0110 correction
	0111 0101 0000
	+0110 correction
	1000 0000 0000

Here the LSD of the sum exceeds 9. The correction of that digit results in the next significant digit exceeding 9. The correction of that digit yields the final correct sum.

It is important to understand the difference between binary and BCD representations of numbers. In BCD, each decimal digit is represented by the corresponding four-bit code word. In binary, the complete number is converted into binary pattern with the appropriate number of bits. For example, the binary representation of $(567)_{10}$ is (1000110111), whereas its BCD representation is (0101 0110 0111).

In the 2 4 2 1 code shown in Table 1.8, decimal 6 is represented by 1100. Another valid representation of 6 in this code is 0110. We have chosen the combinations in Table 1.8 to make this code *self-complementing*. A code is said to be self-complementing if the code word of the nines complement of N (i.e., 9-N) can be obtained by complementing each bit of the code word for N. This property of the code makes taking the complement easy to implement in digital hardware. A necessary condition for a code to be self-complementing is that the sum of weight of the code is 9. Thus, both the (2 4 2 1) and (6 4 2 −3) codes are self-complementing, while BCD is not.

1.7.2 Nonweighted Codes

Table 1.9 shows two popular codes. They do not have any weight associated with each bit of the code word.

Excess-3 is a four-bit self-complementing code. The code for each decimal digit is obtained by adding 3 to the corresponding BCD code word. Excess-3 code has been used in some older computer systems. In addition to being self-complementing, thereby making subtraction easier, this code enables simpler arithmetic hardware. This code is included here for completeness and is no longer commonly used.

The *gray code* is a four-bit code in which the 16 code words are assigned so that there is a change in only one bit position as we move from one code word to the subsequent code word. Such codes are called *cyclic codes*. Because only one bit changes from code word to code word, it is easier to detect an error if there is a change in more than one bit. For example, consider the case of a shaft position indicator. Assume that the shaft position is divided into 16 sectors indicated by the gray code. As the shaft rotates, the code words change. If at any time there is a change in two bits of the code word compared with the previous one, there is an error. Several cyclic codes have been devised and are commonly used.

TABLE 1.9   Nonweighted Codes

Decimal	Excess-3 code	Gray code
0	0011	0000
1	0100	0001
2	0101	0011
3	0110	0010
4	0111	0110
5	1000	0111
6	1001	0101
7	1010	0100
8	1011	1100
9	1100	1101
10	Nd	1111
11	Nd	1110
12	Nd	1010
13	Nd	1011
14	Nd	1001
15	Nd	1000

Nd: Not defined.

1.7.3   Error Detection Codes

Errors occur during digital data transmission as a result of the external noise introduced by the medium of transmission. For example, if a digital system uses BCD code for data representation and if an error occurs in the LSB position of the data 0010, the resulting data will be 0011. Because 0011 is a valid code word, the receiving device assumes that the data are not in error. To guard against such erroneous interpretations of data, several error detection and correction schemes have been devised. As the names imply, an error detection scheme simply detects that an error has occurred, while an error correction scheme corrects the errors. We will describe a simple error detection scheme using *parity checking* here. For information on more elaborate schemes, such as Cyclic Redundancy Check (CRC), Check sums, and XModem protocols, see the books by Kohavi and Ercegovac and Lang listed in the reference section at the end of the chapter.

In the simple *parity check* error detection scheme, an extra bit (known as a *parity bit*) is included in the code word. The parity bit is set to 1 or 0, depending on the number of 1s in the original code word, to make the total number of 1s even (in an *even parity* scheme) or odd (in an *odd parity* scheme). The sending device sets the parity bit. The receiving device checks the incoming data for parity. If the system is using an even parity scheme, an error is detected if the receiver detects an odd number of 1s in the incoming data (and vice versa). A parity bit can be included in the code words of each of the codes described above.

Table 1.10 shows two error detection codes. The first code is the even parity BCD. The fifth bit is the parity bit, and it is set for even parity. The second code is known as the 2-out-of-5 code. In this code, five bits are used to represent each decimal digit. Two and only two bits out of five are 1s. Out of the 32 combinations possible using five bits, only 10 are utilized to form this code. This is also an even parity code. If an error occurs in transmission, the even parity is lost and detected by the receiver.

In this simple parity scheme, if two bits are in error, the even parity is maintained and we will not be able to detect that an error has occurred. If the occurrence of more than one error is anticipated, more elaborate parity-checking schemes using more than one parity bit are used. In fact, it is possible to devise codes that not only detect errors but also correct them, by including enough parity bits. For example, if a block of words is being transmitted, each word might include a parity bit and the last word in the block might be a parity word each bit of which checks for an error in the corresponding bit position of each word in the block. (See Problem 1.20.) This scheme is usually referred to as *cross-parity checking* or *vertical and horizontal redundancy check* and is a coding scheme that detects and corrects single errors. R. W. Hamming invented a single-error detecting/correcting scheme using a distance-3 code. That is, any code word in this scheme differs from other code words in at least 3 bit positions.

1.7.4 Alphanumeric Codes

If alphabetic characters, numeric digits, and special characters are used to represent information processed by the digital system, an alphanumeric code is needed. Two popular alphanumeric codes are Extended BCD Interchange Code (EBCDIC) and American Standard Code for Information Interchange (ASCII). Table 1.11 shows these codes. ASCII is more commonly used, and EBCDIC is used primarily in large IBM computer systems. Both EBCDIC and ASCII are eight-bit codes and hence can represent up to 256 elements.

In a general computer system, each and every component of the system need not use the same code for data representation. For example, in the simple calculator system shown in Figure 1.4, the keyboard produces ASCII-coded characters corresponding to each keystroke. These ASCII-coded numeric data are then converted by the processor into BCD for processing. The processed data is then reconverted into ASCII for the printer. Such code conversion is common, particularly when devices from various vendors are integrated to form a system.

TABLE 1.10 Error Detection Codes

Decimal	Even parity BCD	2-out-of-5
0	0000 0←parity	11000
1	0001 1 bit	00011 No specific
2	0010 1	00101 parity bit
3	0011 0	00110
4	0100 1	01001
5	0101 0	01010
6	0110 0	01100
7	0111 1	10001
8	1000 1	10010
9	1001 0	10100

1.8 Data Storage and Register Transfer

Let us now examine the operation of a digital computer in more detail, to better understand the data representation and manipulation schemes. The binary information is stored in digital systems, in devices such as flip-flops and magnetic cores. (Magnetic cores are now obsolete.) We will call such storage devices *storage cells*. Each cell can store one bit of data. The *content* (or state) of the cell can be changed from 1 to 0 or 0 to 1 by the signals on its inputs, whereas the content of the cell is determined by sensing its outputs. A collection of storage cells is called a *register*. An n-bit register can thus store n-bit data. The number of bits in the most often manipulated data unit in the system determines the *word* size of the system. That is, a 16-bit computer system manipulates 16-bit numbers most often and its word size is 16 bits. Computer systems with 8-, 16-, and 32-bit words are common. Other machines have word sizes of 36, 64, and larger number of bits. An 8-bit unit of data is commonly called a *byte*, and a four-bit unit is a *nibble*. Once the word size of the machine is defined, *half-word* and *double-word* designations are also used to designate data with half or twice the number of bits in the word.

A digital system manipulates the data through a set of *register transfer* operations. A register transfer is the operation of moving the contents of one register (i.e., the source) to another register (i.e., the destination). The source register contents remain unchanged after the register transfer, while the contents of the destination register are replaced by those of the source register.

Let us now examine the set of register transfer operations needed to bring about the addition of two numbers in a digital computer. The memory unit of the digital computer is composed of several *words*, each of which

TABLE 1.11 Alphanumeric Codes

Character	EBCDIC code	ASCII code
blank	0100 0000	0010 0000
.	0100 1011	0010 1110
(	0100 1101	0010 1000
+	0100 1110	0010 1011
$	0101 1011	0010 0100
*	0101 1100	0010 1010
)	0101 1101	0010 1001
−	0110 0000	0010 1101
/	0110 0001	0010 1111
;	0110 1011	0010 0111
'	0111 1101	0010 1100
=	0111 1110	0011 1101
A	1100 0001	0100 0001
B	1100 0010	0100 0010
C	1100 0011	0100 0011
D	1100 0100	0100 0100
E	1100 0101	0100 0101
F	1100 0110	0100 0110
G	1100 0111	0100 0111
H	1100 1000	0100 1000
I	1100 1001	0100 1001
J	1101 0001	0100 1010
K	1101 0010	0100 1011
L	1101 0011	0100 1100
M	1101 0100	0100 1101
N	1101 0101	0100 1110
O	1101 0110	0100 1111
P	1101 0111	0101 0000
Q	1101 1000	0101 0001
R	1101 1001	0101 0010
S	1110 0010	0101 0011
T	1110 0011	0101 0100
U	1110 0100	0101 0101
V	1110 0101	0101 0110
W	1110 0110	0101 0111
X	1110 0111	0101 1000
Y	1110 1000	0101 1001
Z	1110 1001	0101 1010
0	1111 0000	0011 0000
1	1111 0001	0011 0001
2	1111 0010	0011 0010
3	1111 0011	0011 0011
4	1111 0100	0011 0100
5	1111 0101	0011 0101
6	1111 0110	0011 0110
7	1111 0111	0011 0111
8	1111 1000	0011 1000
9	1111 1001	0011 1001

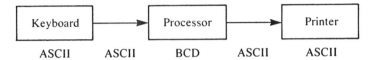

Figure 1.4 Coding in a simple calculator system.

can be viewed as a register. Some of the memory words contain data, and others contain instructions for manipulating the data. Each memory word has an *address* associated with it. In Figure 1.5, we have assumed that the word size is 16 bits and the memory has 30 words. The program is stored in memory locations 0 through 10, and only two data words at addresses 20

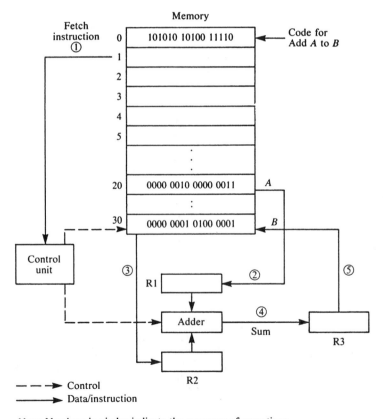

Note: Numbers in circles indicate the sequence of operations.

Figure 1.5 Register transfers for an ADD operation.

and 30 are shown. Word 0 contains the instruction ADD A TO B, where A and B are the operands stored in locations 20 and 30, respectively. The instruction is coded in binary, with the 6 MSBs representing the add operation and the remaining 10 bits used to address the two operands, 5 bits for each operand address.

The control unit fetches the instruction ADD A TO B from the memory word 0. This fetch operand is a register transfer. The instruction is analyzed by the control unit to decode the operation called for and the operand address required. The control unit then sends a series of control signals to the processing unit to bring about the set of register transfers needed.

We will assume that the processing unit has two operand registers, R1 and R2, and a results register, R3. The adder adds the contents of R1, and R2 and stores the result in R3.

To carry out the ADD A TO B instruction, the control unit

1. Transfers the contents of memory word 20 (operand A) to R1,
2. Transfers the contents of memory word 30 (operand B) to R2,
3. Commands the adder to add R1 and R2 and sends the result to R3,
4. Transfers the contents of R3 to memory word 30 (operand B).

This description is obviously very much simplified compared with the actual operations that take place in a digital computer. Nevertheless, it illustrates the data flow capabilities needed in a digital system.

As seen from this example, the binary data stored in a register can be interpreted in various ways. The context in which the register content is examined determines the meaning of the bit pattern stored. For example, if the register is part of the control unit, and if it is accessed during the instruction fetch phase, its contents are interpreted as operation codes. However, if the contents of a register in the processing unit are accessed during the data fetch phase of the instruction execution, they are interpreted as binary numbers. Figure 1.6 shows a 16-bit register and three interpretations of its contents. Note that the contents of this register have no meaning if they are to be considered as BCD digits, since 1010 is not a valid BCD digit.

The data transfer and manipulative capabilities of the digital system are brought about by digital *logic circuits*. In later chapters, we will discuss the analysis and design of the logic circuits and memory subsystems that form the components of a digital system.

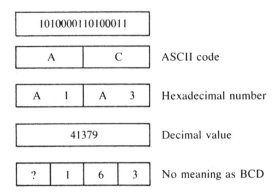

Figure 1.6   Data interpretations.

1.9 Summary

The topics covered in this chapter form the basis for all the data representation schemes that are used in digital systems. We have presented the most common number systems and conversion procedures. The most basic arithmetic schemes in these number systems and popular representation schemes have been examined. Various binary codes encountered in digital systems have been discussed. We have also had a brief introduction to the operation of a digital computer based on the register transfer concept. The floating-point representation and associated accuracy problems have been introduced.

References

Chu, Y. *Computer Organization and Programming*, Englewood Cliffs, NJ: Prentice-Hall, 1972.

Ercegovac, M. D. and Lang, T. *Digital Systems and Hardware/Firmware Algorithms*, New York: John Wiley, 1985.

Hamming, R. W. "Error Detecting and Correcting Codes," *Bell System Technical Journal*, 29 (April 1950):147–160.

Hwang, K. *Computer Arithmetic*, New York: John Wiley, 1979.

Kohavi, Z. *Switching and Automata Theory*, New York: McGraw-Hill, 1970.

Shiva, S. G. *Computer Design and Architecture*, Glenview, IL: HarperCollins, 1991.

Problems

1.1   Convert the following decimal numbers to base 3, base 5, base 8, and base 16: 245, 461, 76.5, 46.45, 231.78, 1023.25.

1.2   List the first 20 decimal numbers in base 7 and base 9.

1.3   Assume that your car's odometer shows the mileage in Octal. If the current reading of the odometer is 24516, how many miles (in decimals) has the car been driven? What will be the new reading if the car is driven 23 (decimal) miles?

1.4   What will be the current odometer reading in the problem above, if the odometer uses a 5 digit Hexadecimal representation? What will be the new reading after driving 23 (decimal) miles?

1.5   Find the ones and twos complements of the following binary numbers

(a) 10010 (b) 110010 (c) 0010101
(d) 10110.0101 (e) 1101.1100 (f) 111010.0011
(g) 1001.0001 (h) 110100.0100 (i) 1010110.111

1.6   Find the nines and tens complement of the following decimal numbers:

(a) 465 (b) 09867 (c) 42678
(d) 8976 (e) 423.76 (f) 561.876
(g) 463.90 (h) 1786.967 (i) 12356.078

1.7   Given the binary numbers below, determine $X + Y$, $X - Y$, $X \times Y$, and X/Y:

(a) $X = 1101010$ (b) $X = 101101$ (c) $X = 1001$
    $Y = 10111$ $Y = 1111$ $Y = 1111$

(d) $X = 110.11$ (e) $X = 1110.101$ (f) $X = 1011.00$
    $Y = 10.11$ $Y = 1011.10$ $Y = 1100$

1.8   Given the octal numbers below, determine $X + Y$, $X - Y$, $X \times Y$, and X/Y:

(a) $X = 533$ (b) $X = 46537$ (c) $X = 26$
    $Y = 234$ $Y = 234$ $Y = 533$

(d) $X = 123.2$ (e) $X = 234.6$
    $Y = 234$ $Y = 156.7$

1.9   Given the hexadecimal numbers below, determine $X + Y$, $X - Y$, $X \times Y$, and X/Y:

(a) $X = 1CF$ (b) $X = 1B59A$ (c) $X = B6$
    $Y = B6$ $Y = C23$ $Y = 1CF$

(d) $X = 2ECD$ (e) $X = 234F.16$
    $Y = 4321$ $Y = 456E$

1.10   Perform the following conversions:

(a) $(234)_{10} = (?)_2$ (b) $(3345)_6 = (?)_2$
(c) $(875)_9 = (?)_{11}$ (d) $(0.3212)_4 = (?)_{10}$
(e) $(87.35)_9 = (?)_{11}$

1.11   Perform the following conversions using base 2^k conversion technique:

(a) $(10110100.00101)_2 = (?)_4$
(b) $(AB143)_{16} = (?)_8$
(c) $(2347.45)_8 = (?)_{16}$
(d) $(110111110.010000011)_2 = (?)_{16}$

1.12   Following the conversion technique of the preceding problem, convert $(2574)_9$ to base 3.

1.13   If $(130)_X = (28)_{10}$, find the value of X (X is a positive integer).

1.14   Perform the following operations:

(a) $11101 + 1111 + 1011$
(b) $111000 - 10101$
(c) $11001101/101$
(d) 11010×11001

1.15   Use (a) twos complement and (b) ones complement arithmetic to perform the following operations:

$$1011010 - 10101$$
$$10101 - 1011010$$

1.16   Use (a) nines complement and (b) tens complement arithmetic to perform the following operations:

$1875 - 924$
$924 - 1875$

1.17   Use an eight-bit twos complement representation (with one sign bit and seven magnitude bits) to perform the following operations:

(a) $113 - 87$ (b) $87 - 113$ (c) $43 + 26$ (d) $96 - 22$
(e) $46 - 77$

1.18   Perform the following operations:

(a) $(7256)_8 \times (23)_8 = (?)_8$
(b) $(56)_8 \times_8 (AF)_{16} = (?)_4$ (base 8 multiplication)

1.19   Represent the following numbers in the IEEE standard floating-point format:

(a) $(11010.010)_2$ (b) $(432.26)_{10}$
(c) $-(10100111.1001)_2$ (d) $-(236.77)_{10}$

1.20 The following four code words were transmitted. The LSB of each code word (row) is a parity bit, and odd parity is used. The last word is a parity word across all the earlier words so that even parity is adopted in each bit position (column). Determine whether an error has occurred. If there is an error, correct the bit in error:

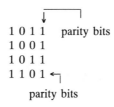

```
1 0 1 1    parity bits
1 0 0 1
1 0 1 1
1 1 0 1
```

parity bits

1.21 Determine the code words for each of the ten decimal digits in the weighted code (6 3 −1 1). The code should be self-complementing.

1.22 Design a four-bit code to represent each of the six digits of the base 6 number system. Make the code self-complementing by selecting appropriate weights.

1.23 A 16-bit register contains the following:

0100100101010111

Interpret the contents as:

(a) a BCD number   (b) a binary number
(c) an excess-3 number (d) two ASCII characters.

1.24 Represent the following in a 16-bit register:

(a) $(356)_{10}$ (b) $(356)_{BCD}$ (c) $(A1)_{ASCII}$

1.25 Express $(746)_{10}$ in (a) BCD, (b) binary, and (c) ASCII, using the minimum number of bits in each case.

1.26 Convert the following IEEE standard floating-point representation into binary:

(a) 0 10000010 10000010 . . . 000
(b) 1 01111000 01000110 . . . 000

1.27 Just as in BCD arithmetic, when two excess-3 digits are added, a correction is needed. Determine the correction factor.

2

Boolean Algebra

2.1 Logic Circuits

A digital system typically consists of several subsystems. For example, the digital computer of Chapter 1 consists of five subsystems. Each subsystem is a hardware component consisting of several *logic circuits*. A logic circuit is an interconnection of several logic devices designed to perform a desired function. It has one or more inputs and one or more outputs. The logic devices used in building circuits are called *gates*. There are two types of logic circuits: *combinational* and *sequential*. The outputs of a combinational circuit at any time are each a function of combination of inputs at that time, without regard to the previous combination of inputs. That is, a combinational circuit does not possess any memory. A circuit with memory is called a sequential circuit. The output of a sequential circuit at any time is a function of not only the input at that time but also the *state* of the circuit at that time. The state of the circuit depends on what has happened to the circuit previously; hence, the state is also a function of the previous inputs and states. The sequential circuits are built out of gates and memory elements such as flip-flops, delay lines, and so forth.

The analysis and design of logic circuits are guided by Boolean algebra. In this section, we will relate Boolean algebraic concepts to logic circuits in an informal manner and then introduce Boolean algebra more formally in subsequent sections. We will also examine the gates commonly used in building combinational logic circuits and show the utility of Boolean algebra in the analysis and design of combinational circuits. A detailed description of combinational circuits will be undertaken in Chapter 4, and flip-flops and sequential circuits will be discussed in Chapter 6. The following examples relate Boolean algebra to logic circuits.

Example 2.1

Consider the addition of two bits, producing a SUM bit and a CARRY bit, as shown in the following table:

		A plus B	
A	B	SUM	CARRY
0	0	0	0
0	1	1	0
1	0	1	0
1	1	0	1

A and B are binary variables that can each take a value of either 0 or 1. The first two columns list the four combinations of values that are possible for the two variables, and the two columns at right show the result of the addition.

Note that the CARRY is 1 only when A is 1 *AND* B is 1, while the SUM is 1 when *either* of the following two conditions is satisfied: (1) A is 0 AND B is 1, (2) A is 1 AND B is 0. That is, SUM is 1 if: (A is 0 AND B is 1) OR (A is 1 AND B is 0), where the operator *OR* is the equivalent of "either . . . or" and the operator *AND* corresponds to "both." Let us say, A' (pronounced "A bar," and meaning "not A") represents the *COMPLEMENT* of A. That is, A is 0 if A' is 1, and vice versa. Similarly, B' represents the complement of B. We can then say that SUM is 1 if (A' is 1 and B is 1) OR (A is 1 AND B' is 1). Therefore,

$$\text{SUM} = (A' \cdot B) + (A \cdot B')$$
$$\text{CARRY} = A \cdot B \tag{2.1}$$

where "+" represents the OR operation (not arithmetic addition); "·" represents AND, and "′" represents NOT. A and B are binary variables. Each can take a value of either 0 or 1. SUM and CARRY are functions of the variables A and B. Together these functions express

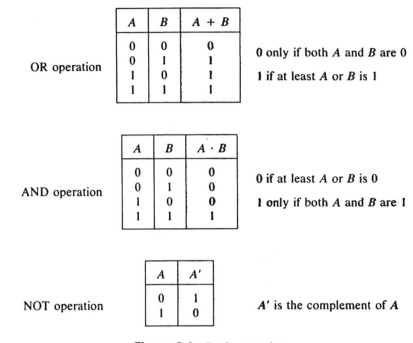

OR operation

A	B	A + B
0	0	0
0	1	1
1	0	1
1	1	1

0 only if both A and B are 0

1 if at least A or B is 1

AND operation

A	B	A · B
0	0	0
0	1	0
1	0	0
1	1	1

0 if at least A or B is 0

1 only if both A and B are 1

NOT operation

A	A'
0	1
1	0

A' is the complement of A

Figure 2.1   Basic operations.

the operation of a combinational circuit called "adder" with two inputs (A and B) and two outputs (SUM and CARRY). The details of the circuit are given in the next subsection.

The AND operation yields a value of 1 only when both the operands have a value of 1; the OR operation yields 1 when either of its operands is a 1 and the NOT operation is a unary operator that complements the value of the operand. Figure 2.1 shows the definitions of these Boolean operators.

Example 2.2 illustrates the conversion of word definition of the function to be performed by a circuit into Boolean functions.

Example 2.2

Consider the following statement:

Subtract if and only if an add instruction is given and the signs are different or a subtract instruction is given and the signs are alike.

Let

S = the subtract action
A = "Add instruction given" condition
B = "Signs are different" condition
C = "Subtract instruction given" condition

The preceding statement can be expressed as

$$S = (A \cdot B) + (C \cdot B') \tag{2.2}$$

Usually the "·" and the parentheses are removed from the expressions when there is no ambiguity. Thus, this function can be written as

$$S = AB + CB' \tag{2.3}$$

As shown by these examples, a Boolean function can be derived from the description of the processing function to be performed by a logic module. We will now examine certain primitive logic devices to illustrate the design of logic circuits from the Boolean function description.

2.1.1 Signals and Gates

The binary variables A, B, SUM, and CARRY in functions (2.1) are each represented in hardware by either two distinct voltage levels or two distinct current magnitudes, as mentioned earlier. Figure 2.2 shows a typical representation. Here, logic-0 corresponds to 0.2 volt, while logic-1 corresponds to +3.5 volts. In practice, logic devices accept a range of voltage levels, rather than a fixed voltage value for each of the two logic values (or states). The output from these devices will be either a 0 or a 1, which in turn corresponds to the two ranges of voltages. The voltages shown in Figure 2.2 are the ranges accepted by a popular logic family known as Transistor-

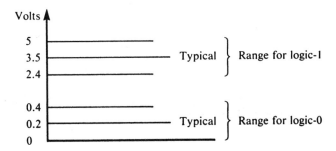

Figure 2.2 Binary signal.

Transistor Logic (TTL). Here, logic 0 corresponds to the range of 0 to 0.4 volt, with the typical value of 0.2 volt, and logic 1 corresponds to 2.4 to 5 volts, with the typical value of 3.5 volts. This assignment of a higher voltage level to logic-1 and a lower voltage level to logic-0 is called *positive logic*. The opposite assignment known as *negative logic* is also used. The choice is usually arbitrary. These logic assignments will be described in more detail in Section 2.9.

The AND, OR, and NOT operators are implemented by logic devices known as gates. Figure 2.3 shows the symbols used to represent such gates. For example, the AND gate accepts the binary signals on its inputs A and B and produces the signal F as its output. The value of F is the AND of the values of A and B, as shown in Figure 2.3. Similarly, the OR gate implements the OR operator and the NOT gate (or the "inverter") complements its input signal. The technology and the fabrication details of these gates are described in Chapter 10.

Figure 2.4 shows the implementation of the SUM and CARRY functions of equations (2.1) using these logic gates. As we trace these implementations (i.e., logic circuits) from inputs to their outputs, note that the signals A and B are first complemented (inverted) to form A' and B', respectively. A and B' are then fed into an AND gate to form $(A \cdot B')$. Similarly, A' and B are fed into an AND gate to form $(A' \cdot B)$. The outputs of the two

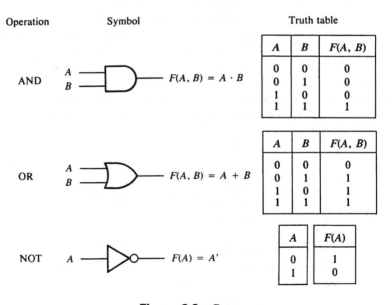

Figure 2.3   Gates.

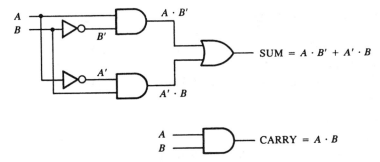

Figure 2.4 Circuits to generate SUM and CARRY.

AND gates then form the inputs to an OR gate, whose output is the SUM. The CARRY is implemented by an AND gate. In these circuits, the connection of two wires is denoted by a dot (·) at their junction. The function of these circuits is to switch the output from one logic value to the other, based on their inputs. Hence, these circuits are also called *switching circuits*. The outputs of these circuits are functions of their inputs. This functionality can be expressed easily in terms of Boolean functions. The implementation of function (2.3) is left as an exercise.

The next section will offer a formal definition of Boolean algebra and the rest of the chapter will describe the terminology and notation useful in digital design. Because of the advances in digital hardware technology, it is now possible to design logic circuits from off-the-shelf components, without formal understanding of Boolean algebra. Nevertheless, Boolean algebra is a useful tool, and familiarity with it makes one a better designer of logic circuits.

2.2 Boolean Algebra

In 1854 George Boole introduced a symbolic notation to deal with logic statements that take a value of "true" or "false." This notation has come to be known as *Boolean algebra*. Claude Shannon adopted Boolean algebra to the analysis of two-valued switching functions. This two-valued Boolean algebra is now called *switching algebra*. Boolean algebra, like any other algebraic system, can be described with a set of elements, a set of operators, and a set of axioms or postulates. These postulates are the assumptions from which the other rules and theorems that govern the algebra are deduced. To define Boolean algebra, we will use a set of postulates developed by E. Huntington in 1904.

Definition: A Boolean algebra is a closed algebraic system containing a set of two or more elements and two binary operators—"+" (OR) and "·" (AND). That is, for every X and Y in set K, $X + Y$ belongs to K and $X \cdot Y$ belongs to K. This is the "closure" postulate of Huntington. In addition to closure, the following postulates must be satisfied by a Boolean algebra.

Postulates

P1. Existence of an identity element 0 with respect to the operator "+," and of an identity element 1 with respect to the operator "·".

(a) $X + 0 = X$
(b) $X \cdot 1 = X$

P2. Commutativity of the operators "+" and "·".

(a) $X + Y = Y + X$
(b) $X \cdot Y = Y \cdot X$

P3. Associativity of the operators "+" and "·".

(a) $X + (Y + Z) = (X + Y) + Z$
(b) $X \cdot (Y \cdot Z) = (X \cdot Y) \cdot Z$

The associative law was not part of the Huntington postulates, but it is included here for convenience. It can be deduced from the other postulates.

P4. Distributivity of:

(a) "+" over "·": $X + (Y \cdot Z) = (X + Y) \cdot (X + Z)$
(b) "·" over "+": $X \cdot (Y + Z) = (X \cdot Y) + (X \cdot Z)$

P5. Existence of the complement X' for every element X in K, so that:

(a) $X + X' = 1$
(b) $X \cdot X' = 0$

2.3 Two-Valued Boolean Algebra

The previous section examined a general Boolean algebraic system. In this book, we will deal only with the application of Boolean algebra to logic circuits that manipulate two-valued logic variables. We will now verify that the set of two elements $K = (0, 1)$, along with the operators AND (·) and OR (+) as defined in Figure 2.1, form a two-valued Boolean algebra.

Closure is obvious since the resulting value of the operations is always 1 or 0. There are two elements 1 and 0 such that $1 \neq 0$, thereby satisfying postulate 1, since (a) $0 + 0 = 0$, $1 + 0 = 1$ and (b) $0.0 = 0$, $1.0 = 0$. The operators are commutative as seen by the symmetry of the tables in Figure 2.1. Associativity and distributivity are verified by Table 2.1. In Table 2.1(a) the first column lists all the possible combinations of values for X, Y, and

TABLE 2.1 Associativity and Distributivity

(a) Associativity

$X\,Y\,Z$	$Y + Z$	$X + (Y + Z)$	$X + Y$	$(X + Y) + Z$
0 0 0	0	0	0	0
0 0 1	1	1	0	1
0 1 0	1	1	1	1
0 1 1	1	1	1	1
1 0 0	0	1	1	1
1 0 1	1	1	1	1
1 1 0	1	1	1	1
1 1 1	1	1	1	1

(b) Distributivity

$X\,Y\,Z$	$Y{\cdot}Z$	$X + (Y{\cdot}Z)$	$X + Y$	$X + Z$	$(X + Y)(X + Z)$
0 0 0	0	0	0	0	0
0 0 1	0	0	0	1	0
0 1 0	0	0	1	0	0
0 1 1	1	1	1	1	1
1 0 0	0	1	1	1	1
1 0 1	0	1	1	1	1
1 1 0	0	1	1	1	1
1 1 1	1	1	1	1	1

Z. The second column lists the values for $(Y + Z)$, and the third column lists the values for the left-hand side of the associativity postulate P3(a). The last column lists the corresponding values for the right-hand side of P3(a). Since the values of both sides of P3(a) are identical for each combination of values of X, Y, and Z (as shown by this table), the two sides of P3(a) are equal, thereby verifying its validity. The (b) parts of these postulates can similarly be verified.

In Table 2.1(b) the distributive property is verified. These tables are called truth tables. Truth tables are described more formally in the next section. Since $0 + 0' = 0 + 1 = 1$, $1 + 1' = 1 + 0 = 1$ and $0{\cdot}0' = 0{\cdot}1 = 0$, $1{\cdot}1' = 1{\cdot}0 = 0$, $0{\cdot}0' = 0{\cdot}1 = 0$, the complement exists.

Thus, the two-valued algebra consisting of elements 1 and 0 and operators "$\cdot$" and "$+$" is a Boolean algebra. We will now describe various properties of this algebra, ones that are useful in the analysis and design of logic circuits.

2.4 Properties of Boolean Algebra

2.4.1 Operator Hierarchy

The right-hand sides of the Boolean equations in the preceding section are Boolean expressions. An expression is formed by combining operators ($+$, $\cdot$, $'$) with operands that are either Boolean variables or constants (1, 0). Each variable can take a value of 0 or 1. Knowing the value of the component variables of an expression, one can find the value of the expression itself. The hierarchy of operations is important in the evaluation of expressions. We always perform the NOT operation first, followed by AND and then OR, in the absence of parentheses. If there are parentheses, the expressions within them are evaluated first (observing the above hierarchy of operations), and then the remaining expression is evaluated.

Example 2.3

Evaluate the function $Z = AB'C + (A'B)(B + C')$, given $A = 0$, $B = 1$, and $C = 1$.

$$Z = (A \cdot B' \cdot C) + (A' \cdot B) \cdot (B + C') \quad \text{Insert ``$\cdot$''}$$
$$= (0 \cdot 1' \cdot 1) + (0' \cdot 1) \cdot (1 + 1') \quad \text{Substitue values}$$
$$= (0 \cdot 0 \cdot 1) + (1 \cdot 1) \cdot (1 + 0) \quad \text{Evaluate NOT}$$
$$= (0) + (1) \cdot (1) \quad \text{Evaluate expressions}$$
$$\text{within parentheses}$$
$$\text{AND operation}$$
$$= 0 + 1 \quad \text{OR operation}$$
$$= 1 \quad \text{(value of } Z \text{ is 1)}$$

2.4.2 Equality of Expressions

Two expressions are said to be *equivalent* if one can be replaced by the other. That is, for two expressions to be equivalent, they should attain identical values for all the combinations of values of their component variables.

2.4.3 Duality

The *dual* of an expression is obtained by replacing each "$+$" in the expression by "$\cdot$", each "$\cdot$" by "$+$", each 1 by 0, and each 0 by 1, and by preserving the existence of all parentheses, whether present or implied.

The *principle of duality* states that if an equation is always valid in Boolean algebra, its dual is also valid.

Example 2.4

Given $X + YZ = (X + Y) \cdot (X + Z)$, its dual is
$$X \cdot (Y + Z) = (X \cdot Y) + (X \cdot Z)$$

Note that part (b) of each of the postulates given earlier is the dual of the corresponding part (a).

2.5 Functions and Their Representations

Boolean functions are represented in various forms. The two popular forms are *truth tables* and *Venn diagrams*. Truth tables represent functions in a tabular form, while Venn diagrams provide a graphic representation. In addition, there are two algebraic representations known as the *standard* (or *normal*) *form* and the *canonical form*.

2.5.1 Truth Tables

In Figure 2.1 we saw the truth tables for the three primitive operations AND, OR, and NOT. In these tables, one column corresponds to each component variable of the function and one column corresponds to the value of the function itself. Because each component variable can take either of the two values 0 and 1 and there can be 2^n combinations of values for a set of n variables, there will be 2^n rows and $n + 1$ columns in a truth table for a function with n component variables. If the expression on the right-hand side of the function is complex, the truth table usually is developed in several steps, as illustrated by Example 2.5.

Example 2.5

Draw the truth table for $Z = AB' + A'C + A'B'C$.

There are three component variables: A, B, and C. Hence, there will be $2^3 = 8$ combinations of values. These eight combinations are shown in the first three columns of Table 2.2. These combinations correspond to binary numbers 000 through 111, or $(0)_{10}$ through $(7)_{10}$.

To evaluate Z in the example function, knowing the values for A, B, and C at each row of the truth table (Table 2.2), we would first generate values for A' and B' and then generate values for AB', $A'C$ and $A'B'C$ by ANDing the values in the appropriate columns for each row. Finally, we would derive the value of Z by ORing the values in the last three columns for each row. Note that evaluating $A'B'C$ corresponds to ANDing A' and B' values, followed by ANDing the value of C. Similarly, if more than two values are to be ORed, they are ORed

TABLE 2.2   Truth Table for $Z = AB' + A'C + A'B'C$

A	B	C	A'	B'	AB'	$A'C$	$A'B'C$	Z
0	0	0	1	1	0	0	0	0
0	0	1	1	1	0	1	1	1
0	1	0	1	0	0	0	0	0
0	1	1	1	0	0	1	0	1
1	0	0	0	1	1	0	0	1
1	0	1	0	1	1	0	0	1
1	1	0	0	0	0	0	0	0
1	1	1	0	0	0	0	0	0

two at a time. The columns corresponding to A', B', AB', $A'C$, and $A'B'C$ are not usually shown in the final truth table.

2.5.2   Venn Diagrams

Venn diagrams provide a graphic representation of functions. In a Venn diagram, each variable is represented by a circle and a rectangle encloses all the circles. The circles may overlap. Figure 2.5(a) shows a Venn diagram with a single variable A. The area inside the circle "belongs to" A and the area outside the circle "does not belong to" A. That is, A is 1 inside the circle and 0 outside the circle. The fact that A is 1 is shown by the shading inside the circle. Figure 2.5(b) shows the Venn diagram for A'. In Figure 2.5(c), the two circles corresponding to variables A and B overlap. $A \cdot B$ corresponds to the area that is common to both variables (i.e., the *intersection* of A and B). $A + B$ corresponds to the area that is either in A or in B, and hence is the *union* of the two areas, as shown in Figure 2.5(d). Similarly, the area belonging to A and B, but not to C, is shown shaded in Figure 2.5(e). This area corresponds to the term ABC'. Boolean constants 1 and 0 are shown in parts (f) and (g), respectively.

Figure 2.6 shows the representation of a three-variable function on a Venn diagram. The areas corresponding to A and B' intersect to give AB'. Similarly, the areas corresponding to B and C' intersect to give BC'. The union of the two diagrams is the diagram for $AB' + BC'$.

All the combinations of values for component variables of a function that can be shown in a truth table can also be shown in Venn diagrams. Figure 2.7 shows such combinations for two- and three-variable functions. Thus Venn diagrams are equivalent to truth-table representation and can be used, for instance, to evaluate an expression when the value of the com-

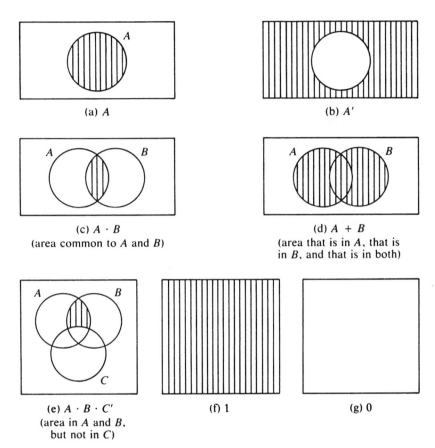

Figure 2.5   Some operations on Venn diagrams.

ponent variables is known or to prove the equality of two expressions (i.e., the two shaded areas must be identical).

2.5.3   Algebraic Representations

We will now define various terms leading to the definition of the two popular algebraic representations of Boolean functions.

A *product term* is obtained by ANDing (i.e., conjunction of) two or more variables. For example, $AB'C$, XYZ', and $P'QRS'$ are all product terms.

A *sum term* is obtained by ORing (i.e., disjunction of) two or more variables. For example, $X + Y'$, $A + B' + C'$, and $P' + Q' + R' + S$ are all sum terms.

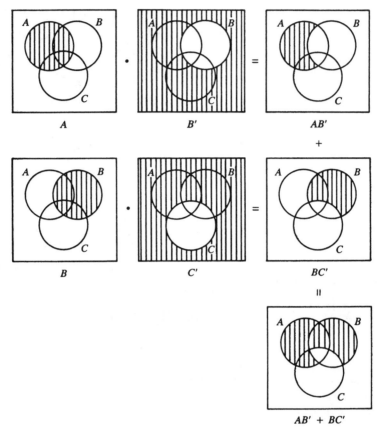

Figure 2.6   Representation of $AB' + BC'$ by Venn diagrams.

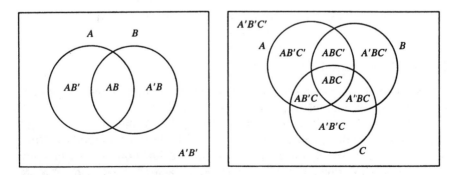

Figure 2.7   Venn diagram designating all possible combinations of variables.

Each occurrence of a variable in either the true form or the complemented form is called a *literal*. The product term $XY'Z$ has three literals; the sum term $A' + B + C' + D$ has four literals.

A product term X is said to be *included* in another product term Y if Y has each literal that is in X. Similarly, a sum term X is included in another sum term Y if Y has each literal that is in X. For example,

XY' is included in $XY'Z$

XY' is included in XY'

$X' + Y$ is included in $X' + Y + Z$

XY' is not included in XY (Why?)

A Boolean function can be expressed either as a sum of several product terms (known as the *sum of products* form) or as a product of several sum terms (known as the *product of sums* form). For example,

$$Q(A, B, C) = AB' + A'C + B'C$$

is in the sum of products form.

$$P(X, Y, Z) = (X + Y')(X' + Y' + Z)$$

is in the product of sums form.

The variables of the functions are explicitly listed on the left-hand sides of the functions above. Such listing is not necessary if we can determine the variables of the functions from the right-hand sides of the functions.

If none of the product terms in the function expressed in the sum of products form are included in the other product terms of the function, the function is said to be in the *normal sum of products* form. For example,

$$Q = AB + AC$$
$$Q = X + Y + Z$$
$$P = AB'C + A'CD + AC'D'$$

are in the normal sum of products form.

Similarly, a function is said to be in the *normal product of sums* form if none of the sum terms of the function are included in the other. For example,

$$P = (X + Y')(X' + Y' + Z')$$
$$Q = (A + B')(A' + B + C')(A + B + C)$$

are in the normal product of sums form. We will abbreviate a normal sum of products form as *SOP form* and a normal product of sums form as *POS form*.

A function in SOP form is said to be in *canonical SOP* form if each of the product terms of the function contains each and every component variable either in complemented or uncomplemented form. That is, if the function is of *n* variables, each product term in the canonical representation will have *n* distinct literals. For example, if *Q* is a function of *A*, *B*, and *C*, then

$$Q = A'B'C + AB'C' + A'B'C'$$

is in canonical SOP form, while

$$Q = A'B + AB'C + A'C'$$

is not, since the first and the last product terms do not contain all three variables.

Similarly, a function in POS form is said to be in *canonical POS* form if each of its sum terms contains all the component variables either in complemented or in uncomplemented form. For example, if *Q* is a function of *A*, *B*, and *C*, then

$$Q = (A' + B + C')(A + B' + C')(A + B + C')$$

is in canonical POS form, while

$$Q = (A' + B)(A + B' + C')(A' + B' + C)$$

is not.

Thus, the only difference between the normal and canonical representations is that in a canonical representation each term contains all the component variables of the function, whereas in the normal form some of the component variables may not be present in one or more terms of the representation.

A truth table can be used to derive the canonical forms for a function, as illustrated by Example 2.6.

Example 2.6

Consider the following truth table for Q, a function of A, B, and C.

A	B	C	Q
0	0	0	0
0	0	1	1
0	1	0	0
0	1	1	1
1	0	0	1
1	0	1	1
1	1	0	0
1	1	1	0

From the truth table it can be seen that Q is 1 when $A = 0$ and $B = 0$ and $C = 1$. That is, Q is 1 when $A' = 1$ and $B' = 1$ and $C = 1$, which means that Q is 1 when $A'B'C$ is 1. Similarly corresponding to other three 1s in the Q column of the table, Q is 1 when either $A'BC$ is 1 or $AB'C'$ is 1 or $AB'C$ is 1. This argument leads to the following representation for Q:

$$Q = A'B'C + A'BC + AB'C' + AB'C$$

which is the canonical SOP form.

In general, to derive the canonical SOP form from the truth table, the following procedure can be used:

1. Generate a product term corresponding to each row in which the value of the function is 1.
2. In each product term, the individual variables are uncomplemented if the value of the variable in that row is 1 and complemented if the value of the variable in that row is 0.

Similarly, the canonical POS form for the function can be derived from the truth table by means of the following procedure:

1. Generate a sum term corresponding to each row in which the value of the function is 0.
2. In each sum term, the individual variables are complemented if the value of the variable in that row is 1 and uncomplemented if the value of the variable in that row is 0.

Thus, the POS form for Q in Example 2.6 is

$$Q = (A + B + C)(A + B' + C)(A' + B' + C)(A' + B' + C')$$

The derivation of canonical forms for another three-variable function P is shown in Example 2.7.

Example 2.7

	A B C	P	Product term (MINTERM)	Sumterm (MAXTERM)
0	0 0 0	1	$A'B'C'$	
1	0 0 1	0		$A + B + C'$
2	0 1 0	0		$A + B' + C$
3	0 1 1	0		$A + B' + C'$
4	1 0 0	1	$AB'C'$	
5	1 0 1	1	$AB'C$	
6	1 1 0	0		$A' + B' + C$
7	1 1 1	0		$A' + B' + C'$

The canonical SOP form is

$$P = A'B'C' + AB'C' + AB'C$$

The canonical POS form is

$$P = (A + B + C')(A + B' + C)(A + B' + C')(A' + B' + C)$$
$$\cdot (A' + B' + C')$$

Note that the SOP form above has 9 literals. Hence, it is less complex than the POS form, which has 15 literals.

Each term in a canonical SOP form is called a *minterm*, and hence the canonical SOP form can be represented as the *sum of minterms*. Similarly, each term in the canonical POS form is a *maxterm*, and the canonical POS form can be expressed as a *product of maxterms*. Minterm and Maxterm are the decimal equivalents of the binary combinations corresponding to function values of 1 and 0, respectively. The minterms of the function in Example 2.7 are 0, 4, and 5, and the maxterms are 1, 2, 3, 6, and 7. The sum of minterms form for the function is

$$P(A, B, C) = \Sigma m\ (0, 4, 5)$$

The product of maxterms form is

$$P(A, B, C) = \Pi M\ (1, 2, 3, 6, 7)$$

These forms are also called minterm and maxterm *list* forms. Here, m and M denote the minterm and maxterm, respectively; Σ denotes the sum and Π denotes the product.

The list forms provide a compact way of representing the function. Note also that for an n variable function,

$$\text{(Number of minterms)} + \text{(number of maxterms)} = 2^n \qquad (2.4)$$

The Venn diagram for the function P in Example 2.7 is:

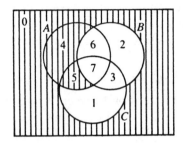

The shaded areas correspond to $P = 1$, and the unshaded areas correspond to $P = 0$. Minterm and maxterm numbers corresponding to all eight areas are also shown.

The following examples illustrate the derivation of one form from the other.

Example 2.8

From the truth table in Example 2.7,

$$P(A, B, C) = A'B'C' + A\ B'C' + AB'C$$

0 0 0	1 0 0	1 0 1 ←Input combinations

(0 for a complemented variable and 1 for an uncomplemented variable)

0   4   5   ←Decimal values

$= \Sigma m\ (0, 4, 5)$ ←Minterm list form

The *minterm list form* is a compact representation for the canonical SOP form. In addition,

$$P(A, B, C) = (A + B + C')(A + B' + C)(A + B' + C')(A' + B' + C)(A' + B' + C')$$
$$\quad\ 0 \quad\ 0 \quad\ 1 \quad 0 \quad\ 1 \quad\ 0 \quad 0 \quad\ 1 \quad\ 1 \quad 1 \quad\ 1 \quad\ 0 \quad 1 \quad\ 1 \quad\ 1$$

↖ Input combinations (1 for a complemented variable and
↖ 0 for an uncomplemented variable)

1   2   3   6   7

↖ Decimal values

$= \Pi M\ (1, 2, 3, 6, 7)$ ←Maxterm list form

The *maxterm list form* is a compact representation for the canonical POS form. If one form is known, the other can be derived, as shown by the following example.

Example 2.9

Given $Q(A, B, C, D) = \Sigma m$ (0, 1, 7, 8, 10, 11, 12, 15).

Q is a four-variable function. Hence, there will be $2^4 = 16$ combinations of input values whose decimal values range from 0 to 15. There are eight minterms. Hence, there should be $16 - 8 = 8$ maxterms; that is,

$$Q(A, B, C, D) = \Pi M \text{ (2, 3, 4, 5, 6, 9, 13, 14)}$$

Figure 2.8 summarizes the relation between the various forms of representation of functions.

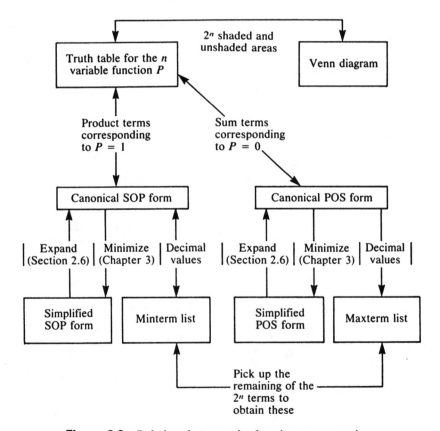

Figure 2.8 Relations between the function representations.

2.6 Theorems

We will now state and prove several theorems of Boolean algebra. These theorems, along with the five postulates, are useful in manipulating switching functions. Some applications of theorems are discussed in the next section. Their application to the minimization of switching functions will be examined in Chapter 3.

Each theorem is listed along with its dual. We will give the algebraic proof of the first part. The dual is proved by the principle of duality. Theorems and postulates used in these algebraic proofs are identified at each step by their numerical designations—for example, T1a, P2b, and so forth. Table 2.3 lists these theorems and postulates for easy reference.

In the following discussion, X, Y, and Z are Boolean variables.

Theorem 1. Idempotency

$$\text{(a) } X + X = X \qquad \text{(b) } X \cdot X = X$$

PROOF:

$$
\begin{aligned}
X \cdot X &= X \cdot X \\
&= X \cdot X + 0 &\quad \text{P1a} \\
&= X \cdot X + X \cdot X' &\quad \text{P5b} \\
&= X \cdot (X + X') &\quad \text{P4b} \\
&= X \cdot 1 &\quad \text{P5a} \\
&= X &\quad \text{P1b}
\end{aligned}
$$

This theorem states that if two sum (product) terms are identical in a SOP (POS) form, one of them can be discarded.

Theorem 2. Properties of 1 and 0

$$\text{(a) } X + 1 = 1 \qquad \text{(b) } X \cdot 0 = 0$$

PROOF:

$$
\begin{aligned}
X + 1 &= X + 1 \\
&= (X + 1) \cdot 1 &\quad \text{P1b} \\
&= 1 \cdot (X + 1) &\quad \text{P2b} \\
&= (X + X') \cdot (X + 1) &\quad \text{P5a} \\
&= X + X' \cdot 1 &\quad \text{P4a} \\
&= X + X' &\quad \text{P1b} \\
&= 1 &\quad \text{P5a}
\end{aligned}
$$

TABLE 2.3 Postulates and Theorems

Postulates

P1. Existence of 0 and 1

(a) $X + 0 = X$

(b) $X \cdot 1 = X$

P2. Commutativity

(a) $X + Y = Y + X$

(b) $X \cdot Y = Y \cdot X$

P3. Associativity

(a) $X + (Y + Z) = (X + Y) + Z$

(b) $X \cdot (Y \cdot Z) = (X \cdot Y) \cdot Z$

P4. Distributivity

(a) $X + (Y \cdot Z) = (X + Y) \cdot (X + Z)$

(b) $X \cdot (Y + Z) = (X \cdot Y) + (X \cdot Z)$

P5. Complement

(a) $X + X' = 1$

(b) $X \cdot X' = 0$

Theorems

T1. Idempotency

(a) $X + X = X$

(b) $X \cdot X = X$

T2. Properties of 1 and 0

(a) $X + 1 = 1$

(b) $X \cdot 0 = 0$

T3. Absorption

(a) $X + XY = X$

(b) $X \cdot (X + Y) = X$

T4. Absorption

(a) $X + X'Y = X + Y$

(b) $X \cdot (X' + Y) = X \cdot Y$

T5. DeMorgan's law

(a) $(X + Y)' = X' \cdot Y'$

(b) $(X \cdot Y)' = X' + Y'$

T6. Consensus

(a) $XY + X'Z + YZ = XY + X'Z$

(b) $(X + Y) \cdot (X' + Z) \cdot (Y + Z)$

$= (X + Y) \cdot (X' + Z)$

T7. Involution

$(X')' = X$

This theorem states that anything ORed with a 1 is always 1 and anything ANDed with a 0 is always 0. We can now collect all the properties of 0 and 1 as follows:

$$X + 0 = X \qquad X \cdot 1 = X$$
$$X \cdot 0 = 0 \qquad X + 1 = 1$$
$$0' = 1 \qquad\quad 1' = 0$$

Theorem 3.   Absorption

(a) $X + XY = X$ \qquad (b) $X \cdot (X + Y) = X$

PROOF:

$$X + XY = X + XY$$
$$= X \cdot 1 + XY \quad \text{P1b}$$
$$= X \cdot (1 + Y) \quad \text{P4b}$$
$$= X \cdot 1 \quad\quad\quad \text{T2a}$$
$$= X \quad\quad\quad\quad \text{P1b}$$

If an isolated variable X is ORed (ANDed) with a product (sum) term containing X, the product (sum) term is absorbed in the isolated variable.

Example 2.10

$$X + XYZ = X$$
$$X' + X'YZ = X'$$
$$X(X + Y + Z) = X$$
$$X + XYZ' + XY'Z' + XY = X$$

Theorem 4.   Absorption

   (a) $X + X'Y = X + Y$ (b) $X \cdot (X' + Y) = X \cdot Y$

isolated variable ———— complement of isolated variable ANDed (or ORed) with others. (Discard it.)

PROOF:

$$X \cdot (X' + Y) = X \cdot (X' + Y)$$
$$= X \cdot X' + X \cdot Y \quad \text{P4b}$$
$$= 0 + X \cdot Y \quad\quad\quad \text{P5b}$$
$$= X \cdot Y + 0 \quad\quad\quad \text{P2a}$$
$$= X \cdot Y \quad\quad\quad\quad\; \text{P1a}$$

Example 2.11

$$X + X'YZ = X + YZ$$
$$XY + (XY)'Z = XY + Z$$
$$X' \cdot (X + Y + Z) = X' \cdot (Y + Z)$$

Theorem 5.   DeMorgan's law

   (a) $(X + Y)' = X' \cdot Y'$ (b) $(X \cdot Y)' = X' + Y'$

PROOF:

See the Venn diagrams of Figure 2.9. The diagrams for $(X + Y)'$ and $X' \cdot Y'$ are alike, thereby proving (a). This theorem provides the procedures for complementing an expression.

To complement an expression, change each variable to its complement, and change each 0 to 1, 1 to 0, "$+$" to "$\cdot$", and "$\cdot$" to "$+$".

An algebraic proof of Theorem 5 follows. Let

$$P = X + Y \qquad \text{and} \qquad Q = X' \cdot Y'$$

If we prove that $P \cdot Q = 0$ and $P + Q = 1$, then by postulate 5, P is a complement of Q, thereby proving the theorem.

$$
\begin{aligned}
P \cdot Q &= (X + Y) \cdot (X'Y') \\
&= (X'Y') \cdot (X + Y) && \text{P2b} \\
&= (X'Y')X + (X'Y')Y && \text{P4b} \\
&= (XX')Y' + X'(YY') && \text{P2b} \\
&= 0 \cdot Y' + X' \cdot 0 && \text{P5b} \\
&= 0 + 0 && \text{T2b} \\
&= 0 && \text{P1a}
\end{aligned}
$$

$$
\begin{aligned}
P + Q &= (X + Y) + X'Y' \\
&= (Y + X) + X'Y' && \text{P2a} \\
&= Y + (X + X'Y') && \text{P3a} \\
&= Y + (X + Y') && \text{T4a} \\
&= Y + Y' + X && \text{P2a} \\
&= (Y + Y') + X && \text{P3a} \\
&= 1 + X && \text{P5a} \\
&= 1 && \text{T2a}
\end{aligned}
$$

Thus, $P = Q'$ and hence T5a is proved.

Example 2.12

$$
\begin{aligned}
\text{(a)} \;\; (X + Y + Z)' &= X' \cdot Y' \cdot Z' && \text{T5a} \\
\text{(b)} \;\; (X' + XZ)' &= (X')' \cdot (YZ)' && \text{T5a} \\
&= X \cdot (YZ)' && \text{T7} \\
&= X \cdot (Y' + Z') && \text{T5b}
\end{aligned}
$$

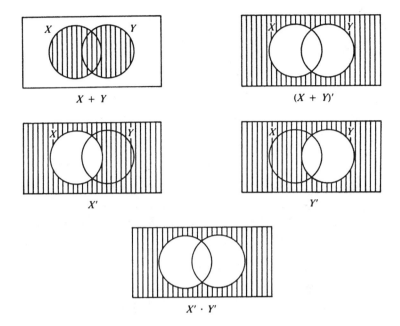

Figure 2.9 Proof of DeMorgan's law.

(c) $(X)(P + Q(X' + L)))' = X' + (P + Q(X' + L))'$ T5b

$= X' + P' \cdot (Q(X' + L))'$ T5b

$= X' + P' \cdot (Q' + (X' + L)')$ T5b

$= X' + P' \cdot (Q' + X'' \cdot L')$ T5a

$= X' + P' \cdot (Q' + X \cdot L')$ T7

$= X' + P'Q' + P'XL'$ P4b

$= X' + XPL' + P'Q'$ P3b, P2a

$= X' + PL' + P'Q'$ T4a

Theorem 6. Consensus

(a) $XY + X'Z + YZ = XY + X'Z$

(b) $(X + Y)(X' + Z)(Y + Z) = (X + Y)(X' + Z)$

PROOF:

$$XY + X'Z + YZ = XY + X'Z + YZ$$

$$= XY + X'Z + 1 \cdot YZ \qquad \text{P2b}$$

$$= XY + X'Z + (X + X')YZ \qquad \text{P5a}$$

$$= XY + X'Z + XYZ + X'YZ \qquad \text{P4b}$$

$$= (XY + XYZ) + (X'Z + X'ZY)$$

$$= XY + X'Z \qquad \text{T3a}$$

To use this theorem, we need to recognize an element (X), its complement (X'), and the included term, which is a product (sum) of the associated elements of X and X'. The included term can then be discarded.

Example 2.13

(a) $A'B + ACD' + BCD' = A'B + ACD'$ T6a

(b) $AB' + A'(P + Q) + B'(P + Q) = AB' + A'(P + Q)$ T6a, b

(c) $(A' + B')Z + ABX + ZX = (AB)'Z + ABX + ZX$ T5b

$$= (AB)'Z + ABX \qquad \text{T6a}$$

Theorem 7. Involution

$$(X')' = X$$

If $X = 0$, $X' = 1$ and $(X')' = 1' = 0$. If $X = 1$, $X' = 0$ and $(X')' = 0' = 1$.

These theorems can be used along with the postulates to manipulate and simplify Boolean expressions. Some of their applications are illustrated in the following sections.

2.6.1 Expansion to Canonical Forms

Any Boolean function can be expanded into one of the canonical forms by using the theorems and postulates. Consider the following product term of a three-variable function of X, Y, Z:

$$XY' \qquad \text{product term}$$

$$= X \cdot Y' \cdot 1 \qquad \text{P1b}$$

$$= XY'(Z + Z') \qquad \text{P5a (since } Z \text{ is missing)}$$

$$= XY'Z + XY'Z' \qquad \text{P4b}$$

This example illustrates the procedure for expanding a product term into its canonical SOP form. We can similarly expand a sum term into a canonical POS form. Consider the following sum term of a three-variable function in X, Y, Z:

$$X + Y' \qquad\qquad \text{sum term}$$
$$= X + Y' + 0 \qquad\qquad \text{P1a}$$
$$= (X + Y') + Z \cdot Z' \qquad\qquad \text{P5b (since } Z \text{ is missing)}$$
$$= (X + Y' + Z)(X + Y' + Z') \quad \text{P4a}$$

The following examples illustrate the complete procedure for expanding Boolean functions into canonical forms.

Example 2.14

$$F(X, Y, Z) = YZ' + X'$$

(a) Include the missing variables:

$$= YZ'(X + X') + X'(Y + Y')(Z + Z')$$

(b) Expand:

$$= XYZ' + X'YZ' + X'YZ + X'YZ' + X'Y'Z + X'Y'Z'$$

(c) Eliminate the repeated terms:

$$= \boxed{XYZ' + X'YZ' + X'YZ + X'Y'Z + X'Y'Z'}$$

(d) Minterm coding:

$$110 \qquad 010 \qquad 011 \qquad 001 \qquad 000$$

(e) Decimal form:

$$6 \qquad 2 \qquad 3 \qquad 1 \qquad 0$$

Therefore,

$$F = \Sigma m \,(0, 1, 2, 3, 6)$$

Example 2.15

$$F(X, Y, Z) = (X + Y')Z'$$
$$= (X + Y')(Z')$$

(a) Include the missing variables:

$$= (X + Y' + Z \cdot Z')(Z' + X \cdot X' + Y \cdot Y')$$

(b) Expand:

$$= (X + Y' + Z)(X + Y' + Z')(Z' + X \cdot X' + Y)$$
$$\cdot (Z' + X \cdot X' + Y')$$

$$= (X + Y' + Z)(X + Y' + Z')(Z' + X + Y)$$
$$\cdot (Z' + X' + Y)(Z' + X + Y')(Z' + X' + Y')$$

(c) Eliminate the repeated terms:

$$= \boxed{(X + Y' + Z)(X + Y' + Z')(X + Y + Z')(X' + Y + Z')(X' + Y' + Z')}$$

Canonical POS form

(d) Maxterm coding:

010   011   001   101   111

(e) Decimal form:

2   3   1   5   7

Therefore,

$$F = \Pi M \ (1, 2, 3, 5, 7)$$

We can summarize the procedure for expanding to canonical forms into the following steps:

1. Include the missing variable or variables:
   (a) as a sum of the variable and its complement, into a product term.
   (b) as a product of the variable and its complement, into a sum term.
2. Expand the resulting expression.
3. Neglect any repeated terms.

If the minterm or maxterm list form is desired:

4. Code the terms:

$$
\begin{aligned}
\text{minterms:} \quad & \text{variable} & = 1 \\
& \text{complement} & = 0 \\
\text{maxterms:} \quad & \text{variable} & = 0 \\
& \text{complement} & = 1
\end{aligned}
$$

5. Express in minterm or maxterm list notation.

Example 2.16 summarizes the relation between the minterm and maxterm forms.

Example 2.16

$$F(X, Y, Z) = \Sigma m \ (0, 1, 3, 5)$$
$$= \Pi M \ (2, 4, 6, 7)$$

If F' is the complement of the function F, then

$$F'(X, Y, Z) = \Sigma m \ (2, 4, 6, 7)$$
$$= \Pi M \ (0, 1, 3, 5)$$

The results are generalizations of the fact that whenever F is 0, F' is 1, and vice versa.

2.6.2 Proving the Equality of Expressions

There are several ways to prove the equality of Boolean expressions. The most direct method is to draw truth tables for both expressions to determine whether they have the same values at all the combinations of the component variable values. We could also expand each expression into the canonical SOP (minterm list) form or canonical POS (maxterm list) form to see whether they have same terms. The other method is to use the theorems just as we did in proving the theorems themselves. The following examples illustrate this last method.

Example 2.17 $AB + AB' = A$

$$
\begin{aligned}
AB + AB' &= AB + AB' \\
&= A(B + B') \quad &\text{P4b} \\
&= A \cdot 1 \quad &\text{P5a} \\
&= A \quad &\text{P1b}
\end{aligned}
$$

Example 2.18 $AB + AB'C = AB + AC$

$$
\begin{aligned}
AB + AB'C &= A(B + B'C) \quad &\text{P4b} \\
&= A(B + C) \quad &\text{T4a} \\
&= AB + AC \quad &\text{P4b}
\end{aligned}
$$

Example 2.19 $AB + A'C = (A + C)(A' + B)$

$$
\begin{aligned}
AB + A'C &= (AB) + A' \cdot C \\
&= (AB + A')(AB + C) \quad &\text{P4a} \\
&= (A + A')(B + A')(A + C)(B + C) \quad &\text{P4a} \\
&= 1 \cdot (A' + B) \cdot (A + C)(B + C) \quad &\text{P5a, P2a} \\
&= (A' + B)(A + C)(B + C) \quad &\text{P1b} \\
&= (A + C)(A' + B) \quad &\text{T6b}
\end{aligned}
$$

2.6.3   Simplifying Boolean Functions

The most important use of the theorems and postulates of Boolean algebra is in the simplification (the minimization of the number of literals) of Boolean expressions. Because this process reduces the expression's complexity, it is useful in reducing the cost of the hardware required to implement the expression. The following examples illustrate the simplification process.

Example 2.20   Simplify

$$XY' + XYZ + \underbrace{(X(Y' + Z'))'}_{\text{T5b}} + (X' + Y')(X' + \underbrace{YX)Z}_{\text{T4a}}$$

$$= XY' + XYZ + X' + \underbrace{(Y' + Z')'}_{\text{T5a}} + \underbrace{(X' + Y')(X' + Y)Z}_{\text{P4a}}$$

$$= XY' + XYZ + X' + \underbrace{Y''}_{\text{T7}}\,\underbrace{Z''}_{\text{T7}} + (X' + \underbrace{YY')Z}_{\text{P5b}}$$

$$= XY' + XYZ + X' + YZ + \underbrace{(X' + 0)Z}_{\text{P1a}}$$
$$\overset{\text{T3a}}{\diagdown}$$

$$= XY' + \underbrace{YZ + X' + X'Z}_{\text{T3a}}$$

$$= XY' + YZ + X'$$
$$\overset{\text{T4a}}{\diagup}$$

$$= Y' + X' + YZ$$
$$\overset{\text{T4a}}{\diagdown}$$

$$= Y' + Z + X'$$

Example 2.21

Simplify

$$F = \underbrace{((Q + Q'R)(Q' + R' + PQ'R)(Q' + R)' + ((Q + R) + (P + Q))')'}_{\text{T7}}$$

$$F' = \underbrace{(Q + Q'R)}_{\text{T4a}}\underbrace{(Q' + R' + PQ'R)}_{\text{T4a}}\underbrace{(Q' + R)'}_{\text{T5a}} + \underbrace{((Q + R) + (P + Q))'}_{\text{T5a}}$$

$$= (Q + R)(Q' + R' + PQ')(Q \cdot R') + \underbrace{(Q + R)'}_{\text{T5a}} \cdot \underbrace{(P + Q)'}_{\text{T5a}}$$
$$\underbrace{}_{\text{T3a}}$$

$$= \underbrace{(Q + R)(Q' + R')(Q \cdot R')}_{\text{P4b}} + \underbrace{(Q' R') \cdot (P' Q')}_{\text{T1b}}$$

$$= \underbrace{((Q + R)Q'}_{\text{P4b}} + \underbrace{(Q + R)R')}_{\text{P4b}}(Q \cdot R') + P'Q'R'$$

$$= (\underbrace{QQ'}_{0} + RQ' + QR' + \underbrace{RR'}_{0})(QR') + P'Q'R'$$

$$= \underbrace{(QR')(RQ' + QR')}_{\text{P4b}} + P'Q'R'$$

$$= \underbrace{QR' \cdot RQ'}_{0} + \underbrace{QR' \cdot QR'}_{\text{T1b}} + P'Q'R'$$

$$= 0 + \underbrace{QR' + P'Q'R'}_{\text{P4b}}$$

$$= R'\underbrace{(Q + P'Q')}_{\text{T4a}}$$

$$= R'(Q + P') = QR' + P'R'$$

Hence,
$$
\begin{aligned}
F &= (QR' + P'R')' & \text{T7} \\
  &= (QR')' \cdot (P'R')' & \text{T5a} \\
  &= (Q' + R)(P + R) & \text{T5b} \\
  &= (R + Q')(R + P) & \text{P2a} \\
  &= R + PQ' & \text{P4a}
\end{aligned}
$$

Note that this algebraic simplification process depends entirely on one's familiarity with the postulates and theorems and one's ability to recognize their application. Of course, this ability varies from individual to individual. Depending on the sequence in which the theorems and postulates are applied, more than one simplified form of the expression may be obtained. Since a minimized function is one with fewer literals than in the original expression, all such minimized forms are valid and acceptable. Thus, *there is generally no single, exclusive minimized form of a Boolean expression.*

Because several mechanical procedures are available for simplifying functions, it is not necessary to be thoroughly familiar with the theorems and postulates of Boolean algebra. Chapter 3 describes some of these procedures. Although function simplification was an important topic in the 1970s, the advances in digital hardware technology, resulting in inexpensive hardware modules, have made function simplification less important. Nevertheless, the theorems and postulates of Boolean algebra form the foundation for all the mechanical procedures used by the practicing logic designer.

2.7 Other Operations

We defined Boolean algebra as consisting of three operators: AND, OR, and NOT. Other operators that are useful in the design and analysis of logic circuits will be discussed in this section.

Let us examine the truth table for the OR function shown below.

X	Y	$F = X + Y$
0	0	0
0	1	1
1	0	1
1	1	1

This table has four rows. Hence, there are four positions in the column corresponding to F. Each of these positions can contain either a 1 or a 0. Thus, the F column can have 2^4, or 16, combinations of values. The OR function shown in the table corresponds to one of the 16 possible combinations. In general, there can be $2^{(2^n)}$ functions for n variables.

Table 2.4 shows the 16 possible combinations, and hence the 16 possible functions, of two variables X and Y. This table is essentially a truth table formatted slightly differently from other truth tables. The four combinations of two variables are listed along the top two rows and the functions

TABLE 2.4 Functions of Two Variables

		Operation	In terms of AND, OR, NOT	Operator symbol
X	0011			
Y	0101			
F_0	0000	Null	0	
F_1	0001	AND	$X \cdot Y$	$X \cdot Y$
F_2	0010	Inhibit Y	$X \cdot Y'$	X/Y
F_3	0011	Transfer X	X	X
F_4	0100	Inhibit X	$X' \cdot Y$	Y/X
F_5	0101	Transfer Y	Y	Y
F_6	0110	EXCLUSIVE-OR	$X \cdot Y' + X' \cdot Y$	$X \oplus Y$
F_7	0111	OR	$X + Y$	$X + Y$
F_8	1000	NOR	$(X + Y)'$	$X \downarrow Y$
F_9	1001	EQUIVALENCE	$X \cdot Y + X' \cdot Y'$	$X \odot Y$
F_{10}	1010	COMPLEMENT	Y'	Y'
F_{11}	1011	Implication	$X + Y'$	$Y \rightarrow X$
F_{12}	1100	COMPLEMENT	X'	X'
F_{13}	1101	Implication	$X' + Y$	$X \rightarrow Y$
F_{14}	1110	NAND	$(X \cdot Y)'$	$X \uparrow Y$
F_{15}	1111	Identity	1	

are listed as the next 16 rows, rather than the usual column format used in truth tables.

Here, F_0 is the null function and F_{15} is the identity function. These functions realize constants 0 and 1, respectively. F_1 corresponds to AND and F_7 corresponds to OR. F_{10} and F_{12} correspond to NOT functions (of Y and X, respectively). F_2 inhibits Y (i.e., X but not Y), and F_4 inhibits X. F_3 and F_5 transfer X and Y, respectively. F_{11} and F_{13} are implications (if Y then X; if X then Y, respectively). F_8 is the complement of OR, and hence it is called NOT-OR or, more commonly, NOR. Similarly, F_{14} is NOT-AND, or NAND. F_6 is a special case of OR in which the case of both variables being 1 is excluded. It is thus called EXCLUSIVE-OR. F_9 is the complement of F_6 and is called EXCLUSIVE-NOR or EQUIVALENCE, since it implies the equivalence of the two variables. That is, F_9 is 1 only when the values of X and Y are equal. The EXCLUSIVE-OR function F_6 is 1 only when the values of X and Y are not equal.

In addition to AND, OR, and NOT operators, NAND, NOR, and EX-CLUSIVE-OR operators are used extensively by logic designers. Figure 2.10 shows the gate symbols for these operators. The other operators in the table

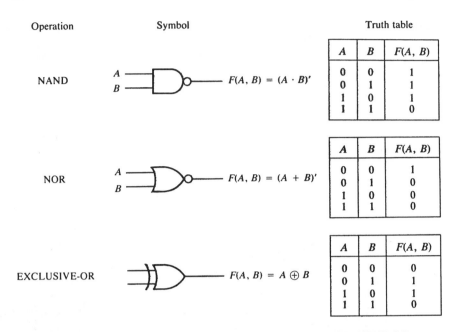

Figure 2.10   Gate symbols for NAND, NOR, and EXCLUSIVE-OR.

are not used. They can be expressed in terms of the common operators, as shown.

2.7.1   A Functionally Complete Set of Operators

A set of operators is said to be functionally complete if any Boolean function can be expressed in terms of the set. Thus, the set containing AND, OR, and NOT is functionally complete. If it can be shown that a set of operators can realize AND, OR, and NOT operations, then that set is also functionally complete. It can be shown that NAND and NOR are each functionally complete. Figure 2.11 shows how the three basic operators are realized by NAND gates only. Proving that NOR is functionally complete will be left as an exercise. NAND and NOR are also called universal operators. The advantage of universal operators is that only one type of gate is used for realizing the logic circuit. Since integrated circuit (IC) chips contain several gates of the same type, using universal gates in the circuit is advantageous in terms of minimizing the number of chips used in the design. We will examine this further in subsequent chapters.

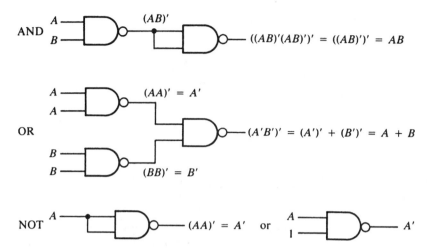

Figure 2.11 Realization of primitive operations using NAND.

2.8 Multiple-Input Gates and Logic Circuits

A NOT gate always has one input and one output. We have shown all the other types of gates to have two inputs and one output. Since each of these binary operators is commutative, the inputs to these gates are interchangeable. This means that the order in which the inputs are connected is immaterial. In practice, gates with more than two inputs are available. Consider, for example, the realization of the function

$$F = X + Y + Z$$

shown in Figure 2.12. This implementation uses a cascade of two OR gates. Because the OR operation is associative, the order in which the cascading is performed is immaterial. This means that the first gate could have Y and Z as inputs, the output of which is fed to the second gate along with X to realize the three-input function above. Because of the commutative and associative properties, the OR operation can be extended to more than two operands. Multiple-input OR gates are thus available. Figure 2.13 shows the symbols for three-input gates. This symbolism can be extended to any number of inputs. The number of inputs to the gate is referred to as its *fanin*. In practice there is a limit to the fanin of a gate.

Like OR, both AND and EXCLUSIVE-OR are commutative and associative. NAND and NOR, however, are not associative. (See Problem 2.14.) Multiple-input NAND and NOR gates realize NOT-AND and NOT-

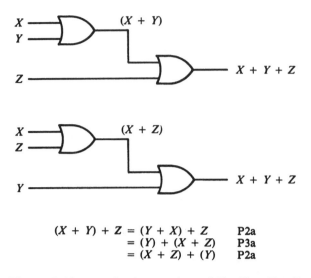

$$(X + Y) + Z = (Y + X) + Z \quad \text{P2a}$$
$$= (Y) + (X + Z) \quad \text{P3a}$$
$$= (X + Z) + (Y) \quad \text{P2a}$$

Figure 2.12   Two implementations of $F = X + Y + Z$.

OR of their inputs, respectively. As such, cascading of two two-input gates will not achieve a three-input realization in the case of NAND and NOR.

A logic or a switching circuit is an interconnection of logic gates, as defined earlier. A logic circuit realizes a Boolean function and provides the appropriate output for each combination of input values, as defined by the truth table of the function. Figure 2.4 showed a logic circuit for realizing the sum. Example 2.22 illustrates the use of multiple-input gates and logic minimization.

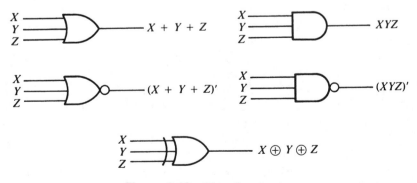

Figure 2.13   Three-input gates.

Example 2.22

Consider the Boolean function

$$F = XYZ' + XYZ + X'Y'Z$$

Figure 2.14(a) shows a logic circuit that realizes this function. Note that a three-input OR gate is used to obtain the OR of the three product terms. Each product term is realized by a three-input AND gate. The complemented variables required are obtained by using NOT gates.

The function shown above can be simplified to

$$F = XY + X'Y'Z$$

and the realization of this function is shown in Figure 2.14(b). By using the simplified function to realize the circuit, we thus saved one three-input gate and could use a two-input rather than three-input AND gate. Minimization procedures will be discussed in Chapter 3, and analysis and synthesis procedures for combinational circuits will be examined in Chapter 4. We will now take a brief look at *integrated circuits* (ICs), the building blocks of modern-day logic circuits.

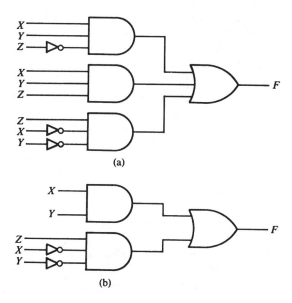

Figure 2.14 A logic circuit. (a) The circuit; (b) simplified circuit.

2.9 Integrated Circuits

So far in this chapter we have concentrated only on the functional aspects of gates and logic circuits in terms of manipulating binary signals. A gate is an electronic circuit made up of transistors, diodes, resistors, capacitors, and other components interconnected to realize a particular function. In this section, we will expand our understanding of gates and circuits to the electronic level of detail.

At the current state of digital hardware technology, the logic designer combines ICs that perform specific functions to realize functional logic units. An IC is a small slice of silicon semiconductor crystal called a *chip*, on which the discrete electronic components mentioned above are chemically fabricated and interconnected to form gates and other circuits. These circuits are accessible only through the pins attached to the chip. There will be one pin for each input signal and one for each output signal of the circuit fabricated on the IC. The chip is mounted in either a metallic or a plastic package. Various types of packages, such as Dual-In-Line Package (DIP) and flat package, are used. DIP is the most widely used package. The number of pins varies from 8 to 64. Each IC is given a numeric designation (printed on the package), and the IC manufacturer's catalogue provides the functional and electronic details on the IC.

Each IC contains one or more gates of the same type. The logic designer combines ICs that perform specific functions to realize functional logic units. As such, the electronic-level details of gates usually are not needed to build efficient logic circuits. But as logic circuit complexity increases, the electronic characteristics become important in solving the timing and loading problems in the circuit.

Figure 2.15 shows the details of an IC that comes from the popular TTL (transistor-transistor logic) family of ICs. It has the numeric designation of 7400 and contains four two-input NAND gates. There are 14 pins. Pin 7 (ground) and pin 14 (supply voltage) are used to power the IC. Three pins are used by each gate (two for the input and one for the output). On all ICs, a ''notch'' on the package is used to reference pin numbers. Pins are numbered counterclockwise starting from the notch. TTL ICs are available in several versions, as designated by the letters S, LS, and so on. We examine the characteristics of these versions in Chapter 10.

Digital and *linear* are two common classifications of ICs. Digital ICs operate with binary signals, while linear ICs operate with continuous signals. In this book, we will be dealing only with digital ICs.

Because of the advances in IC technology, it is now possible to fabricate a large number of gates on a single chip. According to the number of

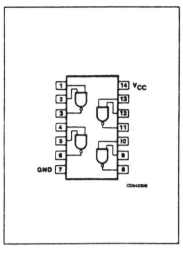

TTL 7400

Figure 2.15   A typical IC (Courtesy of Signetics Corporation).

gates it contains, an IC can be classified as a small-, medium-, large- or very-large-scale circuit. An IC containing a few gates (approximately 10) is called a small-scale integrated (SSI) circuit. A medium-scale integrated (MSI) circuit has a complexity of around 100 gates and typically implements an entire function, such as an adder or a decoder, on a chip. An IC with a complexity of more than 100 gates is a large-scale integrated (LSI) circuit, while a very-large-scale integrated (VLSI) circuit contains thousands of gates.

There are two broad categories of IC technology, one based on *bipolar* transistors (i.e., p-n-p or n-p-n junctions of semiconductors) and the other based on the *unipolar* Metal Oxide Semiconductor Field Effect Transistor (MOSFET). Within each technology, several logic families of ICs are available. The popular bipolar logic families are TTL and Emitter-Coupled Logic (ECL). P-channel MOS (PMOS), N-channel MOS (NMOS), and Complementary MOS (CMOS) are all popular MOS logic families. A new family of ICs based on gallium arsenide has been introduced recently. This technology has the potential of providing ICs that are faster than ICs in silicon technology.

In the following discussion, we will examine functional-level details of ICs. These details are adequate to build circuits using ICs. Var-

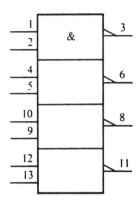

Figure 2.16   IEEE standard symbol for TTL 7400 IC.

ious performance characteristics of ICs will be introduced, without electronic-level justification. Such justification is deferred to Chapter 10, where
electronic-level details of basic circuits in various technologies will be
examined.

In addition to the details of the type provided in Figure 2.15, the IC
manufacturer's catalogue contains such information as voltage ranges for
each logic level, fan-out, propagation delays, and so forth, for each IC. In
this section, we will introduce the major symbols and notation used in the
IC catalogues and describe the most common characteristics in selecting and
using ICs.

Throughout this book we will continue to use the graphic symbols
introduced thus far to represent gates. A new standard set of symbols, based
on *dependency notation*, recently has been adopted by IEEE. Figure 2.16
shows the new symbol for the TTL 7400 IC of Figure 2.15. Here, the ''&''
represents the AND operator, the four partitions of the rectangle correspond
to the four gates, and the half-arrows on the outputs correspond to the ''bubbles'' on the outputs of the NAND gate symbol used earlier. The IEEE
standard symbolism is further described in appendix A at the end of this
book.

The following symbols will be used throughout this book to represent
various units of measure and scale factors:

Characteristic	Unit of measure	Scale factor
Resistance	ohm (Ω)	
Voltage	volts (V)	kilo (10^3): k
Current	amperes (A)	milli (10^{-3}): m
Power	watt (W)	micro (10^{-6}): μ
Capacitance	farad (f)	nano (10^{-9}): n
Energy	joule (J)	pico (10^{-12}): p
Time	second (s)	

2.9.1 Positive and Negative Logic

As mentioned earlier, two distinct voltage levels are used to designate logic values 1 and 0, and in practice these voltage levels are ranges of voltages, rather than fixed values. Figure 2.2 showed the voltage levels used in the TTL technology: the high level corresponds to the range of 2.4 to 5 V and the low level corresponds to 0 to 0.4 V. These two levels are designated *H* and *L*.

In general, once the voltage levels are selected, the assignment of 1 and 0 to those levels can be arbitrary. In the so-called *positive logic* system, the higher of the two voltages denotes logic-1, and the lower value denotes logic-0. In the *negative logic* system, the designations are the opposite. The following table shows the two possible logic value assignments.

	Positive logic	Negative logic
Logic-1	*H*	*L*
Logic-0	*L*	*H*

Note that *H* and *L* can both be positive, as in TTL, or both negative, as in ECL (*H* = −0.7 to −0.95 V, *L* = −1.9 to −1.5 V). It is the assignment of the relative magnitudes of voltages to logic-1 and logic-0 that determines the type of logic, rather than the polarity of the voltages.

Because of these "dual" assignments, a logic gate that implements an operation in the positive logic system implements its dual operation in the negative logic system. IC manufacturers describe the function of gates in terms of *H* and *L*, rather than logic-1 and logic-0. As an example, consider the TTL 7408 IC, which contains four two-input AND gates (Figure 2.17a). The function of this IC as described by the manufacturer in terms of *H* and *L* is shown in the voltage table of Figure 2.17(b). Using positive logic, the

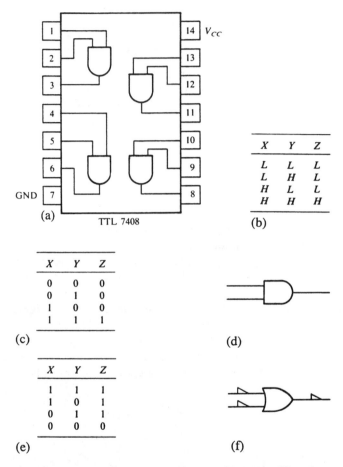

Figure 2.17 Positive and negative logic. (a) TTL 7408, (b) voltage table, (c) positive logic truth table, (d) positive logic AND, (e) negative logic truth table, (f) negative logic OR.

voltage table (b) can be converted into the truth table (c) representing the positive logic AND, as shown by the gate (d). Assuming negative logic, table (b) can be converted into truth table (e), which is the truth table for OR. The negative logic OR gate is shown in (f). The half-arrows on the inputs and the output designate them to be negative logic values. Note that the gates in (d) and (f) both correspond to the same physical gate but function either as positive logic AND or as negative logic OR. Similarly, it can be shown that the following dual operations are valid:

Positive logic	Negative logic
OR	AND
NAND	NOR
NOR	NAND
EXCLUSIVE-OR	EQUIVALENCE
EQUIVALENCE	EXCLUSIVE-OR

We will assume the positive logic system throughout this book and as such will not use negative logic symbolism. In practice, a designer may combine the two logic notations in the same circuit (mixed logic), so long as the signal polarities are interpreted consistently.

2.9.2 Signal Inversion

We have used a "bubble" in NOT, NOR, and NAND gate symbols to denote signal inversion (complementation). This bubble notation can be extended to any logic diagram. Some examples are shown in Figure 2.18. The NOT gate symbol in (a) implies that when the input X is asserted (e.g., at H), the output is low (L). The input is said to be *active-high* and the output is said to be *active-low*. An alternative symbol for a NOT gate, with an active-low input, is shown in (b). An AND gate with active-low inputs is shown in (c). The output of this gate is high only when both inputs are low. Note that this is the INVERT-AND or a NOR gate. A typical IC with four inputs and one output is shown in (d). Input A and output E are active-low, and inputs B, C, and D are active-high. An L on input A would appear as an H internal to the IC, and an H corresponding to E internal to the IC would appear as an L external to the IC. That is, an active-low signal is active when it carries

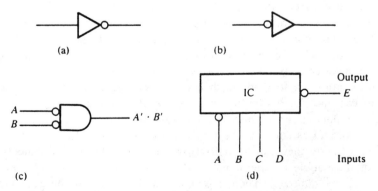

Figure 2.18 Bubble notation. (a) NOT gate; (b) NOT with active-low input; (c) AND with active-low inputs; (d) a typical IC.

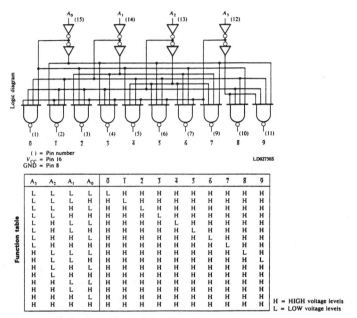

( ) = Pin number
V_{CC} = Pin 16
GND = Pin 8

LD02750S

Function table

A_3	A_2	A_1	A_0	0	1	2	3	4	5	6	7	8	9
L	L	L	L	L	H	H	H	H	H	H	H	H	H
L	L	L	H	H	L	H	H	H	H	H	H	H	H
L	L	H	L	H	H	L	H	H	H	H	H	H	H
L	L	H	H	H	H	H	L	H	H	H	H	H	H
L	H	L	L	H	H	H	H	L	H	H	H	H	H
L	H	L	H	H	H	H	H	H	L	H	H	H	H
L	H	H	L	H	H	H	H	H	H	L	H	H	H
L	H	H	H	H	H	H	H	H	H	H	L	H	H
H	L	L	L	H	H	H	H	H	H	H	H	L	H
H	L	L	H	H	H	H	H	H	H	H	H	H	L
H	L	H	L	H	H	H	H	H	H	H	H	H	H
H	L	H	H	H	H	H	H	H	H	H	H	H	H
H	H	L	L	H	H	H	H	H	H	H	H	H	H
H	H	L	H	H	H	H	H	H	H	H	H	H	H
H	H	H	L	H	H	H	H	H	H	H	H	H	H
H	H	H	H	H	H	H	H	H	H	H	H	H	H

H = HIGH voltage levels
L = LOW voltage levels

Figure 2.19 TTL 7442 BCD-to-decimal decoder (Courtesy of Signetics Corporation).

a low logic value, while an active-high signal is active when it carries a high logic value. A bubble in the logic diagram indicates an active-low input or output. For example, the BCD-to-decimal decoder IC (7442) shown in Figure 2.19 has four active-high inputs corresponding to the four input BCD bits and ten active-low outputs. Only the output corresponding to the decimal value of the input BCD number is active at any time. That is, the output corresponding to the decimal value of the BCD input to the IC will be *L* and all other outputs will be *H*.

When two or more ICs are used in the circuit, the active-low and active-high designations of input and output signals must be observed for the proper operation of the circuit, although ICs of the same logic family generally have compatible signal-active designations.

It is important to distinguish between the negative logic designation (half-arrow) and the active-low designation (bubble) in a logic diagram. These designations are similar in effect, as shown by the gate symbols in Figure 2.20, and as such can be replaced by each other. In fact, as shown in (c) and (d) on the figure a half-arrow following the bubble cancels the effect of the bubble on the signal, and hence both the half-arrow and the bubble can be removed. Then, the inputs and outputs of the gate are of different polarity. For example, if the half-arrow and the bubble are removed from the output of the negative logic NOR gate in (c), the inputs represent negative logic and the outputs represent positive logic polarities.

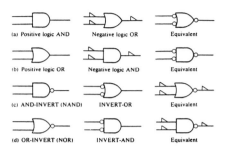

(a) Positive logic AND Negative logic OR Equivalent

(b) Positive logic OR Negative logic AND Equivalent

(c) AND-INVERT (NAND) INVERT-OR Equivalent

(d) OR-INVERT (NOR) INVERT-AND Equivalent

Figure 2.20 Equivalent symbols.

2.10 Current Trends

So far in this chapter we have concentrated on two-valued logic, since this logic system has been the only practical one from an implementation point of view. There have been two other trends in generalizing the concepts presented in this chapter: multivalued and fuzzy logic systems. We provide a brief description of these trends in this section.

2.10.1 Multivalued Logic

Allowing the logic signal to take more than two values can provide several advantages. For instance, in a three-valued logic each signal represents three values: 0, 1 and 2. Thus, a wire can carry three distinct logic values; two wires can carry nine combinations of logic values (compared to only four in a two-valued logic), and so on. In general, if an n-valued logic system is used, p wires can carry n^p logic values. The other advantages of multivalued logic system are that fewer gates would be required to perform certain (especially, arithmetic) operations and they can be performed faster. The major problem with implementation is the difficulty in controlling the voltage levels to the close tolerances needed to distinguish between the multiple values being represented. Due to this problem, multivalued implementations have not yet achieved the reliability, cost and performance advantages of two-valued implementations.

The concept of multivalued logic was first formulated by Post (1921) in his propositional calculus. Over the last twenty years several multivalued algebraic systems have been proposed. One such algebra presented by Irving (1973) used the following operators:

$$\text{AND: } A \cdot B = \text{Min}(A, B)$$
$$\text{OR: } A + B = \text{Max}(A, B)$$
$$\text{CYCLE: } A^{\rightarrow B} = A \text{ plus } B, \text{ Modulo } N$$
$$\text{COMPLEMENT: } A' = P \text{ minus } A$$

where, N = base of the algebra $(N > 2)$, $P = N - 1$, and A and B are the elements of the set L of N logic values $(0, 1, 2, \ldots, P)$.

It can be shown that the above set of operators forms a functionally complete set. A computer oriented minimization procedure for this algebra can be found in Su (1972) and implementation of these operators in hardware is provided in Vranesic (1970).

There is still a very active interest in multivalued logic. Much work has been performed in terms of implementing multivalued systems in hardware. Multivalued implementations have been utilized to a limited extent, in connecting intra-chip signals (where two-valued logic results in very high wiring complexities). But, no practical multivalued systems exists today or is expected to be implemented in the near future. We will not provide further details on the topic in this book. Refer to the proceedings of international symposium on multivalued logic held annually for latest details.

2.10.2   Fuzzy Logic

Note that we defined Boolean algebra as a closed algebraic system containing a set of elements and binary operators. This set of elements is more specifically a 'crisp' set since each element in the set represents a distinct value. In practice, we cannot represent all the real life scenarios in crisp sets. For instance, if the color of an object can only be either black or white, a crisp set representation would be sufficient. But, if the color can be grey, almost black, or almost white, then the crisp set representation is inadequate. Then a "fuzzy" set representation is needed. Other typical examples requiring a fuzzy representation are: lighting in the room is too bright, some radiation levels are unhealthy, the road is somewhat slippery, he is almost tall, etc.

Fuzzy set theory was introduced by Zadeh in early 1960s to propagate the transition from traditional (crisp) mathematical modeling to a much more qualitative "rough" modeling. This theory provided a mechanism to represent vague notions in a mathematically sound way. The most prominent application of fuzzy logic theory has been in the area of automatic control theory. It has been applied to several practical systems including the control of consumer goods such as cameras.

More recent applications of fuzzy logic is in applications such as pattern recognition, chemical process modelling, database management and operations research. In these areas fuzzy theory is used to provide "approximate reasoning" capability based on the fuzzy information available in knowledge bases.

The topics above are clearly beyond the scope of this book. For further information, refer to the books and journals listed in the references.

2.11   Summary

This chapter has provided a formal introduction to Boolean Algebra. The terminology and concepts introduced here will be used throughout the book. Applications of Boolean algebra to logic circuit design and analysis will be discussed in greater detail in subsequent chapters. The functional level details of ICs introduced in this chapter are sufficient to understand the material in subsequent chapters. Further electronic level details on these devices are provided in Chapter 10. This chapter also provided a very brief introduction to two current trends: multivalued and fuzzy logic systems.

References

Bandemer, H. and Gottwald, S., *Fuzzy Sets, Fuzzy Logic, Fuzzy Methods with Applications*, New York: John Wiley, 1995.

Ercegovac, M. D. and Lang, T., *Digital Systems and Hardware/Firmware Algorithms*, New York: John Wiley, 1985.

Irving, T. A. and Nagle, H. T., "An approach to multivalued logic," *Conference Record of the International Symposium on Multiple-valued Logic*, pp. 89–105, Toronto, 1973.

Kohavi, Z., *Switching and Automata Theory*, New York: McGraw-Hill, 1970.

Post, E. L., "Introduction to a general theory of elementary propositions," *Amer. J. of Mathematics*, *43*, pp. 163–185, 1921.

Roth, C. H., *Fundamentals of Logic Design*, St. Paul, MN: West, 1985.

Su, S. Y. H. and Cheung, P. T., "Computer minimization of multivalued switching functions," *IEEE Transactions on Computers*, C-21, No. 9, pp. 995–1003, September 1972.

The TTL Data Book Series, Dallas, TX: Texas Instruments, 1984.

TTL Manual, Sunnyvale, CA: Signetics, 1978.

Vranesic, Z. G., Lee, E. S. and Smith, K. C., "A many-valued algebra for switching systems," *IEEE Transactions on Computers*, C-19, No. 10, pp. 904–971, October 1970.

Zadeh, L. A., "Fuzzy sets," *Information and Control*, *8*, pp. 338–353, 1965.

Problems

2.1 Given $A = 0$, $B = 1$, $C = 0$, and $D = 1$, find the value of F in each of the following:

$$F = AB' + C$$
$$F = AB' + C'D + CD$$
$$F = A'B(A + B' + C' \cdot D) + B'D$$
$$F = (A + B')(C' + A)(A + B \cdot C)$$
$$F = ((A + B')C + D')AB' + CD'(C' + A'(B + C'D))$$

2.2 Draw a truth table for each of the following:

$$Q = XY' + X'Z' + XYZ$$
$$Q = (X' + Y)(X' + Z')(X + Z)$$
$$Q = AB'(C' + D) + ABC' + C'D'$$
$$Q = A'BC + AB'D' + A' + B' + CD'$$
$$Q = (X + Y + Z')(Y' + Z)$$

2.3 Draw a Venn diagram for each function in problem 2.2.

2.4 State whether the following identities are true or false (use truth tables):

$$XY' + X'Z + Y'Z = X'Y + X'Z$$
$$(B' + C)(B' + D) = B' + CD$$
$$A'BC + ABC' + A'BD = BD' + ABC'$$
$$X'Z + X'Y + XZ = X'YZ' + X'YZ + X'Z$$
$$(P + Q' + R)(P + Q' + R') = Q' + PR' + RP'$$

2.5   Use the algebraic method to solve problem 2.4.

2.6   Indicate whether the following statements are true or false:
      $X + Y'$ is a conjunction.
      $XY'Z$ is a product term.
      $AB'C'$ is a disjunction.
      $(A + B' + C')$ is a sum term.
      AB' is included in $ABCD$.
      $(A + B')$ is included in $(A + B + C)$.
      $A + B + C'$ is included in ABC'.

2.7   State whether the following identities are in (a) normal POS form, (b) normal
      SOP form, (c) canonical POS form, or (d) canonical SOP form:
      $F(X, Y, Z) = XY' + YZ' + Z'Y'$
      $F(A, B, C, D) = (A + B' + C')(A' + C' + D)(A' + C')$
      $F(P, Q, R) = PQ' + QR'(P + Q') + (R' + Q')$
      $F(A, B, C) = (A + B + C')(A' + B' + C')(A' + B + C')$
      $F(A, B, C, D) = ABC'D + AB'CD' + A'B'CD$
      $F(A, B, C) = (A + B' + C)(A + B')(A + B + C')$
      $F(X, Y, Z) = XY'Z + X'Y'Z + X'Y' + XYZ$

2.8   Find the dual for each of the following:
      $AB'C + A'B + B'C'$
      $(A + B' + C)(A' + B + C')(A' + B')$
      $XY'(Z + XY') + X'Y'$
      $AB(C' + D') + A'B'C' + CD'(A + C')$
      $AB' + CD'(A' + B') = 0$

2.9   Find the complement of each of the following:
      $AB' + A'(A + B')$
      $A'B(C + A') + ((AB)(B' + C'))' + (ABC)'$
      $W'X'(Y' + Z'W') + (XYZ)'(X' + W'X') + (X' + Y')$
      $A'BC' + (ABC)' + AB'(C' + B')$
      $PQ'(R'S' + P'Q')RS' + (P + QR)'SR'$

2.10 Express the following functions in canonical SOP and POS forms. (Hint:
      Draw the truth table for each one first.)
      $F(A, B, C) = (A + B')C' + A'C$
      $F(X, Y, Z) = (X + Y')(X' + Z) + ZY'$
      $F(A, B, C, D) = AB'C + A'BC'D + A'BCD' + B'D'$
      $F(W, X, Y, Z) = WX' + Z'(Y' + W') + W'Z'Y'$

2.11 Express the following in minterm list form:

$$F(A, B, C) = (A + B')C' + A'C$$
$$F(X, Y, Z) = (X + Y')(X + Z)(Z + Y')$$
$$F(P, Q, R) = \Pi M\ (0, 1, 5)$$
$$F(A, B, C, D) = \Pi M\ (1, 2, 3, 7, 9, 10, 15)$$
$$F(W, X, Y, Z) = WZ' + (W' + X')YZ' + W'Z'X'$$
$$F(A, B, C) = 1$$
$$F(A, B, C) = 0$$

2.12 Express the following in maxterm list form:

$$F(A, B, C) = (A + B') + C' + A'C$$
$$F(X, Y, Z) = (X + Y')(X' + Z) + ZY'$$
$$F(P, Q, R, S) = (P + Q')R' + P'S'R' + PQ'(S' + R' + Q')$$
$$F(A, B, C, D) = \Sigma m\ (0, 1, 5, 7, 11, 14, 15)$$
$$F(A, B, C) = 1$$
$$F(A, B, C) = 0$$

2.13 If two functions are equal, they will have the same minterms and maxterms. Use this fact to solve problem 2.4.

2.14 Verify the following algebraically:
   (a) The dual of EXCLUSIVE-OR is equal to its complement.
   (b) $X \uparrow (Y \uparrow Z) \neq (X \uparrow Y) \uparrow Z$, where $\uparrow$ is the NAND operator.
   (c) $(X \oplus Y) \oplus Z = X \oplus (Y \oplus Z)$

2.15 Use only NOR gates to realize AND, OR, and NOT functions.

2.16 Given

$$P(A, B, C) = \Sigma m\ (0, 1, 3, 5)$$
$$Q(A, B, C) = \Sigma m\ (1, 4, 5, 7)$$

Find
   (a) P' in minterm list form,
   (b) P' in maxterm list form,
   (c) $P' \cdot Q$ in minterm list form,
   (d) $P' + Q$ in maxterm list form.

2.17 Simplify the following:
   (a) $P' + PQR + QR'$
   (b) $X'Y(Y' + Z') + W'X'Z' + WX'Y'(Z + Z'X)$
   (c) $AB'C'(D + BC') + C'D + A'B'$
   (d) $((X'Y)' + XY' + Z(X + YZ)' + X'Z')'$
   (e) $XY'Z + XY'Z' + XZ'Y + YZ'X + XZ'Y'$

2.18 Implement the simplified functions from problem 2.17 using AND, OR, and NOT gates.

2.19 Given the function $P(A, B, C) = \Sigma m\ (0, 1, 3, 5)$,
   (a) express P in canonical forms,
   (b) implement simplified P using AND, OR, and NOT gates,
   (c) implement simplified P using AND and NOT gates only,
   (d) implement simplified P in OR and NOT gates only.

2.20 Draw a truth table to show the function of an odd number detection circuit. That is, the output of the circuit will be 1 only if the decimal value of the binary inputs is odd. Assume three inputs. Derive a simplified circuit.

3

Minimization of Boolean Functions

The complexity of a logic circuit is a function of the number of gates in the circuit. The complexity of a gate generally is a function of the number of inputs to it. Because a logic circuit is a realization (implementation) of a Boolean function in hardware, reducing the number of literals in the function should reduce the number of inputs to each gate and the number of gates in the circuit—thus reducing the complexity of the circuit (as shown by Example 2.22).

However, reducing the number of literals in a function is not always the criterion for circuit minimization. Various other criteria are used. For instance, in the design of ICs, various considerations (e.g., minimizing the area used by the circuit on the silicon wafer, the regularity of the circuit structure from a fabrication point of view) are more important. In implementing logic circuits using programmable logic arrays (PLAs) (see Chapter 9), merely minimizing the number of literals does not yield a less complex circuit. But eliminating some product terms from the SOP (sum of products) form of the function reduces the circuit's complexity. In the design of logic circuits using ICs, reducing the number of ICs in the circuit means reducing the cost of the circuit. Since each IC contains a set of gates of the same type, it is important to reduce the IC count by using as many gates of the same type as possible. As long as the IC count is not increased, the circuit complexity will not increase as a result of an extra gate used in the circuit.

Nevertheless, we will concentrate here on the Boolean function-minimization procedures that reduce the literal count in the function. The algebraic method used in Chapter 2 to minimize functions is tedious and error prone. Its success depends on our ability to recognize the application of a theorem or a postulate during the minimization process. Such recognition may not be obvious. Further, there is no general set of rules to aid that recognition.

In this chapter, we will examine two popular minimization techniques. The first is based on a graphical representation of Boolean functions using *Karnaugh maps* (*K-maps*), and the second is a tabular method devised by Quine and McCluskey, called the *Quine-McCluskey procedure* (*QM procedure*). Both of these methods are mechanical in nature. K-maps are useful in minimizing functions with up to five or six variables. The QM procedure is useful for functions of any number of variables and can easily be programmed to run on a digital computer.

Generally, several minimum functions can be obtained for a given function using either method, based on the choices made during the minimization process. All minimum functions with the same number of literals yield circuits of the same complexity; hence, any of them can be selected for implementation.

3.1 K-maps

K-maps are modified Venn diagrams. Consider the Venn diagram for two variables, A and B, shown in Figure 3.1(a). The four areas correspond to minterms m_0, m_1, m_2, and m_3, as shown in (b). Each of these areas is represented by a square (block) in the map (c). Here the union of the two blocks in the second column yields A (i.e., $AB' + AB = A$), and hence the area covered by these two blocks corresponds to A. Similarly, the first column corresponds to A'; the first row corresponds to B', and the second row corresponds to B. The usual forms of the two-variable K-map are shown in (d) and (e). The values 0 and 1 shown at the top of the map—in (e)—correspond to the values of A, and those shown on the left side are the values of B. The block numbers are the same as the minterm numbers.

Figure 3.2 shows a three-variable K-map. There are 2^3, or 8, blocks corresponding to the 8 minterms. The blocks are arranged so that the two right-hand columns correspond to A, the two middle columns correspond to B, and the second row corresponds to C. Thus, block 6 corresponds to the minterm code of 110 or ABC', which is the area in both A and B, but not in C. Note that there are four blocks corresponding to each variable in the K-map. The allocation of variables and the block numbers are shown in (b). The four combinations of values for variables A and B are shown at the top

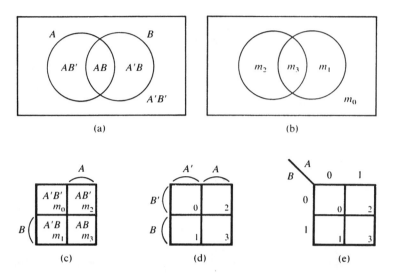

Figure 3.1 Two-variable K-maps.

and the values for C are shown on the left side. Figure 3.3 shows a four-variable K-map.

An n-variable K-map contains 2^n blocks, each corresponding to a minterm of the n-variable function. There will be exactly 2^{n-1} blocks in the K-map, corresponding to each variable in the function.

The unique property of K-maps is that only one variable changes in value as we move from one block on the K-map to an adjacent block. For example, in Figure 3.3, only the value of D changes as we move from block 4 to block 5; only B changes as we move to block 0, and only A changes

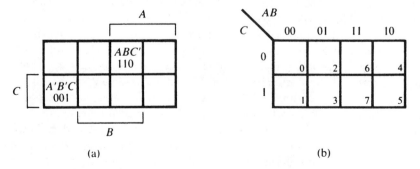

Figure 3.2 Three-variable K-map.

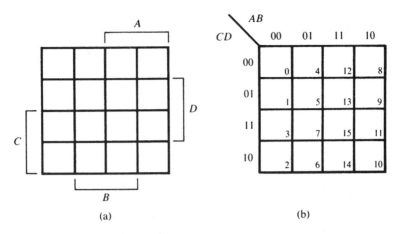

Figure 3.3 Four-variable K-map.

as we move to block 12. Two minterms (maxterms) are said to be logically *adjacent* when they differ in only one variable (i.e., a variable appears in true form in one minterm and complemented form in the other and all other variables appear in the same form in both the minterms). The K-map thus transforms the *logical adjacency* into the *physical adjacency* of minterms. That is, the minterms that are represented physically adjacent to each other on the K-map are also logically adjacent. This property is useful in minimizing functions.

Note that the blocks in the top row of the four-variable K-map (shown in Figure 3.3) are each adjacent to the corresponding blocks in the bottom row, as are the blocks in the first and last columns. Thus, the four-variable K-map should be viewed as a toroid (doughnut) that is formed by bringing the top and bottom edges together to form a cylinder whose ends are brought together to form the toroid.

Block numbering in a K-map follows a simple pattern: starting from the top left corner with a 0, continue numbering the blocks in column 1 sequentially—except that the numbers of last two blocks in the column are interchanged. Continue this numbering scheme with subsequent columns sequentially, except that the last two columns are interchanged. This numbering scheme is valid so long as the assignment of variables is in the order shown in the K-maps in Figures 3.2 and 3.3, wherein the first two variables (A and B) are assigned to columns and the remaining variables (C in Figure 3.2, C and D in Figure 3.3) are assigned to rows.

The following K-maps show another ordering of the variables:

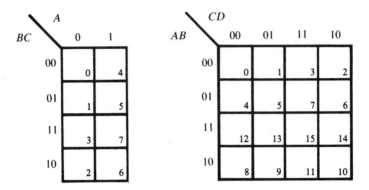

Note that the block numbering also changes accordingly. But these K-maps also maintain the physical adjacency of logically adjacent terms. As such, these maps are also valid forms. (We will retain the ordering shown in Figures 3.2 and 3.3.)

Figure 3.4 shows a five-variable K-map. Note that the left-hand map corresponds to $A = 0$, while the right-hand map corresponds to $A = 1$. Because of this assignment of A, when the two maps (planes) are held one over the other, the two blocks that are in the same position in the planes are also adjacent to each other.

Figure 3.5 shows a six-variable K-map. Among the four planes shown, each is adjacent to the plane to its right (or left) and to the plane below (or

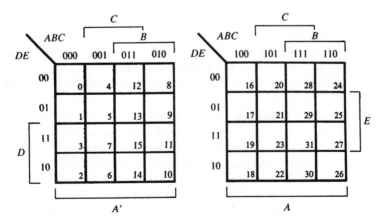

Figure 3.4 Five-variable K-map. The two parts of the map are treated as two planes, one superimposed on the other. The blocks in the same position on each plane are also logically adjacent.

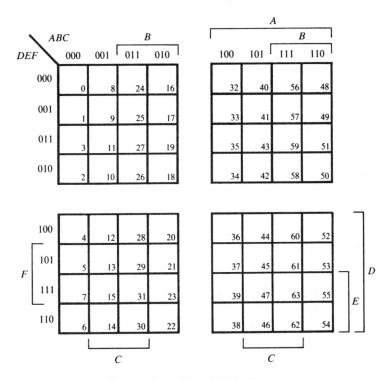

Figure 3.5 Six-variable K-map.

above) it. The planes that are adjacent should be held over each other to bring the logically adjacent blocks into physical adjacency.

As can be seen from Figures 3.2 through 3.5, the number of blocks in a K-map doubles when the number of variables increases by 1. In addition, the degree of adjacency is directly proportional to the number of variables. That is, each block in an *n* variable K-map can have *n* blocks adjacent to it. Because a cube has only six sides, it is difficult to represent an adjacency of more than six on a K-map. Thus, K-maps are useful in representing functions of up to six variables. When the number of variables exceeds 6, tabular methods, such as the QM procedure, are more helpful.

3.1.1 Representation of Functions on K-maps

On Venn diagrams, we represent functions by shading the areas. On K-maps, each block is given a value of 0 or 1, depending on the value of the function.

Each block corresponding to a minterm will have a value of 1; all other blocks will have 0s, as shown in the following examples.

Example 3.1

$F(X, Y, Z) = \Sigma\, m\,(0, 1, 4)$

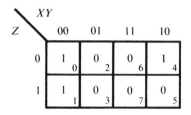

Place a 1 corresponding to each minterm.

Example 3.2

$F(A, B, C, D) = \Pi M\,(1, 4, 9, 10, 14)$

Place a 0 corresponding to each maxterm.

Usually 0s are not shown explicitly on the K-map. Only 1s are shown, with a blank block corresponding to a 0.

3.1.2 Plotting the SOP Form

When the function is given in the SOP form, the equivalent minterm list can be derived by the method of Chapter 2 and the minterms plotted on the K-map. An alternative and faster method is to intersect the areas on the K-map that correspond to each product term, as illustrated in Example 3.3.

Example 3.3

$F(X, Y, Z) = XY' + Y'Z'$

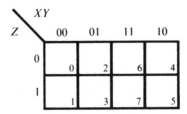

Note: X corresponds to blocks 4, 5, 6, 7 (all the blocks where X is 1). Y' corresponds to blocks 0, 1, 4, 5 (all the blocks where Y' is 1). XY' corresponds to their intersection = 4, 5. Similarly,

$$Y' = 0, 1, 4, 5$$
$$Z' = 0, 2, 4, 6$$
$$\therefore Y'Z' = 0, 4$$

Therefore, the K-map will have 1 in the union of (4, 5) and (0, 4), which is (0, 4, 5), as shown below:

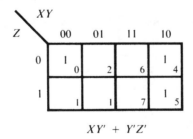

$$XY' + Y'Z'$$

Alternatively, XY' corresponds to the area where $X = 1$ and $Y = 0$, which is the last column; $Y'Z'$ corresponds to the area where both $Y = 0$ and $Z = 0$, which is blocks 0 and 4. Hence the union of the two corresponds to blocks 0, 4, 5.

Note also that in this three-variable K-map, if a product term has two variables missing (as in Y), we use four 1s corresponding to the four minterms that can be generated out of a single-variable product term. In general, a product term with n missing variables will be represented by 2^n 1s on the K-map.

Example 3.4

$$P(A, B, C, D) = AB' + A'BC + C'D'$$

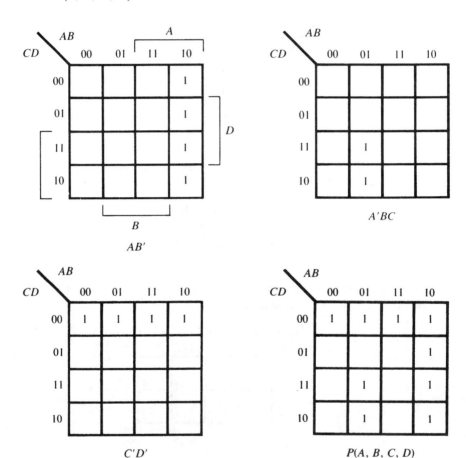

3.1.3   Plotting the POS Form

The procedure for plotting a POS expression is similar to that for the SOP form, except that 0s are used instead of 1s.

Example 3.5

$$F(X, Y, Z) = (X + Y')(Y' + Z')$$

	XY			
Z	00	01	11	10
0		0		
1		0		

$(X + Y') = 0$ only if $X = 0$ and $Y' = 0$. That is, $X = 0$ and $Y = 1$ or the area $(X'Y)$.

	XY			
Z	00	01	11	10
0				
1		0	0	

$(Y' + Z') = 0$ only if $Y' = 0$ and $Z' = 0$; that is, $Y = 1$ and $Z = 1$ or the area (YZ).

	XY			
Z	00	01	11	10
0		0		
1		0	0	

$F(X, Y, Z)$ is 0 when either $(X + Y')$ is 0 or $(Y' + Z') = 0$ or the area $(X'Y) + (YZ)$.

	XY			
Z	00	01	11	10
0	1		1	1
1	1			1

$F(X, Y, Z)$

As can be seen by this example, plotting a POS function on the K-map is somewhat tedious, but straightforward once apparent. An alternative method is simply to convert the POS function into canonical form by the algebraic method of Chapter 2 and insert a 0 into each block of the K-map corresponding to a maxterm in the function. The remaining blocks corresponding to the minterms of the function are then inserted with 1s.

3.1.4 Minimization

Because two terms on a K-map that are physically adjacent are also logically adjacent, and the presence of 1 in a block corresponds to the function taking a value of 1, two adjacent 1s on the K-map can be combined by an OR operation. The result of this operation is to eliminate the variable that changes in value between the two terms and combine the two terms into one, with one fewer literal. Example 3.6 illustrates this process.

Example 3.6

Consider the following K-map for a four-variable function:

	AB 00	01	11	10
CD 00	0	4	1 (12)	1 (8)
01	1	1 (5)	1 (13)	1 (9)
11	3	7	15	11
10	2	6	14	10

Here, blocks 8 and 12 are adjacent. Minterm 8 corresponds to 1000 or $AB'C'D'$ and minterm 12 corresponds to 1100 or $ABC'D'$. Combining these two terms yields

$$AB'C'D' + ABC'D' = AC'D'(B + B') \quad \text{P4b}$$
$$= AC'D'(1) \qquad\qquad \text{P5a}$$
$$= AC'D' \qquad\qquad\quad \text{P1b}$$

This combination is shown below by the grouping of 1s on the K-map. By this grouping, we eliminated one variable, B, from the product term, since B changes its value between the blocks in this grouping.

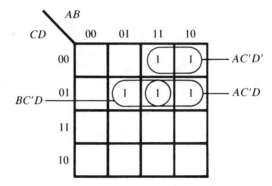

Similarly, grouping blocks 9 and 13 yields $AC'D$, and grouping blocks 5 and 13 yields $BC'D$.

Now, if we combine $AC'D'$ with $AC'D$, i.e.,

$$AC'D' + AC'D = AC'(D + D') \quad \text{P4b}$$
$$= AC' \cdot (1) \qquad\qquad \text{P5a}$$
$$= AC' \qquad\qquad\quad \text{P1b}$$

In effect, this is equivalent to grouping all four 1s in the top right corner of the K-map:

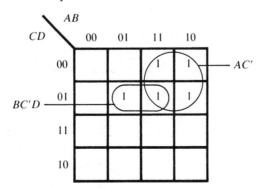

By forming a group of two adjacent 1s, we eliminated one literal from the product term; by grouping four adjacent 1s, we eliminated two literals. In general, if we group 2^n adjacent 1s, we can eliminate n literals. Hence, in simplifying functions it is advantageous to form as large a group of 1s as possible. The number of 1s in any group must be a power of 2 (i.e., 1, 2, 4, 8, ...). Once the groups are formed, the product term corresponding to each group can be derived by the following general rules.

1. Start with a product term that contains all the variables in the function in uncompleted form. Move from block to block within the group to observe the variable that changes in value and eliminate it.
2. For the variables that are not eliminated in step 1:
   (a) If the value of the variable is 1 within the group, leave it uncomplemented.
   (b) If the value of the variable is 0 within the group, complement it.

For example, consider the group of four 1s in the K-map shown above:

$ABCD$	Start with all variables.
$ABCD$	A remains 1 in all four blocks.
$A\!B\!CD$	B changes in value.
$A\!BC'D$	C remains 0.
$A\!BC'\!D$	D changes in value.

Therefore, the product term corresponding to this grouping is AC'.

We can summarize these observations into the following procedure for simplifying functions:

1. Form groups of adjacent pairs of 1s. If a 1 does not have an adjacent 1, it forms a group by itself.
2. Combine the groups formed in the first step into larger groups, if possible. The number of 1s in each group must be a power of 2. Two groups can be combined only if each 1 in the first group has an adjacent 1 in the second group.
3. Cover each 1 on the K-map at least once. The same 1 can be included in several groups if necessary, since $X + X = X$ (Theorem 1).
4. Select the smallest number of groups to cover all the 1s on the map.

5. Translate each group into a product term.
6. OR the product terms to obtain the minimized function.

Recall that in recognizing the adjacencies, the right-hand edge on a three-variable map is treated as the same as the left-hand edge, thus making block 0 adjacent to block 4, and block 1 adjacent to block 5. Similarly, on a four-variable map, the top and bottom edges can be brought together to form a cylinder. The two ends of the cylinder are brought together to form a toroid (like a doughnut). The following examples illustrate the groupings on the K-maps and the corresponding simplifications.

Example 3.7

$F(X, Y, Z) = \Sigma m \ (1, 2, 3, 6, 7)$

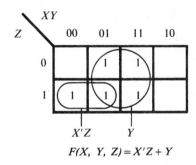

$F(X, \ Y, \ Z) = X'Z + Y$

Example 3.8

$F(A, B, C, D) = \Sigma m \ (2, 4, 8, 9, 10, 11, 13, 15)$

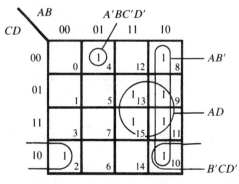

$F(A, \ B, \ C, \ D) = AB' + AD + B'CD' + A'BC'D'$

Example 3.9

$F(X, Y, Z, W) = \Sigma m \ (0, 4, 5, 8, 12, 13)$

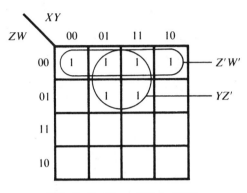

$$F(X, Y, Z, W) = YZ' + Z'W'$$

Example 3.10

$F(A, B, C, D) = \Sigma m \ (0, 1, 2, 7, 8, 9, 10)$

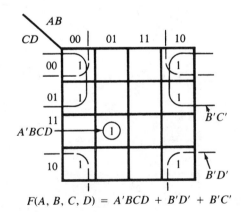

$$F(A, B, C, D) = A'BCD + B'D' + B'C'$$

Example 3.11

$$F(A, B, C, D) = ABC' + ABC + BCD' + BCD + AB'D' + A'B'D'$$
$$+ A'BC'D$$

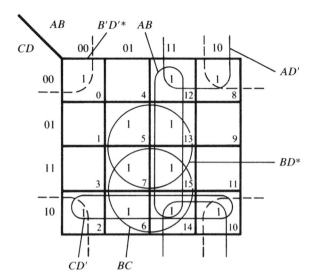

Here, groups marked by an asterisk are "essential," since minterm 0 is covered only by $B'D'$, and minterm 5 is covered only by BD.

Once the essential groups are chosen, the only minterms left uncovered are 6, 12, and 14. To cover 6, we can choose either BC or CD'. Either of these choices covers minterm 14 in addition to minterm 6. To cover the remaining minterm 12, we can choose either AD' or AB. Hence, there are four simplified forms:

$$F(A, B, C, D) = BD + B'D' + BC + AD'$$
$$\text{or} \qquad \text{or}$$
$$CD' \qquad AB$$

Definition: Each of the product terms obtained by the above grouping of 1s in the K-map is called a *Prime Implicant* (PI). A PI is a product term that is not contained by any other product term of the function. It is a candidate to be a term in the minimized function.

Thus, there are six PIs in Example 3.11 and only four are needed in the minimum function, since all the minterms (i.e., all 1s) are *covered* by these four PIs. Note that two of the four PIs are essential since they cover a 1 not covered by any other PI.

Definition: An *essential PI* is a PI that covers a minterm that is not covered by any other PI and hence must be selected to form the minimum function.

If some minterms are left uncovered after the selection of all essential PIs, one or more *nonessential PIs* must be selected to cover those minterms. In the selection of nonessential PIs, the PIs corresponding to larger groupings on the K-map must be selected over those corresponding to smaller groups, since a larger grouping yields fewer literals.

This discussion applies equally well to functions in POS form when "sum" and "maxterm" are substituted for "product" and "minterm."

Obtaining the Minimized Function in POS Form   In Chapter 4, we will examine various circuit realizations of Boolean functions. For some realizations, the function must be in POS form. Minimum POS form can be obtained by combining 0s rather than 1s on the K-map for the given function. The minimization procedure is similar to that for the SOP form, except that in the derivation of sum terms from the grouping of 0s, a variable is complemented if its value in the group is 1. Otherwise, it is uncomplemented. The following alternate procedure can also be adopted to obtain the minimized function in POS form:

1. Plot the function F on the K-map.
2. Derive the K-map for F' by changing 1 to 0 and 0 to 1 on the K-map.
3. Minimize the K-map for F' to obtain F' in SOP form.
4. Use DeMorgan's Law to derive F from F'.

Example 3.12 illustrates both of these procedures.

Example 3.12

(a)   Minimization by grouping 0s

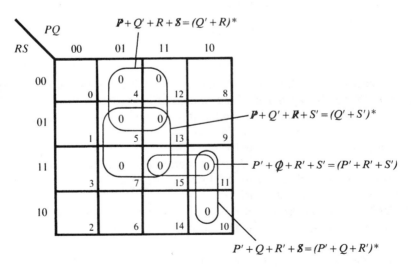

$$F = (Q' + R)(Q' + S')(P' + Q + R')$$

(b)   Alternative procedure

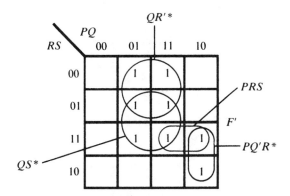

$$F' = QR' + QS + PQ'R$$
$$F = F'' = (QR' + QS + PQ'R)'$$
$$= (QR')' \cdot (QS)' \cdot (PQ'R)' \qquad \text{T5a}$$
$$= (Q' + R)(Q' + S')(P' + Q + R') \qquad \text{T5b}$$

3.2   Incompletely Specified Functions

So far we have assumed that all the combinations of values for variables of a function are valid and that the logic circuit implementing the function will receive all those combinations as inputs. In practice, some functions will be incompletely specified in the sense that some combinations of input values may not occur, or, if a combination occurs, the output given by the circuit may not affect the system operation. Such input conditions are called *don't-cares*. Don't-cares are represented by a *d* on the K-map and truth tables. A *d* can be treated as either a 1 or a 0 when groups of 1s are formed on the K-map. If a *d* on the K-map helps the formation of a larger group, it can be treated as 1; otherwise, it can be treated as a 0 and ignored. Thus, the don't-cares in a function provide some flexibility in the simplification of functions. The following examples illustrate the utility of don't-cares.

Example 3.13

Imagine that our task is to design a logic circuit that produces an output of 1 when the combination of input values correspond to an even decimal number. It is known that the input is always a BCD digit.

Since BCD is a four-bit code, there will be 16 rows in the truth table for the circuit. Only the first 10 combinations are used by the BCD; hence, the last 6 combinations do not occur on the input of the circuit. These 6 combinations are thus don't-cares. The following truth table represents the operation of the desired circuit:

Minterm	Inputs W X Y Z	Output F
0	0 0 0 0	1
1	0 0 0 1	0
2	0 0 1 0	1
3	0 0 1 1	0
4	0 1 0 0	1
5	0 1 0 1	0
6	0 1 1 0	1
7	0 1 1 1	0
8	1 0 0 0	1
9	1 0 0 1	0
10	1 0 1 0	*d*
11	1 0 1 1	*d*
12	1 1 0 0	*d*
13	1 1 0 1	*d* Don't-cares
14	1 1 1 0	*d*
15	1 1 1 1	*d*

We will denote this truth table by using the following minterm list form:

$$F(W, X, Y, Z) = \Sigma m \,(0, 2, 4, 6, 8) + d(10, 11, 12, 13, 14, 15)$$

or

$$= \Sigma m \,(0, 2, 4, 6, 8) + d(10-15)$$

where "−" indicates the range of terms.

In the maxterm list form, the proceeding function is:

$$F(W, X, Y, Z) = \Pi M \, (1, 3, 5, 7, 9) \cdot d(10{-}15)$$

The K-maps shown below illustrate the utility of the don't-care conditions. The first K-map shows the groupings without using the don't-cares, and the second one shows them with the don't-cares. As can be seen, by using don't-cares appropriately, we can reduce the function to one literal, though 5 literals were needed otherwise. Note that in the second K-map, don't-cares in blocks 11, 13 and 15 are not used and hence are treated as 0s, while the other don't-cares are treated as 1s.

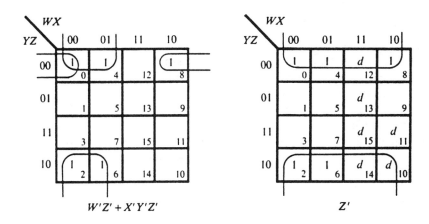

$$W'Z' + X'Y'Z' \qquad\qquad Z'$$

Example 3.14

Let us design a decision-making circuit for an elevator. The function of the circuit is to provide two outputs—UP and DOWN—to indicate the direction in which the elevator is to move. We will assume that the building has three floors. The inputs to the circuit are a direction signal (D), which if 1 indicates that the elevator is moving up (otherwise, the elevator is moving down) and three floor request signals (S_1, S_2, and S_3) corresponding to each floor. The truth table indicating the circuit operation at floor 2 is shown here:

D	S_1	S_2	S_3	UP	DOWN
0	0	0	0	0	0
0	0	0	1	1	0
0	0	1	0	d	d
0	0	1	1	d	d
0	1	0	0	0	1
0	1	0	1	0	1
0	1	1	0	d	d
0	1	1	1	d	d
1	0	0	0	0	0
1	0	0	1	1	0
1	0	1	0	d	d
1	0	1	1	d	d
1	1	0	0	0	1
1	1	0	1	1	0
1	1	1	0	d	d
1	1	1	1	d	d

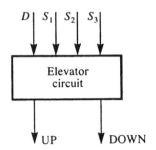

Once the elevator is at floor 2, S_2 will not be on, and hence the input combinations corresponding to $S_2 = 1$ will yield don't-care outputs. Generally the elevator will move up if there is a request for floor 3 and down if there is a request for floor 2. If there are simultaneous requests for both floors, however, the conflict is resolved based on the input signal D, to continue the direction of travel up or down.

The K-maps for UP and DOWN outputs and the corresponding functions are shown below:

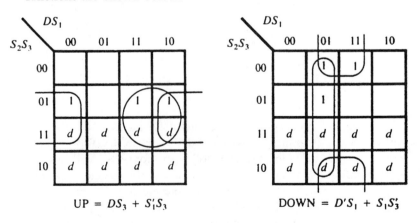

UP = $DS_3 + S_1'S_3$ DOWN = $D'S_1 + S_1S_3'$

This is an overly simplified circuit. Similar circuits are required at the other two floors. As is probably apparent, there will be additional don't care conditions for these circuits, since the elevator can travel in only one direction from floors 1 and 3.

3.3 Quine-McCluskey Procedure

As mentioned earlier, K-maps are useful for functions of up to five or six variables. Since the number of blocks in a K-map doubles with the addition of each variable, minimizing functions with many variables becomes tedious. The QM procedure is a tabular procedure that is useful in such cases.

The QM procedure also uses the logical adjacency property to reduce the Boolean function. Two minterms are logically adjacent if they differ in one and only one position (i.e., one of the variables appears in uncomplemented form in one minterm and in complemented form in the other). Such a variable is eliminated by combining the two minterms. The QM procedure compares each minterm with all the others and combines them if possible. The QM procedure uses the following steps:

1. First, the minterms (and don't-cares) of the function are classified into groups so that each term in a group contains the same number of 1s in the binary representation of the term.
2. Then the groups formed in step 1 are arranged in the increasing order of number of 1s. Let the number of groups be n.
3. Each minterm in the group i ($i = 1$ to $n - 1$) is compared with those in group ($i + 1$); if the two terms are adjacent, a combined term is formed. The variable thus eliminated is represented as "$-$" in the combined term.
4. The matching operation of step 3 is repeated on the combined terms until no more combinations can be done. Each combined term in the final list is a PI.
5. A *PI chart* is then constructed, in which there is one column for each minterm (don't-cares are not listed) and one row for each PI. An $\times$ in a row-column intersection indicates that the PI corresponding to the row covers the minterm corresponding to the column.
6. Then all the essential PIs (i.e., the PIs that cover at least one minterm not covered by any other PI) are located.
7. A minimum number of PIs from the remaining ones are selected to cover those minterms not covered by the essential PIs.
8. The set of PIs thus selected forms the minimum function.

The QM procedure is illustrated in Example 3.15.

Example 3.15

$$F(A, B, C, D) = \Sigma m \underbrace{(0, 2, 4, 5, 6, 9, 10)}_{\text{Minterms}}$$
$$+ \Sigma d \underbrace{(7, 11, 12, 13, 14, 150)}_{\text{Don't-cares}}$$

STEPS 1 AND 2:

Each minterm and don't-care is expanded into binary form, and groups of terms with the same number of 1s are formed. The groups are then arranged in the order of increasing numbers of 1s. Note that the don't-cares are treated as equivalent to minterms until the selection of PIs in step 5.

✓	0	0000	Group 0: terms with no 1s
✓	2	0010	Group 1: terms with one 1
✓	4	0100	
✓	5	0101	
✓	6	0110	
✓	9	1001	Group 2: terms with two 1s
✓	10	1010	
✓	12	1100	
✓	7	0111	
✓	11	1011	Group 3: terms with three 1s
✓	13	1101	
✓	14	1101	
✓	15	1111	Group 4: terms with four 1s

STEP 3:

The minterm in group 0 is compared with each minterm in group 1. Because 0 is adjacent to both 2 and 4, we obtain (0, 2) or 00-0 and (0, 4) or 0-00. Note that the hyphen indicates the missing variable when the adjacent terms are combined. This completes all the possible comparisons between groups 0 and 1. The terms that were combined are shown with a check mark.

We now repeat the comparison process between terms in groups 1 and 2. Minterm 2 combines with only 6 and 10, and minterm 4 combines with only 5, 6, and 12 to yield the five terms shown in group 2 below. This matching process is repeated with groups 2 and 3 and 3 and 4. Note that all the original terms have participated in the combining process, as indicated by the check marks. If there was a term that could not have been combined with any other term, it would have been a PI.

√(0, 2)	00-0	(obtained by matching groups 0 and 1)
√(0, 4)	0-00	
√(2, 6)	0-10	
√(2, 10)	-010	
√(4, 5)	010-	(obtained by matching groups 1 and 2)
√(4, 6)	01-0	
√(4, 12)	-100	
√(5, 7)	01-1	
√(5, 13)	-101	
√(6, 7)	011-	
√(6, 14)	-110	
√(9, 11)	10-1	(obtained by matching groups 2 and 3)
√(9, 13)	1-01	
√(10, 11)	101-	
√(10, 14)	1-10	
√(12, 13)	110-	
√(12, 14)	11-0	
√(7, 15)	-111	
√(11, 15)	1-11	(obtained by matching groups 3 and 4)
√(13, 15)	11-1	
√(14, 15)	111-	

Note: The check mark indicates that the term is used at least once in forming a combined term.

STEP 4:

We now continue with the process of matching the terms obtained in the previous step. In order for two terms to be combined now, they must have a hyphen in the same position and must differ in one position only. Thus, (0, 2) from the first group combines with (4, 6) of the second group, yielding (0, 2, 4, 6) or 0--0. Similarly, (0, 4) can be combined with (2, 6). Since this combination yields the same term as the previous one, it is discarded and now shown.

(0, 2, 4, 6)	0--0	Same as (0, 2, 4, 6)
(2, 6, 10, 14)	--10	Same as (2, 10, 6, 4)
√(4, 5, 6, 7)	01--	Same as (4, 6, 5, 7)
√(4, 5, 12, 13)	-10-	Same as (4, 12, 5, 13)
√(4, 6, 12, 14)	-1-0	Same as (4, 12, 6, 14)
√(5, 7, 13, 15)	-1-1	Same as (5, 13, 7, 15)
√(6, 7, 14, 15)	-11-	Same as (6, 14, 7, 15)
(9, 11, 13, 15)	1--1	Same as (9, 13, 11, 15)
(10, 11, 14, 15)	1-1-	Same as (10, 14, 11, 15)
√(12, 13, 14, 15)	11--	Same as (12, 14, 13, 15)

We now repeat the comparison between the three groups obtained to derive additional combinations. No combinations are possible between the first two groups. Similarly, (2, 6, 10, 14) in the second group cannot be combined with any term in the third group. The other three terms in the second group combine with some terms in the third group. Two terms in the third group do not combine with any term. The following table shows the combinations so far.

(0, 2, 4, 6)		0--0	PI_1
(2, 6, 10, 14)		--10	PI_2
(4, 5, 6, 7, 12, 13, 14, 15)		-1--	PI_3
(9, 11, 13, 15)		1--1	PI_4
(10, 11, 14, 15)		1-1-	PI_5

No other combinations are possible. Hence, the five terms above are the PIs.

STEP 5:

We can now draw a PI chart in which there will be one row corresponding to each PI and one column corresponding to each minterm. Don't-cares are ignored. An × in the row-column intersection on the

PI chart indicates that the minterm corresponding to that column is covered by the PI corresponding to the row. The PI chart for the above function is shown below.

	Minterms						
	✓ 0	✓ 2	✓ 4	✓ 5	✓ 6	✓ 9	10
* PI_1	⊗	×	×		×		
PI_2		×		×		×	
* PI_3			×	⊗	×		
* PI_4						⊗	
PI_5							×

STEP 6:

If there is a single × in any column of the PI chart, the minterm corresponding to that column is covered by only one PI, the PI corresponding to the row in which the × is located. This PI is essential. PI_1, PI_3, and PI_4 are thus essential. Single ×s are circled on the PI chart to highlight essential PIs.

STEP 7:

Once PI_1, PI_3, and PI_4 are selected, all the minterms except 10 are covered. (The minterms covered by the PI are checked as soon as the PI is selected.) To cover minterm 10, we can select either PI_2 or PI_5, since they contribute the same number of literals.

STEP 8:

The reduced function is

$$F(A, B, C, D) = PI_1 + PI_3 + PI_4 + PI_2 \text{ or } PI_5$$
$$= 0\text{--}0 + \text{-}1\text{--} + 1\text{--}1 + \text{--}10 \text{ or } 1\text{-}1\text{-}$$
$$= (A'D' + B + AD + CD')$$
$$\text{or } (A'D' + B + AD + AC)$$

Example 3.16 compares the results obtained by the K-map minimization method with those from the QM procedure.

Example 3.16

Consider the four-variable function of Example 3.11:

$$F(A, B, C, D) = \Sigma m \, (0, 2, 5, 6, 7, 8, 10, 12, 13, 14, 15)$$

minimization by the QM procedure is shown below:

STEPS 1 AND 2:

0	0000	Group 0
2	0010	Group 1
8	1100	
5	0101	Group 2
6	0110	
10	1010	
12	1100	
7	0111	Group 3
13	1101	
14	1110	
15	1111	Group 4

STEP 3:

(0, 2)	00-0
(0, 8)	-000
(2, 6)	0-10
(2, 10)	-010
(8, 10)	10-0
(8, 12)	1-00
(5, 7)	01-1
(5, 13)	-101
(6, 7)	011-
(6, 14)	-110
(10, 14)	1-10
(12, 13)	110-
(12, 14)	11-0
(7, 15)	-111
(13, 15)	11-1
(14, 15)	111-

(This step corresponds to grouping adjacent 1s on the K-map.)

Step 4:

(0, 2, 8, 10)	-0-0	PI_1
(2, 6, 10, 14)	--10	PI_2
(8, 10, 12, 14)	1--0	PI_3
(5, 7, 13, 15)	-1-1	PI_4
(6, 7, 14, 15)	-11-	PI_5
(12, 13, 14, 15)	11--	PI_6

(This step corresponds to forming groups of four 1s on the K-map.)

Step 5:

PI Chart

	Minterms										
	√	√	√		√	√	√		√		
	0	2	5	6	7	8	10	12	13	14	15
* PI_1	⊗	×				×	×				
PI_2		×		×			×			×	
PI_3						×	×	×		×	
* PI_4			⊗		×				×		×
PI_5			×	×					×	×	
PI_6								×	×	×	×

Step 6:

The essential PIs are 1 and 4, since they are the only PIs that cover minterms 0 and 5, respectively. Once these PIs are selected, all minterms but 6, 12, and 14, are covered.

Step 7:

For minterm 6 to be covered, either PI_2 or PI_5 is chosen. Minterm 14 is covered by either choice. For the remaining minterm 12 to be covered, either PI_3 or PI_6 is chosen.

STEP 8:

Thus, there are four possible minimum functions (which can *also be obtained from the K-map method* shown in Example 3.11:

$$F(A, B, C, D) = PI_1 + PI_3 + PI_2 + PI_3$$

or   or

$$PI_5 \quad PI_6$$

The QM procedure is easily programmable on a computer. Because the procedure stores each minterm and all intermediate terms that are generated, large storage capacity is needed as the number of variables in the function grows. For each additional variable, the amount of storage required doubles. Furthermore, the number of comparisons grows exponentially, since the procedure depends on an exhaustive comparison of terms. Several modifications have been used to reduce the storage and computing time requirements. These modified QM procedures were implemented as computer programs during the 1970s. (For a listing of such programs, see the works by Shiva and Nagle, and Dietmeyer that appear in the references at the end of this chapter.)

3.3.1 Covering Problem

The process of selecting a minimum number of PIs to cover a Boolean function (i.e., finding a *minimum cover*) from the PI chart is not always as simple as the one in Example 3.16. The following rules can be used to derive the minimum number of PIs when the PI chart is complex.

Definition: A row (column) i of a PI chart dominates row (column) j if row (column) i contains an $\times$ in each column (row) in which row (column) j contains an $\times$.

Because a row of the PI chart corresponds to a PI, a dominating row covers all the minterms that are covered by a row dominated by it. Hence the following rule is valid.

RULE 1:

A row dominated by another row can be eliminated from the chart, providing the PI corresponding to the dominated row does not have fewer literals compared with the PI corresponding to the dominating row. When several identical rows are present in a chart, all but the one whose PI has the fewest literals can be eliminated.

When a column dominates another column, the PIs that cover the minterm corresponding to the dominated column also cover the minterm corresponding to the dominating column. Hence the following rule.

RULE 2:

A column dominating another column can be eliminated. All but one of the identical columns can be eliminated.

These rules are useful in reducing the complexity of the PI chart. The general procedure for finding a minimum cover for the Boolean function can be summarized as follows.

1. Select all the essential PIs. If these PIs cover all the minterms, stop; otherwise, go to the second step.
2. Apply Rules 1 and 2 to eliminate redundant rows and columns from the PI chart of nonessential PIs. When the chart is thus reduced, some PIs will become essential (i.e., some columns will have a single ×). Go back to the first step.

Rules 1 and 2 do not always help in reducing PI charts. A *cyclic chart* is a chart that contains no essential PIs and cannot be reduced by these rules. In such cases, a row corresponding to a PI with a minimum number of literals is first selected (arbitrarily). The row corresponding to that PI and columns corresponding to the minterms covered by the PI are then removed from the chart. If possible, the two-step procedure shown above is then applied. If the chart remains cyclic after the selection of the first PI, another PI is selected, again arbitrarily. A minimum cover is obtained by repeating the process of selecting a row and applying the two-step procedure.

Examples 3.17 and 3.18 illustrate the application of these procedures.

Example 3.17

Minimize $F(A, B, C) = \Sigma m\ (0, 1, 3, 4, 6, 7)$.

The following lists show the derivation of PIs using the QM procedure:

Minterm ABC		Term	ABC	PI
✓ 0	000	(0,1)	00-	1
✓ 1	001	(0, 4)	-00	2
✓ 4	100	(1, 3)	0-1	3
✓ 3	011	(4, 6)	1-0	4
✓ 6	110	(3, 7)	-11	5
✓ 7	111	(6, 7)	11-	6

The PI chart:

| | Minterms | | | | | |
PI	0	1	3	4	6	7
1	×	×				
2	×			×		
3		×	×			
4				×	×	
5			×			×
6					×	×

As can be seen, the PI chart shown here does not have an essential PI and cannot be reduced by applying Rules 1 and 2. It is cyclic. By arbitrarily selecting PI_1 and removing the minterms covered by PI_1 (i.e., 0 and 1), we obtain the following chart:

PI	3	4	6	7
2		×		
3	×			
4		×	×	
5	×			×
6			×	×

Row 4 dominates row 2. Hence, row 2 can be removed. After this removal, column 6 dominates column 4, and hence column 6 can be removed, yielding the following chart:

| | Minterms | | |
PI	3	4	7
3	×		
* 4		⊗	
5	×		×
6			×

PI$_4$ is now essential and covers minterms 4 and 6. Further, row 5 dominates rows 3 and 6, which can be removed, leaving PI$_5$ to cover minterms 3 and 7.

Thus, the minimum cover consists of PI$_1$, PI$_4$, and PI$_5$. That is,

$$F(A, B, C) = A'B' + AC' + BC$$

Example 3.18

Minimize

$$F(A, B, C, D, E) = \Sigma m\ (0, 1, 2, 5, 14, 16, 18, 24, 26, 30)$$
$$+ d(3, 13, 28).$$

We will leave the derivation of PIs for this function as an exercise but will show the PI chart below:

					Minterms					
					✓					✓
PI		0	1	2	5	14	16	18	24	26	30
1	(24, 26, 28, 30)								×	×	×
2	(16, 18, 24, 26)						×	×	×	×	
3	(0, 2, 16, 18)	×		×			×	×			
4	(0, 1, 2, 3)	×	×	×							
*5	(14, 30)					⊗					×
6	(5, 13)				×						
7	(1, 5)		×		×						

PI$_5$ is essential and covers minterms 14 and 30. After that PI is selected, the chart reduces to the following:

			Minterms					
PI	0	1	2	5	16	18	24	26
1							×	×
2					×	×	×	×
3	×		×		×	×		
4	×	×	×					
6				×				
7		×		×				

Row 2 dominates row 1, and row 7 dominates row 6. Once rows 1 and 6 are removed, PI_2 and PI_7 become essential. The selection of PI_2 and PI_7 covers all minterms except 0 and 2. These two can be covered through the selection of either PI_3 or PI_4, since both of them have the same number of literals.

Hence, the minimum cover for the function consists of PI_5, PI_2, PI_7, and either PI_3 or PI_4.

3.3.2 Multiple-Output Functions

In practice, a logic circuit can have more than one output. Such circuits are represented by multiple functions, one for each output. Although each function can be minimized separately by the QM procedure, simultaneous minimization of functions yields a better minimization, since some terms may be able to be shared between the functions. The QM procedure can be adapted to minimize multiple-output functions of the same set of input variables. To extend the QM procedure to the multiple-output case, the following rules are necessary.

1.   Each minterm must have a flag to show which function it is a minterm of.
2.   Two minterms can be combined only if they possess one or more common flags in addition to being logically adjacent. The resulting term from the combination carries only those flags that are common to both terms that were combined.
3.   A term can be checked off from the list of terms only if all the flags that the term possesses appear in the resulting term when two terms are combined.

Example 3.19 illustrates this minimization procedure.

Example 3.19

Minimize

$$A(P, Q, R, S) = \Sigma\, m\, (0, 2, 10, 12)$$
$$B(P, Q, R, S) = \Sigma\, m\, (2, 4, 6)$$
$$C(P, Q, R, S) = \Sigma\, m\, (0, 2, 10, 14, 15)$$

We will first form the list of minterms from all the functions in the usual order and attach flags to each minterm, as shown below:

Minterm	PQRS	Flag
√ 0	0000	AC
2	0010	ABC
√ 4	0100	B
√ 6	0110	B
√10	1010	AC
12	1100	A
√14	1110	C
√15	1111	C

Note that 0 can be combined only with 2, resulting with (0, 2) with a flag of *AC*. 0 and 4 cannot be combined, since they have no common flag. Furthermore, 2 cannot be checked off, since all the three flags of 2 cannot be attached to (0, 2), but 0 can be checked off. The following list shows the result of combining these terms:

Term	PQRS	Flag
(0, 2)	00-0	AC
(2, 6)	0-10	B
(2, 10)	-010	AC
(4, 6)	01-0	B
(10, 14)	1-10	C
(14, 15)	111-	C

No further reduction of the preceding list is possible. We can then draw the PI chart. In the PI chart, we will have three sections, one for each function.

PI	Flag	A				B			C				
		√ 0	√ 2	√ 10	√ 12	√ 2	√ 4	√ 6	√ 0	√ 2	√ 10	√ 14	√ 15
(2)	ABC		×			×				×			
*(12)	A				⊗								
*(0, 2)	AC	⊗	×						⊗	×			
(2, 6)	B					×		×					
*(2, 10)	AC		×	⊗						×	×		
*(4, 6)	B						⊗	×					
(10, 14)	C										×	×	
*(14, 15)	C											×	⊗

From this chart, we can see that all PIs except (2), (2, 6), and (10, 14) are essential. The following are the minimum covers:

Function A: (0, 2), (12), and (2, 10)
Function B: (4, 6) and select (2, 6) as a nonessential PI
Function C: (0, 2), (2, 10), and (14, 15)

Note that two PIs are common to functions A and C, thereby reducing the cost of implementation.

3.4   CAD Tools for Simplification

As we have seen in this chapter, the K-map technique for simplification becomes impractical for functions with more than six variables. The QM procedure is a better technique when the number of variables exceeds six. But, a major problem is that the number of PIs grows rapidly as the number of variables in the function increases. Finding a minimum cover in such cases is also very difficult. Several heuristic techniques have been devised over the years. These techniques concentrate on generation of a subset of PIs that can cover the function, rather than finding all PIs. They also attempt to find a minimum cover rapidly. There have been several implementations of the QM procedure combined with heuristics techniques, as computer programs. Refer to Bartee [1961] and Shiva and Nagle [1974] for some examples. *Expresso*, available from the University of California, Berkeley, is another two-level minimization program. The book by Brayton, et al [1984] provides further details on various minimization techniques. There are also commercial packages for logic minimization available. CALCAD from SOFCAD Electronics, and SCALD system from Valid Logic Systems are two examples. The field of CAD tools for logic design is still very active. Refer to *IEEE Transactions on CAD of Integrated Circuits and Systems, IEEE*

Design and Test of Computers magazine and commercial publications such as *EDN* and *Computer Design* for the latest details.

3.5 Summary

This chapter has presented two logic minimization techniques. Both are superior to the algebraic technique, since they are mechanical and thus are less error prone. The K-map method is suitable for functions of up to five or six variables, while the QM procedure is suitable for functions with any number of variables, although the complexity increases rapidly, as the number of variables and hence the number of PIs increase. The CAD tools mentioned in the previous section would be useful for minimizing large functions. If the function has a large number of variables, but only a few terms, it might still be easier to use the algebraic method for simplification. The progress in hardware technology has made it possible to fabricate very complex hardware on an IC chip and has brought down the cost of hardware dramatically. It is said that minimization at the gate level implementation is not all that important anymore because of the low cost of hardware; rather the emphasis is on minimizing the number of chips in the system. Thus, the classic minimization techniques described here have lost the limelight, though an exposure to them makes one a better logic designer.

References

Bartee, T. C., "Computer Design of Multiple-Output Logical Networks," *IRE Transactions on Electronic Computers*, EC-10 (2):21–30, 1961.

Brayton, R. K., G. D. Hachtel, C. T. McMullan, and A. L. Sangiovanni-Vincentelli, *Logic Minimization Algorithms for VLSI Synthesis*, Boston: Kluwer, 1984.

Karnaugh, M., "The Map Method for Synthesis of Combinational Logic Circuits." *AIEE Communications on Electronics* (November 1953):593–599.

Kohavi, Z., *Switching and Finite Automata Theory*. New York: McGraw-Hill, 1970.

Mano, M., *Digital Design*. Englewood Cliffs, N.J.: Prentice-Hall, 1991.

McCluskey, E. J., *Introduction to the Theory of Switching Circuits*. New York: McGraw-Hill, 1965.

Prather, R. E., *Introduction to Switching Theory: A Mathematical Approach*. Boston: Allyn and Bacon, 1967.

Quine, W. V., "The Problem of Simplifying Truth Functions." *American Mathematical Monthly* 59 (October 1952):521–531.

Roth, C. H., *Fundamentals of Logic Design*. St. Paul, Minn.: West Publishing, 1985.

Shiva, S. G. and Nagle, H. T., A series of three articles on computer-aided logic design. *Electronic Design* 22 (October 11, October 25, and November 8, 1974).

Problems

3.1   Given $P(A, B, C) = \Sigma m\ (0, 1, 3) + d(2, 7)$ and $Q(A, B, C) = \Sigma m\ (1, 3, 5, 7)$
       $+ d(6)$, use K-maps to find:
       (a) P' in minterm list form.
       (b) P' in maxterm list form.
       (c) $(P' \cdot Q)$ in minterm list form.
       (d) $(P' + Q)$ in maxterm list form.

3.2   Solve problem 2.4 using K-maps.

3.3   Solve problem 2.9 using K-maps.

3.4   Solve problem 2.10 using K-maps.

3.5   Solve problem 2.11 using K-maps.

3.6   Find the minimum SOP forms for each of the following functions using K-map:
       (a) $F = \Sigma m\ (0, 2, 3, 4, 6)$
       (b) $F = \Pi M\ (0, 1, 4)$
       (c) $F = BC'D' + BC'D + A'C'D' + BCD' + A'B'CD'$

3.7   Find the minimum POS forms for the functions in problem 3.6.

3.8   Given $P(A, B, C, D) = \Sigma m\ (0, 2, 4, 7, 8, 10)$ and $Q(A, B, C, D) = ABD +$
       $B'C'D$, use K-maps to find $(P \oplus Q)$ in the minimum (a) SOP form, and (b)
       POS form.

3.9   For the following functions:
        (i) $F = \Sigma m\ (1, 4, 5, 6, 8, 9, 11) + d(7, 15)$
       (ii) $F = \Sigma m\ (2, 3, 6, 8, 9, 11, 13) + d(1, 12, 14)$
      (iii) $F = \Sigma m\ (3, 6, 7, 8, 9, 10, 18, 21, 22, 23, 26, 29, 30)$

       find, using K-maps:
       (a) All PIs.
       (b) Essential PIs.
       (c) The minimum SOP form.
       (d) The minimum SOP form for F'.
       (e) The minimum POS form for F.
       (f) The minimum POS form for F'.

3.10   Use K-maps to simplify the following:
       (a) $F(A, B, C, D) = \Sigma m\ (2, 3, 4, 10, 12, 13) + d(11, 14, 15)$
       (b) $F(A, B, C, D, E) = \Sigma m\ (0, 7, 11, 13 - 16, 23, 28 - 31)$
                               $+ d(1, 2, 17, 19, 25)$

3.11   Solve problem 3.9 using the QM procedure.

3.12   Verify your solutions for problem 3.10 using the QM procedure.

3.13 A logic circuit is to be designed with two control inputs ($C1$ and $C2$) and one data input (D). The output $Z = 1$, when $C1 = C2 = 1$; $Z = 0$ when $C1 = C2 = 0$; $Z = D$ when $C1 = 1$ and $C2 = 0$; $Z = D'$ when $C1 = 0$ and $C2 = 1$.
   (a) Draw the truth table for the circuit.
   (b) Derive the minimum SOP form for Z.
   (c) Derive the minimum POS form for Z.
   (d) Implement the minimum circuit from (b) and (c).

3.14 Given

$$F(P, Q, R, S, T, V) = \Sigma m \ (1, 2, 3, 15, 17, 18, 19, 26, 32, 48, 63)$$
$$+ \ d(16, 28, 29, 30)$$

   (a) Find the minimum SOP form using the QM procedure.
   (b) Find the minimum SOP form assuming there are no don't-cares.
   (c) Find the minimum SOP form assuming all the don't-cares are changed to minterms.
   Answer (b) and (c) by reworking the PI chart obtained in (a).

3.15 Minimize the following functions using the QM procedure:
   (a) $P(A, B, C, D, E, F) = \Sigma m \ (16, 28, 53, 60, 63)$
   (b) $P(A, B, C, D, E, F) = \Sigma m \ (7, 8, 9, 13, 17, 41, 45, 57)$
       $+ \ d(1, 15, 44, 56)$
   (c) $P(A, B, C, D, E, F, G) = \Sigma m \ (28, 39, 52, 65, 102, 103, 120)$

3.16 Minimize the following multiple-output functions using the QM procedure:
   (a) $P(A, B, C, D) = \Sigma m \ (1, 2, 7, 14, 15)$
       $Q(A, B, C, D) = \Sigma m \ (1, 2, 9, 12, 13, 15)$
       $R(A, B, C, D) = \Sigma m \ (2, 7, 11, 15)$
   (b) $X(A, B, C, D) = \Sigma m \ (6, 9, 14) + d(3, 4, 11)$
       $Y(A, B, C, D) = \Sigma m \ (3, 7, 14) + d(9, 12)$

3.17 Assume that a combinational circuit has four inputs and one output. The output is the ODD parity bit for the 4-bit data on the input. That is, the output is 1 if the number of nonzero bits in the input is even; otherwise, it is 0. Draw the truth table and implement a minimum circuit.

3.18 Assume that the inputs to the circuit in problem 3.17 are BCD values. That is, only patterns corresponding to 0 through 9 are valid. Design the circuit.

3.19 Design a minimum circuit to convert a BCD number to an excess-3 number. That is, the circuit will have four inputs and four outputs. Use four K-maps to simplify the circuit.

3.20 Use the multiple-output QM procedure to simplify the functions in problem 3.19. Compare the complexity of this implementation to the one in problem 3.19.

4

Combinational Circuits

4.1 Introduction

A combinational circuit generally consists of several input signals, several output signals, and an interconnection of gates. Each output is a function of the combination of inputs to the circuit. The input and output signals correspond to binary variables. We examined the binary system of representation in Chapter 1. The Boolean algebraic concepts introduced in Chapter 2 are used in representing and manipulating the functions describing the combinational circuit. The minimization techniques introduced in Chapter 3 are used in reducing the complexity of these circuits. (Combinational circuit design was introduced informally in Chapter 2, in Examples 2.1 and 2.2.) We will formalize the design procedure in this chapter and introduce various circuit implementations of a given set of functions describing a combinational circuit.

Combinational circuits are the basic components of any digital system. In fact, a sequential circuit consists of a combinational logic portion and a set of memory elements, which in turn are most often an interconnection of gates (see the discussion of flip-flops in Chapter 6). As such, the analysis and design of combinational circuits are the most important aspects of logic design. We will introduce the analysis and design procedures in this chapter. At this point, we will not be concerned with the electronic level of detail;

rather, we will assume that the gates are primitive circuit components and that all signals are binary-valued variables. Chapter 10 will provide electronic-level details of gates and introduce various IC technologies. In Chapter 5 we will return to combinational circuits to examine the most popular circuits available in the form of ICs and how to design using them as the circuit components. Chapter 9 will take up the design of combinational circuits using programmable IC devices.

Several types of analysis can be performed on a logic circuit. We performed the complexity analysis of the circuits in Chapter 3, before logic minimization. There are other popular analysis methods as well.

Functional analysis involves deriving the functionality or the relation between the circuit's outputs and inputs. The result of this analysis is either the truth table for the circuit or the expression of each output as a function of the inputs to the circuit.

Each gate in the circuit introduces a *propagation delay* into the flow of a signal through it. The circuit response time is thus a function of delay characteristics of the circuit's gates. *Timing analysis* determines the response time (and hence speed) and other timing characteristics that may contribute to erroneous operation of the circuit.

Load analysis enables the determination of excess loads (if there are any) in the circuit. Each gate connected to the output of another gate *loads* the latter gate. Electronic constraints limit the load on a gate and hence the number of gates connected to its output.

In the next section, we will discuss functional analysis. Later sections will discuss the combinational logic design procedure and various implementation schemes, as well as timing and loading analysis methods.

4.2 Functional Analysis

Figure 4.1 shows the model of an n-input, m-output combinational circuit. Since there are n inputs, there will be 2^n combinations of input values. For each of these combinations of inputs, there will be only one output combination. The truth table for this circuit will have 2^n rows and $n + m$ columns. Alternatively, the function performed by the circuit can be described by m Boolean functions.

Example 4.1

Consider the circuit shown in Figure 4.2. There are four input signals (W, X, Y, and Z) and two output signals (P and Q). We can trace the signal propagation through the gates from the inputs to the outputs of the circuit, as shown in the figure, to derive P and Q as functions of W, X, Y, and Z. The functions are

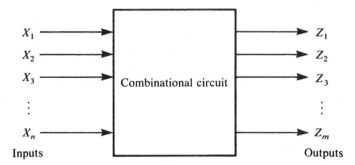

Figure 4.1   Combinational circuit.

$$P = XY' + X'Y + (WZ)'$$
$$Q = (XY(WZ)' + XY' + X'Y + (WZ)')'$$

These functions can be converted to standard or canonical forms, or a truth table can be derived using the procedures from Chapter 2. The truth table can also be derived by tracing the signal values from the inputs to the outputs for each combination of values possible. Because there are four input variables, we first draw the truth table with 16 combinations of input values (see Figure 4.3(b)). We then impose each combination of values on the input lines and note the values for the outputs tracing through the circuit. For example, Figure 4.3(a) shows the condition corresponding to $W = 0$, $X = 0$, $Y = 0$, and $Z = 0$. Tracing through the circuit, we note that $P = 1$ and $Q = 0$. This process is

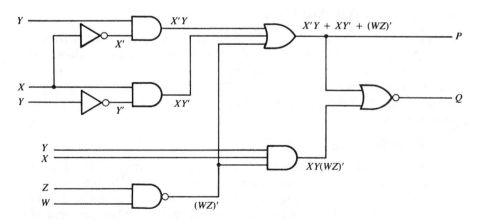

Figure 4.2   A sample circuit.

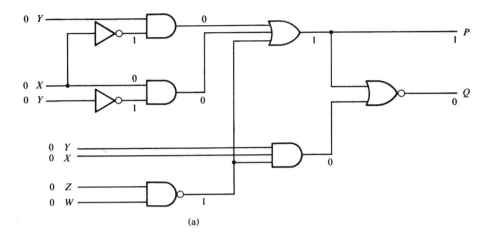

(a)

W X Y Z	P Q
0 0 0 0	1 0
0 0 0 1	1 0
0 0 1 0	1 0
0 0 1 1	1 0
0 1 0 0	1 0
0 1 0 1	1 0
0 1 1 0	1 0
0 1 1 1	1 0
1 0 0 0	1 0
1 0 0 1	0 1
1 0 1 0	1 0
1 0 1 1	1 0
1 1 0 0	1 0
1 1 0 1	1 0
1 1 1 0	1 0
1 1 1 1	0 1

(b)

Figure 4.3 Analysis example. (a) Circuit with 0000 inputs; (b) truth table.

repeated for the other 15 input combinations to determine the complete truth table shown in (b).

Note that the above functions can be simplified to

$$P = XY' + X'Y + W' + Z'$$
$$Q = WX'Y'Z + WXYZ$$

Boolean functions are first minimized before realizing the circuit using gates. But if a circuit is available, it is first analyzed to derive its output functions, which can then be checked for minimization potential.

4.3   Design

Design is the process of converting the functional specification into a logic circuit. The combinational logic design procedure consists of the following steps:

1. From the word statement of the process for which the logic circuit is to be built, derive the input and the output signals required.
2. Analyze the word statement and derive the relation between each output and the inputs and express those relations using a truth table.
3. Obtain the minimum function for each output.
4. Implement the circuit from the minimum functions.

The first two steps are probably the most difficult, since the interpretation of the word statement of the process is not always straightforward. The word statement may not be precise and may not even be complete. But these two steps are critical, since any error made at this point is propagated through the remaining steps in the design procedure. It is important to identify any don't-care conditions that exist at this level. It may not always be necessary to derive the truth table. In some cases, it is possible to derive the output functions directly from the word statement.

The output functions are minimized either individually or by the multiple-output minimization technique shown in Chapter 3.

Various forms of circuits can then be built to realize each function. The four popular realizations are

1. AND-OR.
2. NAND-NAND.
3. OR-AND.
4. NOR-NOR.

The first two forms are derived from the SOP form of the function, and the last two are derived from the POS form. We will illustrate the realization of all four forms in Example 4.2.

Example 4.2

We will start with the following truth table description of the circuit:

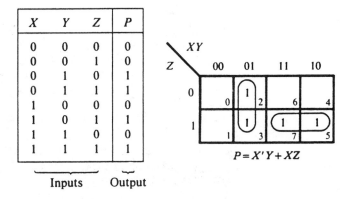

X	Y	Z	P
0	0	0	0
0	0	1	0
0	1	0	1
0	1	1	1
1	0	0	0
1	0	1	1
1	1	0	0
1	1	1	1

Inputs   Output

$P = X'Y + XZ$

4.3.1 AND-OR Circuits

From the K-map above, the minimized SOP form for P is

$$P = X'Y + XZ$$

P is a sum of two product terms. Hence, we can use a two-input OR gate to generate P, as shown below:

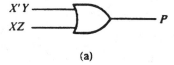

(a)

Each of the inputs to this OR gate is a product of two variables. A two-input AND gate can be used to realize each product term. The outputs of these AND gates are connected to the inputs of the OR gate, as shown below.

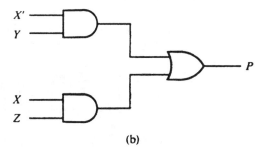

(b)

Variable X is required both in complemented and uncomplemented form. We can use a NOT gate to generate the complemented form and the complete circuit, as shown below:

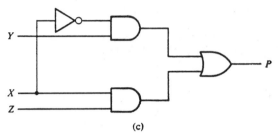

(c)

This circuit is a *three-level* circuit, since the maximum number of gates a signal propagates through, from the input to the output of the circuit, is three. The first level consists of NOT gates, the second level consists of AND gates, and the third level consists of the OR gate.

In the circuit shown above, we assumed that each input signal was available on a wire in uncomplemented form. The complemented signal was derived using a NOT gate. In practice, both complemented and uncomplemented forms of input signals are often available (e.g., two wires for each input signal). Hence, the first level of NOT gates is not specifically shown in the circuit. Thus, the above circuit is usually represented as a *two-level* circuit (as in (b) above). This is called a *two-level AND-OR* circuit, since the first level consists of AND gates and the second level consists of an OR gate.

4.3.2 OR-AND Circuits

An OR-AND circuit can be realized, starting from the POS form of the function. The first level of the circuit will consist of OR gates (corresponding to the sum terms), and the second level will consist of an AND gate (corresponding to the product). The K-map below shows the derivation of the minimum POS form for the function P above:

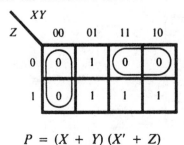

$$P = (X + Y)(X' + Z)$$

We will again assume that the input signals are available in both complemented and uncomplemented forms and demonstrate the two-level OR-AND
circuit:

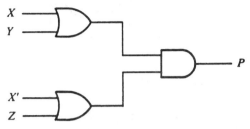

4.3.3 NAND-NAND Circuits

In Chapter 2, the NAND and NOR operations were shown to be universal
operations, since each of the three primitive operations AND, OR, and NOT
can be realized using either only NAND operations or only NOR operations.
This universal characteristic of NAND and NOR operations enables us to
build logic circuits with only one type of gate.

Figure 2.11 involved the realization of primitive operators using
NAND gates only. In light of the figure, consider the AND-OR circuit for
the example function P. Figure 4.4 shows the transformation of the AND-
OR circuit in (a) into a NAND-NAND circuit. Each AND gate in the circuit
is replaced with two NAND gates, as shown in (b). Each OR gate is replaced
with an equivalent circuit consisting of three NAND gates, as shown in (c).
The circuit (c) now has only NAND gates. There are some redundant gates
in (c). Gates 5 and 8 are not needed, since these gates simply complement
the input signal $(X'Y)'$ twice. Similarly, gates 7 and 9 can be removed. The
circuit in (d) then is the minimum two-level NAND-NAND circuit. The
circuits in (a) and (d) are equivalent, since both realize the same function
$X'Y + XZ$.

A NAND-NAND circuit thus can be derived from an AND-OR circuit,
simply by replacing each gate in the AND-OR circuit with a NAND gate
with the same number of inputs as that of the gate it replaces.

4.3.4 NOR-NOR Circuits

A NOR-NOR implementation can be derived by replacing each gate in the
OR-AND circuit with a NOR gate with the same number of inputs as that
of the gate it replaces. Figure 4.5 shows the NOR-NOR circuit for P.

The procedures for transforming AND-OR and OR-AND circuits to
NAND-NAND and NOR-NOR circuits are applicable to all two-level circuits as long as there are no direct inputs to the second level of the circuit.

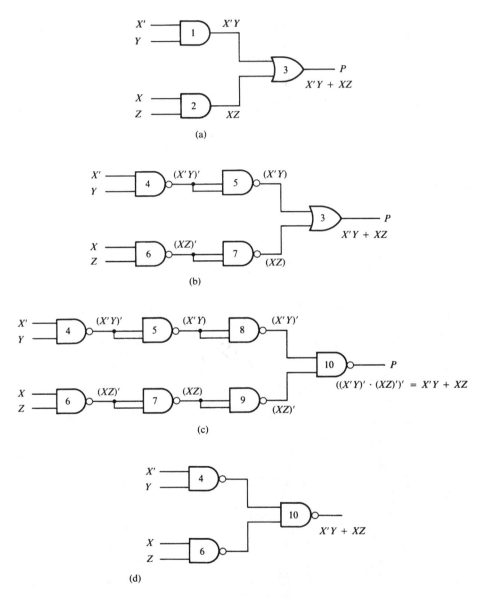

Figure 4.4 NAND-NAND transformation. (a) AND-OR circuit; (b) replace AND gates; (c) replace OR gate; (d) remove redundant gates (NAND-NAND circuit).

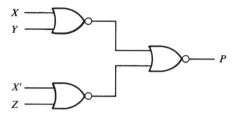

Figure 4.5 NOR-NOR circuit.

If an input is fed directly to the second level in the original AND-OR or OR-AND circuit, it must be inverted before it is fed into the second level of the corresponding NAND-NAND or NOR-NOR circuit. Figure 4.6 illustrates this process.

If the circuit to be transformed consists of more than two levels (i.e., it is a multiple-level circuit), the gate substitution process shown above cannot be adopted. In such cases, each gate in the circuit should be replaced by the equivalent NAND or NOR circuit. The circuit should then be analyzed to remove any redundant gates. Figure 4.7 shows an example. Gates 2, 3, and 5 form a two-level AND-OR circuit, which is replaced by a cor-

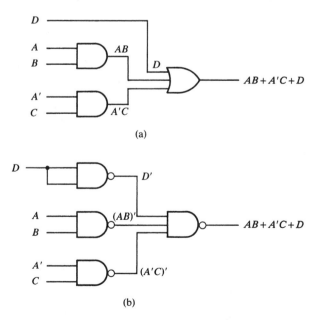

Figure 4.6 NAND-NAND conversion. (a) AND-OR; (b) NAND-NAND.

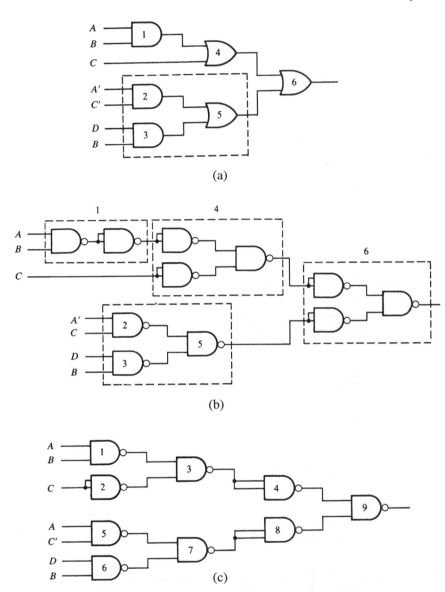

(a)

(b)

(c)

Figure 4.7 NAND transformation of a multiple-level circuit.(a) Original circuit; (b) NAND gate substitution; (c) circuit with only NAND gates.

responding NAND circuit in (b). Gates 1, 4, and 6 are replaced by equivalent NAND circuits. The redundant gates in (b) are then removed to obtain the NAND circuit in (c). Note that the circuit in (c) requires 9 gates, compared with 6 in (a).

Implementing the circuit of Figure 4.7(a) (using TTL 7400 series of ICs) requires the following ICs: one 7408 (quad two-input AND gates) and one 7432 (quad two-input OR gates). Because each IC contains four gates, one AND gate and one OR gate are left unused. The circuit in Figure 4.7(c), however, requires three 7400s (Quad two-input NAND gates). Three NAND gates are left unused. All the ICs in the implementation of (c) are of the same type, while those in (a) are of two types. In addition, the number of levels has increased by one in the transformation to (c) from (a), thereby making circuit (c) slower, compared with circuit (a). The only advantage here is that the ICs required for (c) are of the same type. This advantage may outweigh the disadvantages of increased numbers of ICs and levels, especially if this circuit is part of the larger system built only out of NAND gates. Since the implementation uses only one type of IC, the unused gates in one circuit implementation can be used more easily to implement other circuits in the system.

We will now examine an alternate procedure for transforming circuits into NAND or NOR circuits. Figure 4.8 shows alternate symbols for NAND and NOR gates. Here, a "bubble" indicates the inversion (NOT) of the signal. As we can see, the NAND is equivalent to NOT-OR, and the NOR is equivalent to NOT-AND. These forms allow a more direct transformation from AND-OR and OR-AND circuits to NAND and NOR forms. Figure 4.9(a) shows the AND-OR circuit of Figure 4.4(a). In order to transform this circuit into a NAND-NAND circuit, we first convert OR gate 3 into a NOT-OR by inserting bubbles at its inputs. To compensate for this signal inversion, we can also insert bubbles at the outputs of gates 1 and 2, as shown in (b). Gate 3 then can be replaced by a two-input NAND gate,

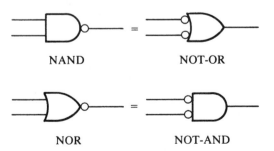

NAND   NOT-OR

NOR   NOT-AND

Figure 4.8   NAND and NOR equivalent symbols.

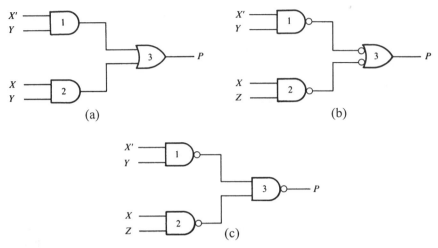

Figure 4.9   Alternate NAND-NAND transformation. (a) AND-OR; (b) insert bubbles; (c) NAND-NAND.

thereby converting the circuit into a NAND-NAND circuit, as shown in (c). This procedure can also be used to transform multiple-output circuits into NAND circuits. Adopting this procedure to the circuit of Figure 4.7 is left as an exercise.

As mentioned earlier, because each IC package contains several gates of the same type, NAND-NAND and NOR-NOR implementations are preferred over the other two implementations for the design of circuits using small-scale (SSI) and medium-scale (MSI) ICs. Thus, using a single gate type in a logic circuit does not require the use of different types of ICs and enables better use of gates on the IC packages—thereby reducing the IC count in the circuit. Further, the NAND and NOR circuits are primitive circuit configurations in major IC technologies, and the AND and OR gates are realized by complementing the outputs of NAND and NOR respectively. Thus, NAND and NOR gates tend to be easier to fabricate as ICs, although at large-scale-integrated (LSI) level and above NANDs and NORs and indeed more complex gates can be intermixed freely for direct implementation on silicon.

All four implementations of P in Example 4.2 have the same number of gates. This is not always the case. The minimized SOP and POS forms generally have different number of literals, and thus the number of gates in the circuits realized by each form will be different, as shown by Example 4.3.

Example 4.3

Figure 4.10 shows the K-maps for P and P', where

$$P(A, B, C, D) = \Sigma m \ (0, 1, 5, 10, 13, 14) + d(4, 7, 11, 15)$$

As can be seen, the minimized SOP form has six literals, and an AND-OR or a NAND-NAND implementation requires three two-input AND gates and one three-input OR gate. The minimized POS form has nine literals, and the implementation requires four three-input gates. This circuit is more complex than the AND-OR circuit.

Various other two-level circuit realizations are possible. We will illustrate these implementations in the next section.

Examples 4.4–4.6 further illustrate the combinational logic design procedure. We have just shown the AND-OR circuit for these examples. Other forms can be derived from the procedures of this section.

Example 4.4

Design a combinational circuit that accepts decimal digits in the form of BCD and produces an output of 1 if the input corresponds to an odd number.

Since the input is BCD, there are four inputs to the circuit. There will be one output. The truth table and the K-map are shown below:

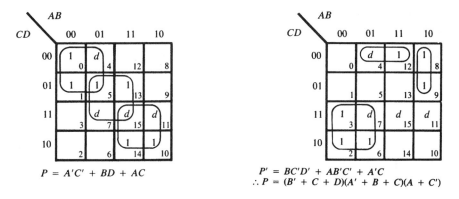

Figure 4.10 A four-variable function.

Inputs *P Q R S*	Output *Z*
0 0 0 0	0
0 0 0 1	1
0 0 1 0	0
0 0 1 1	1
0 1 0 0	0
0 1 0 1	1
0 1 1 0	0
0 1 1 1	1
1 0 0 0	0
1 0 0 1	1
1 0 1 0 through 1 1 1 1	Don't-cares

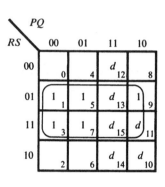

Since the last six combinations are not allowed in BCD, they do not appear on the circuit's input and hence are treated as don't-cares. From the K-map, the minimized function is

$$Z = S$$

No gates are required. The output Z is simply connected to the input S, as shown below:

$$S \longrightarrow Z$$

In this case, a close inspection of the truth table would have been sufficient to infer the minimized output function. Nevertheless, minimization by such observation is not always possible.

Example 4.5

Here, we are required to build a code converter that converts the BCD input to Excess-3. The circuit will have four inputs and four outputs. The following truth table shows the operation of the circuit. Note that the last six combinations are again don't-cares.

Inputs (BCD)				Outputs (Excess-3)			
A	B	C	D	W	X	Y	Z
0	0	0	0	0	0	1	1
0	0	0	1	0	1	0	0
0	0	1	0	0	1	0	1
0	0	1	1	0	1	1	0
0	1	0	0	0	1	1	1
0	1	0	1	1	0	0	0
0	1	1	0	1	0	0	1
0	1	1	1	1	0	1	0
1	0	0	0	1	0	1	1
1	0	0	1	1	1	0	0
1	0	1	0	*d*	*d*	*d*	*d*
1	0	1	1	*d*	*d*	*d*	*d*
1	1	0	0	*d*	*d*	*d*	*d*
1	1	0	1	*d*	*d*	*d*	*d*
1	1	1	0	*d*	*d*	*d*	*d*
1	1	1	1	*d*	*d*	*d*	*d*

(don't-cares)

Figure 4.11 shows the four K-maps, the minimized output functions, and the circuit diagram required to implement the code converter. Note that

$$Y = C \cdot D + C' \cdot D'$$

and

$$Z = D'$$

Again, these minimized functions can be inferred from a close observation of the truth table.

Example 4.6

Design a combinational circuit that produces the product of two two-bit numbers.

Since the product of two two-bit numbers can be at most four bits long, there will be four outputs from the circuit. There will also be four inputs. In the truth table below, the most significant two bits (P and Q) on the input correspond to the two-bit number A, and the least significant two bits (R and S) correspond to the number B.

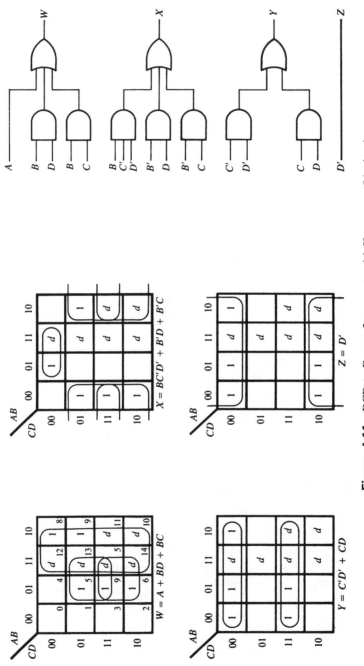

Figure 4.11 BCD-to-Excess-3 converter. (a) K = maps; (b) circuit.

Inputs				Outputs			
A		B					
P	Q	R	S	W	X	Y	Z
0	0	0	0	0	0	0	0
0	0	0	1	0	0	0	0
0	0	1	0	0	0	0	0
0	0	1	1	0	0	0	0
0	1	0	0	0	0	0	0
0	1	0	1	0	0	0	1
0	1	1	0	0	0	1	0
0	1	1	1	0	0	1	1
1	0	0	0	0	0	0	0
1	0	0	1	0	0	1	0
1	0	1	0	0	1	0	0
1	0	1	1	0	1	1	0
1	1	0	0	0	0	0	0
1	1	0	1	0	0	1	1
1	1	1	0	0	1	1	0
1	1	1	1	1	0	0	1

Figure 4.12 shows the K-maps, the minimum function for each of the four outputs, and the circuit diagram.

4.4 Other Two-Level Circuits

As mentioned earlier, several other two-level circuit realizations are possible. Consider, for instance, the four realizations of P shown in Figure 4.13. We can derive four circuits for P' simply by changing the second-level gate to a gate that realizes the complement of what the original gate realized. That is, the second-level gate is changed

1. From OR to a NOR to form an AND-NOR.
2. From NAND to an AND to form a NAND-AND.
3. From AND to a NAND to form an OR-NAND.
4. From NOR to an OR to form a NOR-OR.

Figure 4.14 shows these circuits. Similarly, we can obtain the four circuit forms for P shown in Figure 4.13 by starting from the four circuit forms for P' and applying the above transformation.

Because there are two levels in each of the circuits we have designed so far—and each level can be made up of one of the four types of gates— it is generally possible to realize 2^4, or 16, two-level circuits. Table 4.1 shows these 16 possibilities. Out of these, 8 forms degenerate to a single operation of either OR, AND, NAND (AND-COMPLEMENT), or NOR (OR-COMPLEMENT). These degenerate forms are identified with an as-

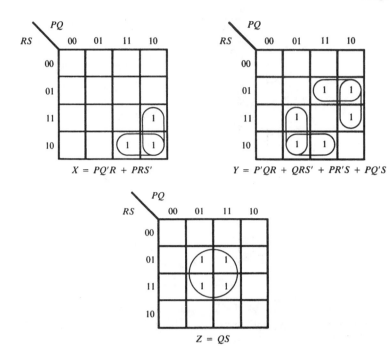

$X = PQ'R + PRS'$

$Y = P'QR + QRS' + PR'S + PQ'S$

$Z = QS$

$W = PQRS$ (from the truth table)

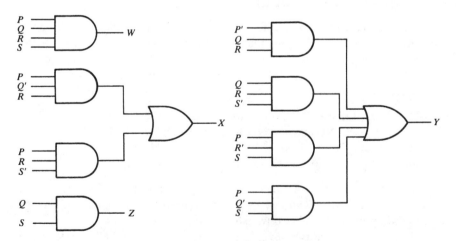

Figure 4.12 Multiplier.

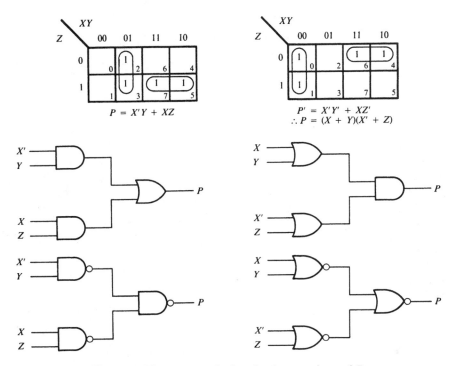

Figure 4.13 Four equivalent implementations of P.

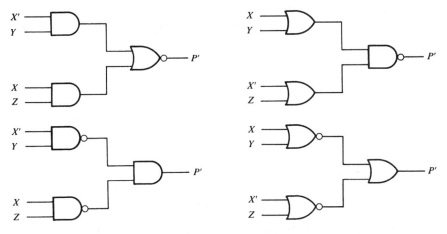

Figure 4.14 Four equivalent implementations of P'.

TABLE 4.1   Two-Level Circuit Forms

Level 1 2	Function realized
AND–AND	AND*
AND–OR	AND–OR
AND–NAND	NAND (AND-COMPLEMENT)
AND–NOR	AND-OR-INVERT
OR–AND	OR–AND
OR–OR	OR*
OR–NAND	OR-AND-INVERT
OR–NOR	NOR (OR-COMPLEMENT)*
NAND–AND	AND-OR-INVERT
NAND–OR	OR*
NAND–NAND	AND-OR
NAND–NOR	AND*
NOR–AND	AND*
NOR–OR	OR-AND-INVERT
NOR–NAND	OR*
NOR–NOR	OR-AND

*Indicates degenerate forms.

terisk in Table 4.1. The remaining 8 forms correspond to the 8 circuit forms shown in Figures 4.13 and 4.14. Table 4.2 classifies the 8 nondegenerate forms into two sets: the first derived from the SOP form and the second derived from the POS form of the function. The circuit forms in each set are equivalent. In addition, the circuit forms listed in the same row of Table 4.2 are duals of each other.

TABLE 4.2   Nondegenerate Circuit Forms

Realized from:	
SOP form	POS form
AND–OR	OR–AND
NAND–NAND	NOR–NOR
NOR–OR	NAND–NAND
OR–NAND	AND–NOR

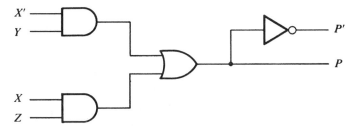

Figure 4.15 Realization of P and P'.

It is often necessary to realize both the given function and its comple-ment. Figure 4.15 shows a method for realizing P and P'. Although this realization converts the circuit into a three-level circuit, it is the least com-plex method, since only one extra gate is needed to realize the complement function.

4.5 EXCLUSIVE-OR and EQUIVALENCE

A gate that is used extensively in building computation-oriented logic cir-cuits (e.g., in arithmetic/logic units) and error detection/correction circuits is the EXCLUSIVE-OR. Figure 4.16 shows the truth table, logic symbol, and implementation of EXCLUSIVE-OR and its complement, EQUIVA-LENCE operations. These operations are both associative and commutative. Two-input EXCLUSIVE-OR gates are commonly available. These gates are more complex than the other types of gates we have used thus far in this chapter, as can be seen in Figure 4.16(c). Implementation of EXCLUSIVE-OR using only NAND gates is shown in (d). This implementation is more complex than the AND-OR implementation in (c). Multiple-input EXCLU-SIVE-OR and EQUIVALENCE operations are usually realized by using two-input gates, as shown in Figure 4.17.

EQUIVALENCE yields an output of 1 when its inputs are equal, and EXCLUSIVE-OR yields an output of 1 when its inputs are unequal. Use of these gates depends on the recognition of such equality conditions during the circuit design. For example, the SUM output of the binary adder circuit of Chapter 2 is the EXCLUSIVE-OR of the input bits (see Figure 4.18).

Figure 4.19 shows another application for EXCLUSIVE-OR gates. Here, the two four-bit numbers A and B are compared, and the output Z is 0 if the two numbers are equal; otherwise, it is 1.

Figure 4.20 shows the truth table for $X \oplus Y \oplus Z$ and $X \odot Y \odot Z$. Note that these two functions are equal. It can be shown that these two functions are always equal when the number of operands is odd. When the

EXCLUSIVE-OR   EQUIVALENCE

X Y	$X \oplus Y$	$X \odot Y$
0 0	0	1
0 1	1	0
1 0	1	0
1 1	0	1

(a)

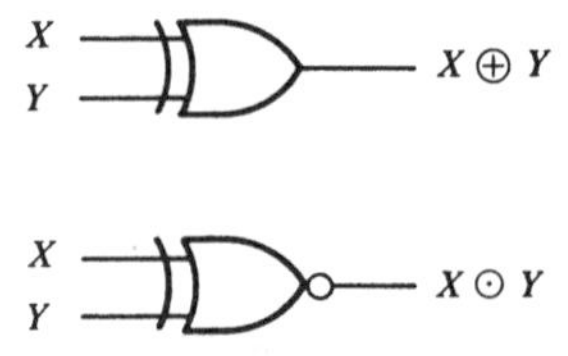

(b)

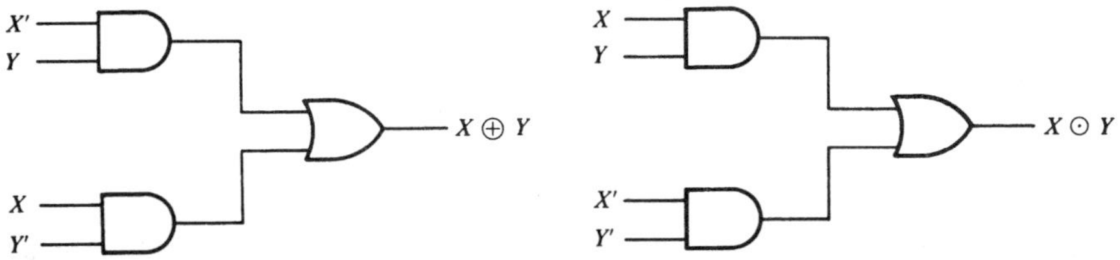

(c)

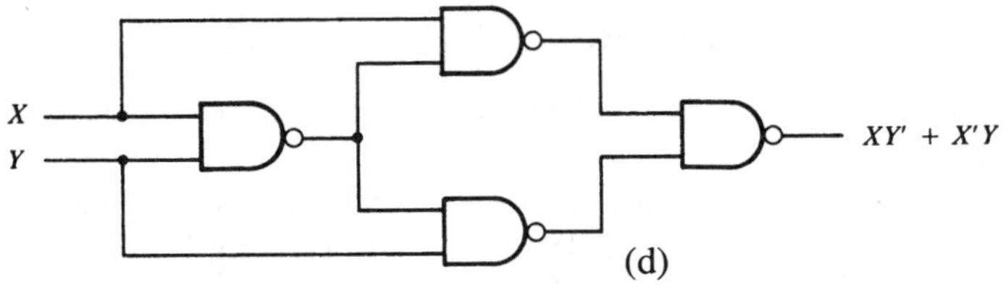

(d)

Figure 4.16 EXCLUSIVE-OR and EQUIVALENCE. (a) Truth table; (b) logic symbol; (c) circuits; (d) EXCLUSIVE-OR using NAND gates.

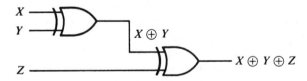

Figure 4.17   Three-input EXCLUSIVE-OR.

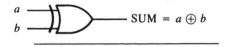

Figure 4.18   SUM circuit.

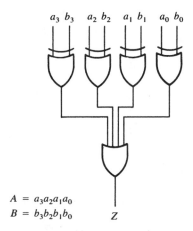

$$A = a_3 a_2 a_1 a_0$$
$$B = b_3 b_2 b_1 b_0 \qquad Z$$

Figure 4.19 Comparator.

X Y Z	$X \oplus Y \oplus Z$	$X \odot Y \odot Z$
0 0 0	0	0
0 0 1	1	1
0 1 0	1	1
0 1 1	0	0
1 0 0	1	1
1 0 1	0	0
1 1 0	0	0
1 1 1	1	1

Figure 4.20 EXCLUSIVE-OR and EQUIVALENCE with an odd number of operands.

number of operands is even, a multiple-operand EXCLUSIVE-OR operation yields a value of 1 when the number of inputs with a value of 1 is odd. However, the multiple-operand EQUIVALENCE denotes the condition in which the number of nonzero inputs is even. The following four-variable K-maps further illustrate these properties.

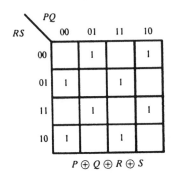

$$P \oplus Q \oplus R \oplus S$$

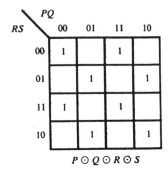

$$P \odot Q \odot R \odot S$$

As we can see, the n-variable EXCLUSIVE-OR function is equivalent to a Boolean function with $2^{(n-1)}$ minterms, whose binary equivalents have an odd number of 1s. Similarly, the n-variable EQUIVALENCE is equivalent to a Boolean function with $2^{(n-1)}$ minterms, whose binary equivalents have an even number of 0s. When the number of variables in the function is odd, the minterms with an even number of 0s are the same as those with an odd number of 1s. Hence, the EXCLUSIVE-OR and EQUIVALENCE functions are equal. When the number of variables is even, these two functions are complements of each other.

Note the pattern of 1s on the K-maps shown above. They are always on alternate diagonals of the K-map. Recognizing this pattern is useful in the design of circuits with EXCLUSIVE-OR gates.

The following examples illustrate the utility of these properties.

Example 4.7

Implement the function represented by the following K-map using EXCLUSIVE-OR and OR gates:

	AB			
CD	00	01	11	10
00		1		1
01	1	1	1	
11		1	1	1
10	1		1	

The K-map is equivalent to the following two maps ORed together:

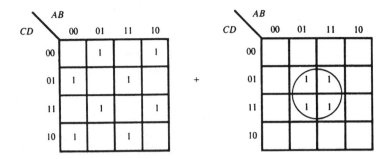

The function is thus

$$P = (A \oplus B \oplus C \oplus D) + BD$$

The circuit is shown below:

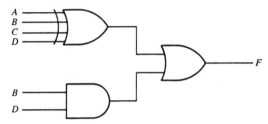

(Is this the minimum circuit? How does it compare with an AND-OR circuit in complexity and speed?)

Example 4.8

Figure 4.21 shows a transmitting station that sends a three-bit message to the receiving station. The transmitter module generates the three-bit message. The parity generator module cascades the three-bit message with a *parity bit*. The parity bit will be 1 if the number of 1s

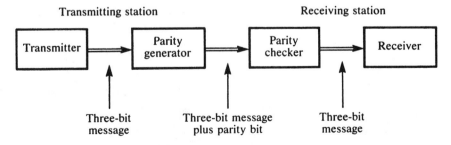

Figure 4.21 A parity scheme.

in the message is odd; otherwise, it is 0. Thus, the four-bit message sent from the transmitting station will always have an even number of 1s. This is an even parity scheme. (If the parity bit is such that the total number of 1s in the complete message is odd, it becomes an odd parity scheme.)

The parity checker in the receiving station checks the incoming four-bit message for its parity. If the number of 1s there is even, then the message received is not in error; if the number of 1s is odd, there has been an error. Once the error is detected, the transmitting station can be requested to retransmit the message. This is a simple, single error-detecting scheme.

Figure 4.22 shows the truth table for the parity generator and the parity checker. Note that the parity bit P is generated by EXCLUSIVE-

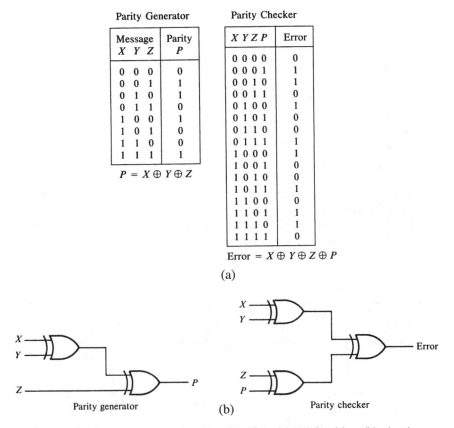

Parity Generator

Message			Parity
X	Y	Z	P
0	0	0	0
0	0	1	1
0	1	0	1
0	1	1	0
1	0	0	1
1	0	1	0
1	1	0	0
1	1	1	1

$$P = X \oplus Y \oplus Z$$

Parity Checker

$X\ Y\ Z\ P$	Error
0 0 0 0	0
0 0 0 1	1
0 0 1 0	1
0 0 1 1	0
0 1 0 0	1
0 1 0 1	0
0 1 1 0	0
0 1 1 1	1
1 0 0 0	1
1 0 0 1	0
1 0 1 0	0
1 0 1 1	1
1 1 0 0	0
1 1 0 1	1
1 1 1 0	1
1 1 1 1	0

$$\text{Error} = X \oplus Y \oplus Z \oplus P$$

(a)

Parity generator   (b)   Parity checker

Figure 4.22 Parity generation and checking. (a) Truth tables; (b) circuits.

ORing the three message bits *X*, *Y*, and *Z*, since *P* = 1 when the message contains an odd number of 1s. The parity-checker circuit generates an error if the number of bits in the four-bit message is odd. Thus, the parity-checker circuit can be realized by the EXCLUSIVE-OR of the four bits of the message. The logic circuits for both of these operations are shown in Figure 4.22. Note also that if *P* is set to 0 in the parity-checker circuit, the circuit generates the EXCLUSIVE-OR of the other three inputs, which is the function of the parity generator. As such, the same circuit can be used for both the parity checker and the parity generator.

4.6 Multiple-Level Circuits Revisited

In Figure 4.7, we first saw a multiple-level circuit. As mentioned earlier, each gate in the circuit introduces a delay into the signal flow through it. These delays accumulate as the signal flows from the input to the output of the circuit. The maximum delay introduced by a two-level circuit is equivalent to two gate delays. As the number of levels increases, the sum of gate delays increases and the circuit takes longer to produce the response. Thus, if circuit speed is a design consideration, it is advantageous to reduce the number of levels in the circuit. Two-level circuits thus are the fastest circuits.

As long as the design follows the procedures discussed in this chapter, the circuit will always be a two-level circuit. When the system being designed is complex and requires a large number of inputs and outputs, the truth table becomes large and minimization becomes difficult. In such cases,

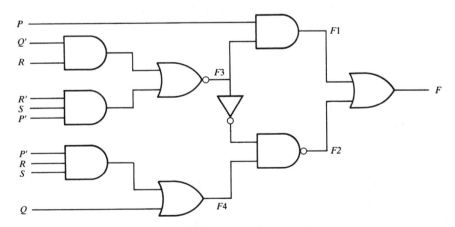

Figure 4.23 A multiple-level circuit.

two-level circuit design becomes impractical. In that case, the system is first partitioned into simpler subsystems, and each subsystem is designed independently. Although each subsystem is realized as a two-level circuit, the overall circuit for the system, which is an interconnection of subsystem circuits, may not be a two-level circuit. In such cases, the functional analysis procedure of this chapter may be adopted to reduce the number of levels in the circuit (if possible), as demonstrated by Example 4.9.

Example 4.9

Consider the five-level circuit shown in Figure 4.23. The functional analysis of the circuit uses the intermediate outputs $F1$, $F2$, $F3$, and $F4$. By tracing through the circuit, we derive the following functions:

$$F3 = (Q'R + P'R'S)' \qquad\qquad\qquad \text{T5a}$$
$$= (Q'R)' \cdot (P'R'S)' \qquad\qquad\qquad \text{T5b}$$
$$= (Q + R')(P + R + S') \qquad\qquad\qquad \text{P4b}$$
$$= \underbrace{PQ + PR' + QR}_{\text{T6a}} + \underbrace{RR'}_{0} + QS' + R'S'$$
$$= PR' + \underbrace{QR + QS' + R'S'}_{\text{T6a}}$$
$$= PR' + QR + R'S'$$
$$F4 = Q + P'RS$$
$$F2 = (F3' \cdot F4)' \qquad\qquad\qquad \text{T5a}$$
$$= F3 + F4'$$
$$= PR' + QR + R'S' + Q'(P + R' + S')$$
$$= PR' + QR + R'S' + Q'P + Q'R' + Q'S'$$
$$= \underbrace{P(R' + Q') + (R' + Q')'}_{\text{T4a}} + \underbrace{S'(R' + Q')}_{\text{T4a}} + Q'R'$$
$$= P + \underbrace{RQ}_{\text{T1a}} + S' + \underbrace{RQ}_{\text{T1a}} + Q'R'$$
$$= P + S' + QR + Q'R'$$
$$F1 = P \cdot F3$$
$$= PR' + PQR + PR'S' \qquad\qquad\qquad \text{T3a}$$
$$= PR' + PQR$$
$$= P(R' + RQ) \qquad\qquad\qquad \text{T4a}$$
$$= PR' + PQ$$
$$F = F1 + F2$$
$$= PR' + PQ + S' + P + QR + Q'R' \qquad\qquad\qquad \text{T3a}$$
$$= P + S' + QR + Q'R'$$

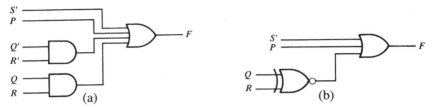

Figure 4.24 Equivalent circuits of the circuit in Figure 4.23. (a) Two-level circuit; (b) three-level circuit.

Figure 4.24 shows two minimum circuits. The circuit in (b) is a three-level circuit because of the two levels required for the EQUIVALENCE implementation; the circuit in (a) is a two-level implementation.

We will return to combinational logic design in Chapter 5, where the popular combinational circuits are introduced along with the IC packages that realize those circuits. We will now examine the timing and loading analysis of combinational circuits.

4.7 Timing Analysis

Timing problems in general are not that severe in simple combinational circuits. However, a timing analysis is usually necessary for any complex circuit. *Timing diagrams* are useful in modeling the delay characteristics of logic circuits. In this section, we will restrict our discussion to the timing problems at the gate level. That is, gates are the primitive components of the circuit and are treated as black boxes with inputs and outputs performing the operation. The electronic details of the gate will not be considered here. In Chapter 10, we will extend the timing analysis concepts to the more detailed electronic circuit level.

Consider the characteristics of the NOT gate shown in Figure 4.25(a). If we plot the time along the X axis and the voltage magnitude (i.e., logic values) along the Y axis, we can obtain the input-output characteristics of the NOT gate shown in (b). Here, the input changes from 0 to 1 at time t_1 and changes back to 0 at t_2. The output correspondingly changes from 1 to 0 at t_1 and changes back to 1 at t_2. This timing diagram represents the ideal characteristics of a NOT gate and does not indicate the delay introduced by the gate.

The *propagation delay* (Δt) introduced by the NOT gate is depicted in (c). Here, the output of the gate is delayed by Δt time units from the input

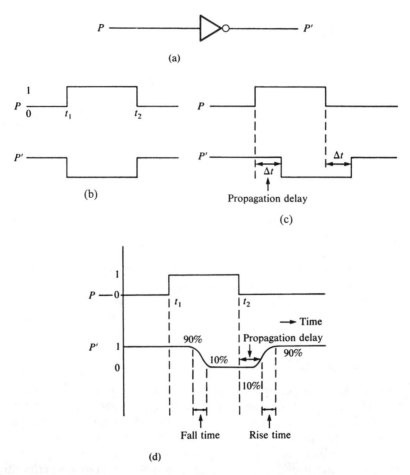

Figure 4.25 Timing characteristics and models of a NOT gate. (a) A NOT gate; (b) ideal gate characteristics; (c) gate with delay model; (d) timing characteristics.

change. With current technology, typical propagation delays are of the order of several nanoseconds.

In diagrams (b) and (c), the transition of a signal from 1 to 0 or 0 to 1 is shown to be instantaneous. In an actual circuit, this is not the case. It takes a finite amount of time for a signal to attain the new value when a transition occurs. The time needed to attain the high value from a lower value is the *rise time*; the time required to attain the low value from a higher value is the *fall time*. The rise and fall timing characteristics of the NOT gate are shown in (d). *Rise time* is defined as the time required to attain the 90% of the final value from 10%. Similarly, the fall time is defined as the time required to attain the 10% of the high value from the 90%, as shown in (d). For our purposes here, we will adopt the timing characteristics of the type shown in (b) and (c) and will not be concerned with the detailed timing diagrams of the type shown in (d).

The diagrams in Figure 4.25(b), (c), and (d) are the timing diagrams for the NOT gate. We will now extend the concept of timing diagrams to logic circuits.

Example 4.10

Figure 4.26 shows the timing diagram for a simple combinational circuit. At t_0, all three inputs (A, B, and C) are at 0. Hence, Z_1, Z_2, and Z are all 0. At t_1, B changes to 1. Assuming that the gates are ideal (i.e., there is no delay), Z_1 changes to 1 at t_1, and hence Z changes to 1. At t_2, C changes to 1. This change does not change either Z_1, Z_2, or Z. At t_3, A changes to 1, pulling A' to 0, Z_1 to 0, and Z_2 to 1; Z remains at 1. The timing diagram now shows the output of the circuit for four input combinations. It can be similarly extended to show the remaining four input combinations.

We will now analyze the effect of propagation delays. Figure 4.27 shows the timing diagram for the above circuit, taking into account the gate delays T_1, T_2, T_3, and T_4, of gates 1, 2, 3, and 4, respectively. Assume that the circuit inputs are at 0, at time t_0. At t_1, B changes to 1, resulting in a change in Z_1 at time $(t_1 + T_2)$ rather than at t_1, because of the propagation delay T_2 of the upper AND gate. This change in Z_1 causes Z to change T_4 later (i.e., at $t_1 + T_2 + T_4$). Changing C at t_2 does not change any other signal values. When A is changed to 1 at t_3, A' falls to 0 at $(t_3 + T_1)$, Z_1 falls to 0 at $(t_3 + T_1 + T_2)$, and Z_2 raises to 1 at $(t_3 + T_3)$. If $T_3 > (T_1 + T_2)$, there is a time period in which both Z_1 and Z_2 are 0, contributing a momentary 0 value for Z. Z raises back to 1, T_4 after Z_2 raises to 1. This momentary 0 value of Z is not revealed by the functional analysis of the circuit, zince Z is 1 for both the input combinations 011 and 111. The glitch (or hazard) is the result of unequal delays in the signal paths of the circuit. Such glitches might result in undesirable circuit operation, especially when the circuit is complex.

4.7.1 Hazards

As stated, hazards are caused by the unequal delay times in the signal paths from the input of the circuit to the inputs of the gate causing the hazard.

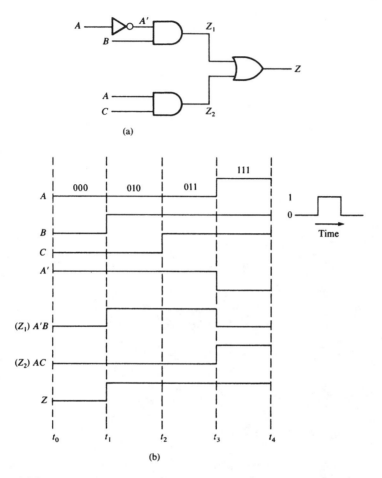

Figure 4.26 Timing analysis (ideal). (a) Circuit diagram (ideal gates); (b) timing diagram.

The delay introduced by gates is very small. Nevertheless, the delay characteristics of two devices are seldom the same, even if the devices are of the same type.

The glitch introduced by the circuit in the example above is called a *static hazard*. A static hazard is one in which a single variable change *may* produce a momentary output change when no such change should occur. In particular, the example shows a *static-1* hazard, since the signal Z was to be at 1 but went to 0 momentarily. We can similarly define a *static-0* hazard

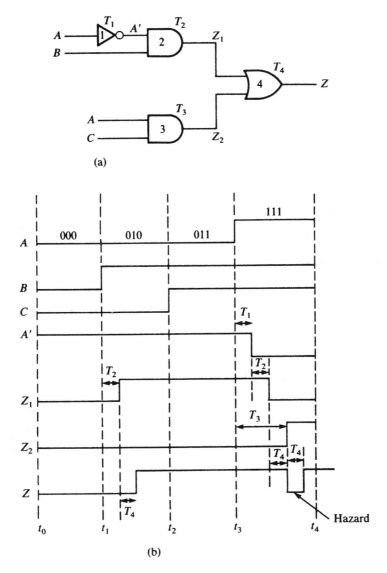

(a)

(b)

Figure 4.27 Timing analysis (actual). (a) Circuit diagram—gates with delays. Assume that $T_3 > (T_1 + T_2)$. (b) Timing diagram.

in which the signal momentarily goes to a 1, while it should have been at 0.

Note that the hazard would not have occurred if T_2 and T_3 were equal. Static hazards can be prevented by adding additional circuitry. In the example above, the hazard occurred when the inputs changed from $ABC = 011$ to 111—that is, when the value of A changed from 0 to 1. This transition is shown by the arrow in the following K-map for the circuit:

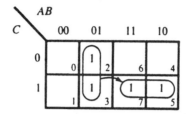

The hazard can be prevented by including a product term to cover this transition, as shown below:

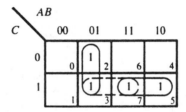

The hazard-free circuit obtained by this method contains some redundant gates and hence is not the minimum circuit.

The second type of hazard is known as the *dynamic hazard*, which also occurs as a result of the differences in delay characteristics of the circuit. These hazards occur when the output of a gate changes (from 1 to 0 or 0 to 1) because of a change in its inputs. Instead of changing from 0 to 1, the output changes from 0 to 1 to 0 to 1—or from 1 to 0 to 1 to 0 when the desired change was 1 to 0. Circuits that are free of static hazards are also free of dynamic hazards.

4.8 Loading Analysis

One other type of analysis that is usually needed on a complex combinational circuit is the *loading analysis*. The output of a gate can only be connected to a limited number of inputs of other gates. In other words, there is a limit to the number of other inputs a gate can "drive." this limit is termed the *fan-out*. The load on a gate must be within the fan-out limit for the proper operation of the circuit. The number of inputs to a gate is termed *fan-in*. Fan-out and fan-in are specified by the gate manufacturer. Chapter 5 gives further details on the fan-in and fan-out computation of various devices. At the logic circuit design level, it is usually sufficient to count the number of other gates to be driven by a gate to make sure that the fan-out constraint is observed. Special types of gates, called "buffers," provide a higher fan-out capability and are available. Either a gate can be designed with buffered outputs or special buffer gates can be inserted at the point in the circuit at which high fan-out is needed. We will discuss buffers and loading analysis further in Section 4.9 and Chapter 10.

4.9 Designing with ICs

As discussed earlier, such functions as adders, decoders, multiplexers, and processors are available as MSI- and LSI-level components. As such, the digital system designer seldom implements the digital system using gate-level (i.e., SSI) ICs only. Rather, the system to be implemented is first partitioned into subsystems that can be implemented using the available MSI and LSI components. Only those subsystems that are not available off-the-shelf are designed with SSI-level components. These subsystems are then interconnected to complete the system implementation.

Several unconventional design approaches are possible with ICs. After introducing various popular ICs, we will describe such approaches in subsequent chapters.

Two common problems need to be resolved if ICs are to be used in the design: *loading* and *timing*. Loading problems occur as a result of fan-out limits of gates and can be solved by using buffers in the circuit. Timing diagrams can be used to analyze timing problems. In Chapter 10, we will describe the circuit characteristics that contribute to the propagation delays and signal rise and fall time. Unequal propagation delays through the various signal paths in the circuit also contribute to timing problems (such as the hazards described in this chapter). Additional gates are used in the appropriate paths to compensate for unequal delays. Chapter 5 gives some examples.

Other important characteristics to be noted while designing with ICs are the active-low and active-high designations of signals, voltage polarities, and low and high voltage values, especially when ICs of different logic families are used in the circuit. ICs of the same family are usually compatible with respect to these characteristics. Special ICs to interface circuits built out of different IC technologies are also available.

Designers usually select a logic family on the basis of the following characteristics:

1. Speed
2. Power dissipation
3. Fan-out
4. Availability
5. Noise immunity (noise margin)
6. Temperature range
7. Cost

The first two characteristics are evaluated by the speed-power product. Another common measure of speed is the maximum clock rate of the flip-flops in the logic family (flip-flops will be discussed in Chapter 6).

Power dissipation is proportional to the current that is drawn from the power supply. The current is inversely proportional to the equivalent resistance of the circuit, which depends on the values of individual resistors in the circuit, the load resistance, and the operating point of the transistors in the circuit. To reduce the power dissipation, the resistance should be increased. However, increasing the resistance increases the rise times of the output signal. The longer the rise time, the longer it takes for the circuit output to "switch" from one level to the other. That is, the circuit is slower. Thus, a compromise between the speed (i.e., switching time) and power dissipation is necessary. The availability of various versions of TTL, for instance, is the result of such compromises.

A measure used to evaluate the performance of ICs is the *speed-power product*. The smaller this product, the better the performance. The speed-power product of a standard TTL with a power dissipation of 10 mW and a propagation delay of 10 ns is 100 pJ.

The *noise margin* of a logic family is the deviation in the H and L ranges of the signal that is tolerated by the gates in the logic family. The circuit function is not affected so long as the noise margins are obeyed. Figure 4.28 shows the noise margins of TTL. The output voltage level of the gate stays either below $V_{L\,max}$ or above $V_{H\,min}$. When this output is connected to the input of another gate, the input treats any voltage above $V_{IH\,min}$ (2.0 V) as high and any voltage below $V_{IL\,min}$ (0.8 V) as low. Thus, TTL provides a *guaranteed noise margin* of 0.4 V. A supply voltage V_{CC} of 5 V is required. Depending on the IC technology, as the load on the gate is increased (i.e., as the number of inputs connected to the output is increased), the output voltage may enter the *forbidden region*, thereby contributing to the improper operation. Care must be taken to ensure that the output levels are maintained by obeying the fan-out constraint of the gate or by using gates with special outputs or with higher fan-out capabilities (such as *buffers*, discussed in Chapter 10).

The *fan-out* of a gate is the maximum number of inputs of other gates that can be connected to its output without degrading the operation of the

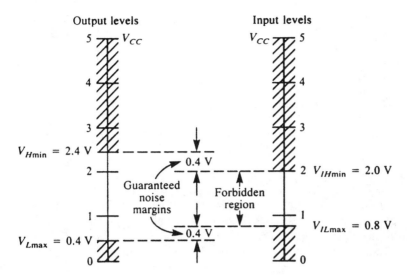

Figure 4.28 TTL noise margins.

gate. The fan-out is a function of the current sourcing and sinking capability of the gate. When a gate provides the driving current to gate inputs connected to its output, the gate is a current *source*, while it is a current *sink* when the current flows from the gates connected to it into its output. It is customary to assume that the driving and driven gates are of the same IC technology. Fan-out of gates in one technology with respect to gates of another technology can be determined according to the current sourcing and sinking capabilities of the gates in both technologies.

When the output of a gate is high, it acts as a current source for the inputs it is driving (see Figure 4.29). As the current increases, the output voltage decreases and may enter the forbidden region. The driving capability is thus limited by the voltage drop. When the output is low, the driving gate acts as a current sink for the inputs. The maximum current that the output transistor can sink is limited by the *heat dissipation* limit of the transistor. Thus, the fan-out is the minimum of these two driving capabilities. (Fan-out will be discussed further in Chapter 10.)

A standard TTL gate has a fan-out of 10. A standard TTL buffer can drive up to 30 standard TTL gates. Various versions of TTL will be described later in Chapter 10. Depending on the version, the fan-out ranges between 10 and 20. Typical fan-out of ECL gates is 25, and that of CMOS is 50.

The popularity of the IC family helps the availability. The cost of ICs comes down when they are produced in large quantities. Therefore, very

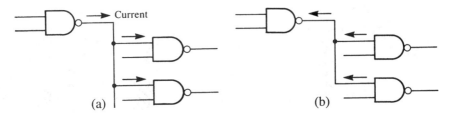

Figure 4.29   Fan-out. (a) Output high, current source; (b) output low, current sink.

popular ICs are generally available. The other availability measure is the number of types of ICs in the family. The availability of a large number of ICs in the family makes for more design flexibility.

Temperature range of operation is an important consideration, especially in such environments as military applications, automobiles, and so forth, where temperature variations are severe. Commercial ICs operate in the temperature range of 0° to 70°C, while ICs for military applications can operate in the temperature range of −55° to +125°C.

The cost of an IC depends on the quantity produced. Popular off-the-shelf ICs have become very inexpensive. As long as off-the-shelf ICs are used in a circuit, the cost of the circuit's other components (e.g., the circuit board, connectors, and interconnections) currently is higher than that of the ICs themselves.

Circuits that are required in very large quantities can be *custom-designed* and fabricated as ICs. Small quantities do not justify the cost of custom design and fabrication. There are several types of *programmable* ICs that allow a *semicustom* design of ICs for special applications. (Designing with programmable ICs is the topic of Chapter 9.)

Bipolar and metal oxide semiconductor technologies have been the two popular IC technologies based on silicon. Another technology based on gallium arsenide (GaAs) is becoming practical, and several GaAs ICs are now commercially available. It is projected that the GaAs will be the IC technology of the 1990s.

Table 4.3 summarizes the characteristics of the popular IC technologies.

4.9.1 Special ICs

Several ICs with special characteristics are available and useful in building logic circuits. We will briefly examine these special ICs in this section, at a functional level of detail. The electronic circuit-level details and character-

TABLE 4.3 Characteristics of Some Popular Logic Families

Characteristic	TTL	ECL	CMOS
Supply voltage (V)	5	-5.2	3 to 18
High-level voltage (V)	2 to 5	-0.95 to -0.7	3 to 18
Low-level voltage (V)	0 to 0.4	-1.9 to -1.5	0 to 0.5
Propagation delay (ns)	5 to 10	1 to 2	25
Fan-out	10 to 20	25	50
Power dissipation (mW) per gate	2 to 10	25	0.1

Note: See Chapter 10 for further details.

istics of these ICs, leading to their special mode of operation, and the need for such ICs in logic circuit design are deferred to Chapter 10.

A gate in the logic circuit is said to be "loaded" when it is required to drive more inputs than its fan-out. Either the loaded gate is replaced with a gate of the same functionality but of a higher fan-out (if available in the IC family) or a *buffer* is connected to its output. The ICs designated as buffers (or "drivers") provide a higher fan-out than a regular IC in the logic family. For example, the following are some of the TTL 7400 series of ICs that are designated as buffers:

    7406   Hex inverter buffer/driver
    7407   Hex buffer/driver (noninverting)
    7433   Quad two-input NOR buffer
    7437   Quad two-input NAND buffer

These buffers can drive approximately 30 standard TTL loads, compared to a fan-out of 10 for nonbuffer ICs.

In general, the outputs of two gates cannot be connected without damaging those gates. Gates with two special types of outputs are available that, under certain conditions, allow their outputs to be connected to realize an AND or an OR function of the output signals. The use of such gates thus results in reduced complexity of the circuit. The need for such gates is illustrated by Example 4.11.

Example 4.11

We need to design a circuit that connects one of the four inputs A, B, C, and D to the output Z. There are four control inputs (C_1, C_2, C_3, and C_4) that determine whether A, B, C, or D is connected to Z, respectively. It is also known that only one of the inputs is connected to the output at any given time. That is, only one of the control inputs will be active at any time. The function of this circuit can thus be represented as

$$Z = P \cdot C_1 + Q \cdot C_2 + R \cdot C_3 + S \cdot C_4$$

Figure 4.30(a) shows the AND-OR circuit implementation of this function.

If each of the AND gates in (a) are such that their outputs can be connected to form an OR function, the four-input OR gate can be eliminated from the circuit. Furthermore, as the number of inputs increases (along with the corresponding increase in control inputs), the circuit can be expanded, simply by connecting the additional AND gate outputs to the common output connection. This way of connecting outputs to realize the OR function is known as the *wired-OR* connection.

In fact, the circuit can be generalized to form a *bus* that transfers the selected source signal to the selected destination. The bus shown in (b) is simply a *common path* that is shared by all the source-to-destination data transfers. W is an additional destination. In this circuit, only one source and one destination can be activated at any given time. For example, to transfer the data from R to Z, the control signals C_3 and C_5 are activated simultaneously; similarly, Q is connected to W when C_2 and C_6 are activated, and so on. All the sources are wired-OR to the bus, and only one of them will be active at any given time. However, more than one destination can be activated simultaneously if the same source signal must be transferred to several destinations. To transfer P to both Z and W, control signals C_1, C_5, and C_6 are activated simultaneously. Buses are commonly used in digital systems when a large number of source and destinations must be interconnected. Using gates whose outputs can be connected to form the wired-OR reduces the complexity of the bus interconnection.

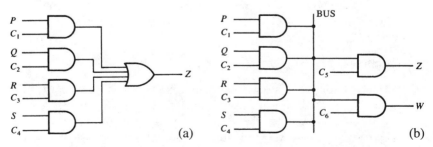

Figure 4.30 (a) Data transfer circuits. (b) Bus.

Two types of gates are available with special outputs that can be used in this mode: (a) gates with *open collector* (or free collector) outputs and (b) gates with *tristate outputs*.

Open Collector Outputs   Figure 4.31 shows two circuits. When the outputs of the TTL open collector NAND gates are tied together, an AND is realized, as shown in (a). The second level will not have a gate. This fact is illustrated with the dotted gate symbol. Note that this *wired-AND* capability results in the realization of an AND-OR-INVERT circuit (i.e., a circuit with a first level of AND gates and an OR-INVERT or NOR gate in the second level) with only one level of gates. Similarly, when open collector ECL NOR gates are used in the first level, we realize an OR when the outputs are tied together, as shown in (b). This *wired-OR* capability results in the realization of an OR-AND-INVERT circuit (i.e., a circuit with a first level of OR gates and a second level consisting of one AND-INVERT or NAND gate) with just one level of gates. (Further details on these open collector gates appear in Chapter 10.)

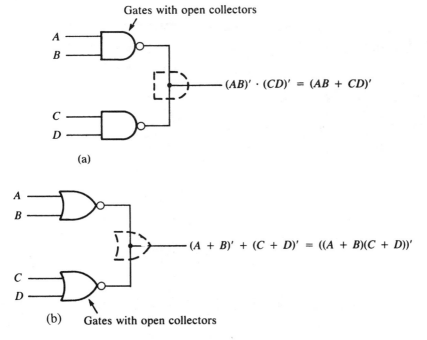

Figure 4.31 Open collector circuits. (a) TTL NANDs, WIRED-AND, AND-OR-INVERT; (b) ECL NORs, WIRED-OR, OR-AND-INVERT.

There is a limit to the number of outputs (typically about 10 in TTL) that can be tied together. When this limit is exceeded, ICs with tristate outputs can be used in place of open collector ICs.

Tristate Outputs In addition to providing two logic levels (0 and 1), the output of these ICs can be made to stay at a *high-impedance* state. An *enable* input signal is used for this purpose. The output of the IC is either at logic 1 or 0 when it is enabled (i.e., when the enable signal is active). When it is not enabled, the output will be at the high-impedance state and is equivalent in effect to the IC not being in the circuit. It should be noted that the high impedance state is not one of the logic levels, but rather is a state in which the gate is not electrically connected to the rest of the circuit.

The outputs of tristate ICs can also be tied together to form a wired-OR as long as *only one* IC is enabled at any time. Note that in the case of a wired-OR (or wired-AND) formed using open collector gates, more than one output can be active simultaneously.

Figure 4.32 shows the bus circuit of Example 4.11 using tristate gates. The control inputs now form the enable inputs of the tristate gates. Tristate outputs allow more outputs to be connected together than the open collector ICs do.

Figure 4.33 shows a tristate IC (TTL 74241). This IC has eight tristate buffers, four of which are enabled by the signal on pin 1 and the other four

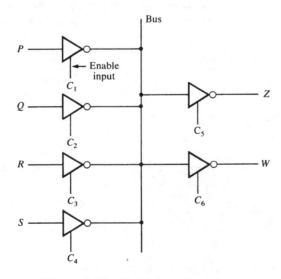

Figure 4.32 Bus using tristate gates.

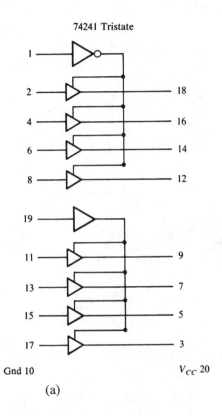

(a)

Control	Input	Output	Control	Input	Output
Pin 1	2	18	Pin 19	11	9
L	L	L	L	L	Z
L	H	H	L	H	Z
H	L	Z	H	L	L
H	H	Z	H	H	H

(b)   Z = High impedance

Figure 4.33 TTL 74241 tristate buffer. (a) Circuit; (b) representative operating values.

by the signal on pin 19. Pin 1 is active-low enable, while pin 19 is active-high enable. Representative operating values are shown in (b).

ICs come with several other special features. Some ICs provide both the true and complement outputs (e.g., ECL 10107); some have a STROBE input signal that needs to be active in order for the gate output to be active (e.g., TTL 7425). Thus, STROBE is an enable input. Figure 4.34 illustrates these ICs. For further details, consult the IC manufacturer manuals listed at the end of this chapter; they provide a complete listing of the ICs and their characteristics.

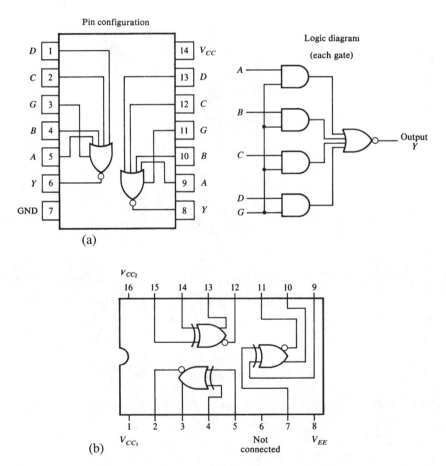

Figure 4.34 Some special ICs. (a) TTL 7425 positive NOR gates with strobe (G); (b) ECL 10107 Triple EXCLUSIVE-OR/NOR.

4.10 Summary

Three analysis procedures for combinational circuits were introduced in this chapter. Functional analysis is the type of analysis that is performed most often. Timing and loading analysis are usually straightforward, except for very complex circuits. The design procedure for combinational circuits was also described in this chapter. Various forms of two-level circuits were introduced, illustrating the flexibility that the designer has in implementing combinational circuits. This chapter also discussed characteristics to be considered in selecting ICs, along with special features to be noted while designing with ICs.

One other logic circuit implementation technique now popular is by using Programmable Logic Devices (PLD). PLDs contain various types of logic elements already fabricated in the package. Circuits are implemented by establishing the interconnections between these logic elements, using special programming equipment. Chapter 9 provides details on this logic implementation technique.

Before the final implementation and production of any logic circuit, it is essential to check its operation to make sure it meets the requirements. Chapter 11 outlines the equipment needed for such experimentation and verification. This mode of verification requires actual ICs. One popular verification technique that does not require the actual hardware elements at this stage is 'simulation.' In this technique, the characteristics of hardware elements are simulated using a computer program. The circuit operation can be simulated by interconnecting these simulated hardware elements appropriately. Several computer aided design (CAD) tools are now available that allow such a simulation. Appendix C provides the details of one such tool.

References

Ercegovac, M. D. and Lang, T. *Digital Systems and Hardware/Firmware Algorithms*, New York: John Wiley, 1985.
Greenfield, J. D. *Practical Digital Design Using ICs*, New York: John Wiley, 1983.
Mano, M. M. *Digital Design*, Englewood Cliffs, NJ: Prentice-Hall, 1991.
MECL System Design Handbook, Phoenix, AZ: Motorola, 1972.
Roth, C. H. *Fundamentals of Logic Design*, St. Paul, MN: West, 1985.
The TTL Data Book Series, Dallas, TX: Texas Instruments, 1984.
TTL Manual, Sunnyvale, CA: Signetics, 1986.

Problems

4.1   Use only NOR gates to realize EXCLUSIVE-OR operations.

4.2   Transform the circuit in Figure 4.7 into a NOR circuit.

4.3   For the circuit shown below:

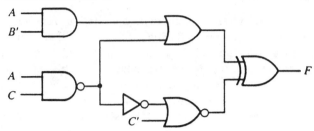

(a) Find minimum *F*.

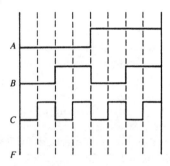

(b) Complete the above timing diagram, assuming ideal gates.
(c) If each gate introduces a delay of Δ*t*, determine the minimum time needed between two input changes, for the circuit to operate properly.
(d) Show the timing diagram with the delay Δ*t*.

4.4   For the circuit shown below, express *P* and *Q* as minimum functions of *A*, *B*, and *C*.

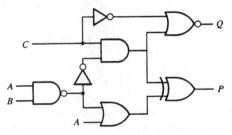

4.5   Derive all four forms of minimum two-level circuits for *P* and *Q* in problem 4.4.

4.6   Draw the timing diagram to show the operation of the circuit in problem 4.4. Assume that the inputs start with *ABC* = 000 and make all the transitions up to *ABC* = 111 (as in problem 4.3). Ignore any gate delays.

4.7 Derive the truth table for the circuit in problem 4.4.

4.8 Given that $F(A, B, C) = \Pi M\ (1, 5, 6, 7)$, find the minimum two-level NAND and NOR circuits.

4.9 Given that $F(A, B, C, D) = A'C'D + BD' + AC'D$, realize the function using the fewest gates possible.

4.10 Design a combinational circuit that has three inputs and produces as its output a three-bit number that is the twos complement of the input. Assume that all input bits are magnitude bits.

4.11 Design a combinational circuit that has four inputs and that produces the twos complement of the input as its output. Treat one of the input bits as a sign bit.

4.12 Design a comparator circuit whose inputs are two two-bit numbers (A and B) and which produces three outputs corresponding to the conditions $A = B$, $A > B$, and $A < B$. An output is 1 if the corresponding condition is satisfied by the inputs; otherwise, it is 0. Use only NAND gates in your design. Minimize the circuit.

4.13 Design the comparator of problem 4.12 using EXCLUSIVE-OR gates. Compare the complexity of the two designs.

4.14 Design a BCD-to-7-segment decoder circuit whose input consists of the four-bit BCD. The circuit should produce 7 outputs, each of which is connected to a segment of the seven-segment display shown below. Whenever the output connected to a segment carries a logic value of 1, the corresponding segment will be on; otherwise, it will be off. Appropriate segments should be on to form the digits 0 through 9 on the display, as shown below.

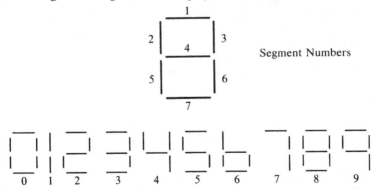

4.15 Two two-bit numbers are to be added. Design a circuit with the two numbers as inputs and their sum as the output.

4.16 A half-adder is a circuit that has two inputs and produces two outputs: one corresponding to the sum of the two input bits and the other corresponding to the carry. Design a minimum NAND circuit for the half-adder.

4.17 A full adder is a circuit that has three inputs and produces two outputs: one corresponding to the sum and the other corresponding to the carry of the three input bits. Design the full adder using the minimum number of gates.

4.18 Design the full adder using two half-adders and a minimum number of additional gates.

4.19 Connect a full adder and a half-adder to form the sum of two two-bit numbers. Note that one of the inputs to the full adder is treated as the CARRY-IN along with the other two input bits to be added to produce the two outputs: sum and carry-out.

4.20 Compare the complexity of the circuits in problems 4.15 and 4.19. Which of these circuits is faster?

4.21 Design a circuit to convert a decimal digit from (2, 4, 2, 1) code to BCD.

4.22 Design a two-level circuit that adds 3 to each BCD digit input to it. Implement this circuit using half- and full adders. Compare the complexity and speed of these implementations.

4.23 Realize the two-input EQUIVALENCE function using

(a) NAND gates.
(b) NOR gates.

4.24 For the following circuit, derive:

(a) The truth table.
(b) The output functions.
(c) The minimum AND-OR circuit.
(d) The minimum OR-AND circuit.

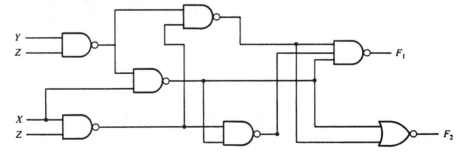

5

Popular Combinational Circuits

5.1 Introduction

With current advances in IC technology, several commonly used functions, such as adders, decoders, and multiplexers, are now available as MSI-level ICs, in addition to the gates available as SSI-level ICs. LSI-level ICs contain such complex functions as an arithmetic/logic unit, a memory unit, or a digital processor, while VLSI-level ICs contain a complete digital system. In this chapter, we will examine the popular MSI-level modules and illustrate the design of combinational circuits using these modules.

It is interesting to trace the evolution of standard modules and the corresponding evolution of hardware design methodology. In the beginning, there were no standard modules. The circuits were implemented using discrete electronic components. As designers realized that some of the circuits were used more frequently than others, such circuits were built on *cards* (*printed circuit boards*) with their inputs and outputs available on a *connector*. The electrical and mechanical characteristics of these cards were then standardized for compatibility. These standard cards were then made available for all designs. Such cards were connected through a *backplane* to complete the system design.

As the technology evolved and the components became smaller, such complex functions as adders, decoders, and multiplexers were implemented as standard modules (cards).

With the advent of IC technology, SSI modules containing several gates and flip-flops were introduced, resulting in reductions in power consumption, cost, and size of the standard modules. As the technology moved into MSI era, functions that previously were implemented on standard cards were converted into ICs. Fabrication of such ICs, however, is cost-effective only when they are mass-produced. Therefore, the major problem was to identify the common functions used most frequently. Further, since the external connections to an IC package are costly, the number of pins—and hence the total number of inputs to and outputs from the chip—had to be minimized.

The cost-effectiveness, and thus the need for mass production, became a dominant issue as IC technology evolved into the LSI era. Two modes of circuit implementation came into being: *custom design* and *programmed design*. Custom design is used for circuits that are needed in large quantities, where the extent of use justifies the cost of designing and producing the IC for that specific circuit. In programmed design, *programmable modules* such as Read Only Memory (ROM), Programmable Logic Arrays (PLAs), and so forth are used. These are standard modules on which a large class of functions can be realized by programming them in the last stage of fabrication. Since the programming stage is the only unique requirement for each new circuit, this mode of circuit implementation is less expensive than the custom design mode.

With state-of-the-art IC technology, we now have the capability to fabricate complex digital systems on a VLSI chip. Very few standard modules of this complexity are used often enough to justify their mass production. Hence, the extensive use of the technology depends on the possibility of designing and fabricating these complex circuits in a cost-effective manner. Innovative design tools are now being devised to conquer the complexity of these chips and reduce the design cost by automating as much of the design process as possible.

The combinational logic design procedure of Chapter 4 used logic gates as the primitive components of the circuit, and hence it is suitable for designing combinational circuits using SSI-level ICs. However, the modern-day logic designer seldom follows the classical combinational logic design procedure. Instead, he or she tries to partition the task at hand to see how much of it can be realized using off-the-shelf ICs. It may not always be possible to find an IC to suit the design at hand. In such cases, the designer tries to use a combination of ICs to realize the circuit. The circuit design thus consists of an interconnection of selected MSI or LSI components, along with some SSI chips as required. When several ICs are used in the circuit, the cost of external hardware, such as interconnections between the ICs and the circuit boards, usually exceeds the cost of the ICs themselves. Thus, a reduction in the number of external connections between the pins of ICs will reduce the cost of the circuit. The circuit function, then, should be concentrated within the IC and as few ICs as possible should be used to implement the circuit. That is, as many MSI (and LSI) components as possible should be used in the circuit, and every effort should be made to utilize as much of the capability of each IC as possible (e.g., as many gates as possible from each SSI-level IC should be used in the circuit).

The off-the-shelf MSI modules used in combinational logic design are usually classified into the following groups:

1. Arithmetic functions and operators (e.g., adders, multipliers, comparators, shifters, parity generators).
2. Decoders/demultiplexers.
3. Code converters.
4. Multiplexers.

We will examine the functional-level details of representative ICs from each of these groups. Alternative design procedures that are possible using these circuits as components will also be illustrated. We will concentrate on the so-called *random logic design*, in which a set of selected chips is interconnected, with no regularity attempted in the design. (Chapter 9 will illustrate *programmable logic design*.)

5.2 Adders

Addition is the most common arithmetic operation performed by a digital computer. If the computer hardware can add two binary numbers, the other three primitive arithmetic operations can be performed with the additional hardware: subtraction is performed by adding the negative of the subtrahend expressed in either twos or ones complement form to the minuend; multiplication involves the repeated addition of the multiplicand to itself; and division involves the repeated subtraction of the divisor from the dividend. Because of the cost of hardware, earlier computers implemented addition in the hardware and used this hardware to realize the other three arithmetic operations by software means (i.e., programming the addition hardware). It is now cost-effective to use hardware to realize all the arithmetic operations. In this section, we will examine some popular adder circuits.

5.2.1 Parallel Binary Adder (PBA)

Consider the addition of two four-bit numbers (A and B):

$$
\begin{array}{rl}
 & c_2 \ c_1 \ c_0 \\
A: & a_3 \ a_2 \ a_1 \ a_0 \\
B: & \underline{b_3 \ b_2 \ b_1 \ b_0} \\
\text{SUM:} & c_3 \ s_3 \ s_2 \ s_1 \ s_0
\end{array}
$$

Here, a_0 and b_0 are the least significant bits (LSBs), and a_3 and b_3 are the most significant bits (MSBs). The addition is performed starting with the LSB position. Addition of a_0 and b_0 yields the sum bit s_0 and a carry of c_0. The carry c_0 is then used in the addition of the bits in the next most significant stage. That is, c_0, a_1, and b_1 are added, yielding s_1 and c_1. This addition process is repeated through the MSB position and applies to the addition of numbers with any number of bits.

A Parallel Binary Adder (BPA) implements this addition process by using two types of adder circuits: a *half-adder* and a *full adder*. In particular, for adding two *n*-bit numbers, the PBA uses one half-adder at the LSB stage and $n - 1$ full adders. The addition of bits is done in parallel since *n* adders are available, one for each bit position.

Figure 5.1 shows the PBA circuit for the four-bit addition. A half-adder is a device that can add two bits producing a SUM and a CARRY as outputs. A full adder adds three input bits and produces a SUM output and a CARRY-OUT output. One of the inputs to the full adder is called CARRY-IN. (Each of these devices is a two-level circuit, as will be shown later in this section.)

In the PBA circuit of Figure 5.1, the carry ripples through the stages of the circuit from the LSB to the MSB. The sum of the two numbers is of the correct value only after the carry has rippled through the MSB. Each stage introduces a two-gate propagation delay. If the number of bits is large,

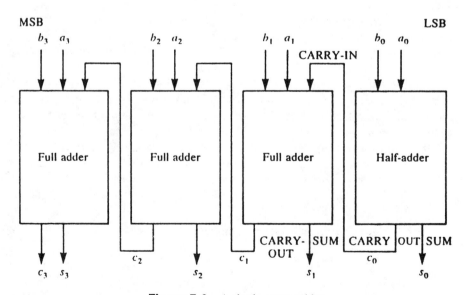

Figure 5.1 A ripple-carry adder.

the carry takes longer to ripple through the MSB and hence the PBA circuit will be slow. Several schemes reduce the carry propagation delay, thereby increasing addition speed. We will examine one such scheme in this section.

Addition in PBA is not strictly parallel since the results at each bit position is not correct until the carry propagates through that stage. As such, the PBA is also called a *pseudoparallel adder*. Because the carry ripples through the stages, it is also called a *ripple-carry adder*.

Figure 5.2 shows the block diagram representation and the truth tables for full and half-adders. From the truth tables, the following functions can be derived.

Half-adder:

$$\text{SUM} = a'b + ab'$$
$$= a \oplus b$$
$$\text{CARRY} = ab \tag{5.1}$$

(Figure 5.3 shows the half-adder circuit.)

Full-adder:

$$\text{SUM} = C'_{in}a'b + C'_{in}ab' + C_{in}ab + C_{in}a'b'$$
$$\text{CARRY-OUT} = C'_{in}ab + C_{in}a'b + C_{in}ab' + C_{in}ab \tag{5.2}$$

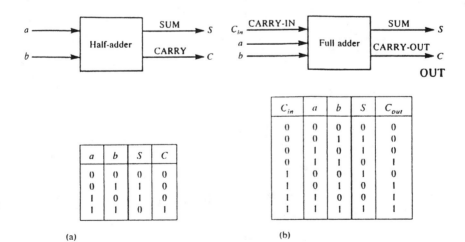

(a)

a	b	S	C
0	0	0	0
0	1	1	0
1	0	1	0
1	1	0	1

(b)

C_{in}	a	b	S	C_{out}
0	0	0	0	0
0	0	1	1	0
0	1	0	1	0
0	1	1	0	1
1	0	0	1	0
1	0	1	0	1
1	1	0	0	1
1	1	1	1	1

Figure 5.2 Adders. (a) Truth table for a half-adder; (b) truth table for a full-adder.

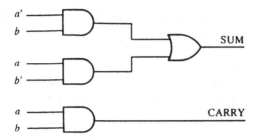

Figure 5.3   Half-adder circuits.

The equations in 5.2 can be reduced to

$$\text{SUM} = C_{in} \oplus a \oplus b$$

$$\text{CARRY-OUT} = ab + C_{in}b + C_{in}a \tag{5.3}$$

(Figure 5.4 shows the full adder circuit.)

5.2.2 Carry Lookahead Adder (CLA)

Consider the ith stage of the pseudoparallel adder with inputs a_i, b_i, and C_{i-1} (CARRY-IN) and the outputs s_i and C_i (CARRY-OUT) shown below:

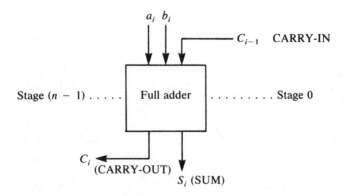

We will define two functions:

$$\text{Generate: } G_i = a_i \cdot b_i \tag{5.4}$$

$$\text{Propagate: } P_i = a_i \oplus b_i \tag{5.5}$$

These functions imply that the stage i generates a carry if $(a_i \cdot b_i) = 1$, and the carry C_{i-1} is propagated to C_i if $(a_i \oplus b_i) = 1$. Substituting G_i and

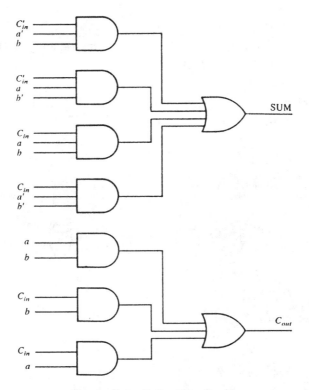

Figure 5.4 Full-adder circuits.

P_i into the equations for the SUM and CARRY functions of the full adder we get

$$\text{SUM: } S_i = a_i \oplus b_i \oplus C_{i-1}$$
$$= P_i \oplus C_{i-1} \qquad (5.6)$$

$$\text{Carry: } C_i = a_i \cdot b_i \cdot C'_{i-1} + a'_i \cdot b_i \cdot C_{i-1} + a_i \cdot b'_i \cdot C_{i-1} + a_i \cdot b_i \cdot C_{i-1}$$
$$= a_i b_i + (a'_i b_i + a_i b'_i)C_{i-1}$$
$$= a_i b_i + (a_i \oplus b_i)C_{i-1}$$
$$= G_i + P_i C_{i-1} \qquad (5.7)$$

G_i and P_i can be generated simultaneously since a_i and b_i are available. Equation 5.7 implies that S_i can be generated simultaneously if all CARRY-IN (C_{i-1}) signals are available. From equation 5.7, we can obtain

$$C_0 = G_0 + P_0 C_{-1}$$

$$C_1 = G_1 + C_0 P_1$$

$$\quad\; = G_1 + G_0 P_1 + C_{-1} P_0 P_1$$

$$\vdots$$

$$C_i = G_i + G_{i-1} P_i + G_{i-2} P_{i-1} P_i + \cdots$$

$$\quad\; + G_0 P_1 P_2 \cdots P_i + C_{-1} P_0 P_1 \cdots P_i \tag{5.8}$$

(C_{-1} is the CARRY-IN to the rightmost bit)

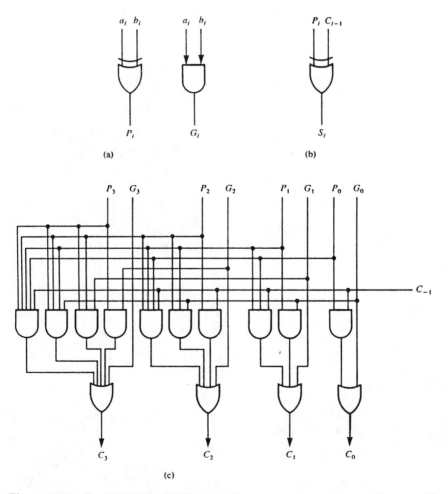

(a)   (b)

(c)

Figure 5.5 Carry lookahead adder. (a) Propagate, carry generate; (b) sum; (c) carry lookahead (four-bit); (d) a four-bit carry lookahead adder.

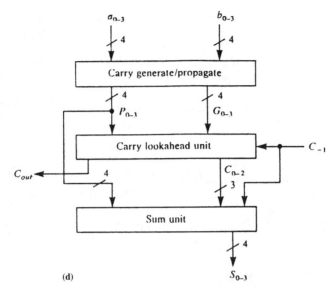

Figure 5.5 *Continued*

From equation 5.8 it can be seen that C_i is a function of only G and P of the ith and earlier stages. All C_is can be generated simultaneously since Gs and Ps are available. Figure 5.5 shows a four-bit carry lookahead adder (CLA) schematic and the detailed circuitry. This adder is faster than the ripple-carry adder since the delay is independent of the number of stages in the adder and is equal to the delay of the three stages of circuitry in Figure 5.5 (i.e., about six gate delays).

5.2.3 Off-the-Shelf Adders

Several adders are available off-the-shelf. TTL 7480 is a full adder IC. TTL 7483 and CMOS 4008 are four-bit PBA circuits with high-speed parallel carry generator circuitry. TTL 74182 is a lookahead carry generator that can be used with the 74181, which is a four-bit arithmetic/logic unit.

Figure 5.6 shows the TTL 74F283A, which can add two four-bit numbers (A and B) plus the incoming carry (C_{in}). This 16-pin IC uses 8 pins for the input of two four-bit numbers, and 4 pins for the four output bits (Σ). Pin 14 is the CARRY-OUT from the MSB and hence is the fifth bit of the sum. The CARRY-OUT is generated by the carry lookahead scheme, thus making the device a fast carry adder. Pin 13 is the CARRY-IN into the adder.

Logic Symbol

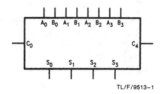

TL/F/9513-1

	C_0	A_0	A_1	A_2	A_3	B_0	B_1	B_2	B_3	S_0	S_1	S_2	S_3	C_4
Logic Levels	L	L	H	L	H	H	L	L	H	H	H	L	L	H
Active HIGH	0	0	1	0	1	1	0	0	1	1	1	0	0	1
Active LOW	1	1	0	1	0	0	1	1	0	0	0	1	1	0

Active HIGH: 0 + 10 + 9 = 3 + 16   Active LOW: 1 + 5 + 6 = 12 + 0

AC Electrical Characteristics:

Symbol	Parameter	74F $T_A = +25°C$ $V_{CC} = +5.0V$ $C_L = 50$ pF			54F $T_A, V_{CC} = $ Mil $C_L = 50$ pF		74F $T_A, V_{CC} = $ Com $C_L = 50$ pF		Units
		Min	Typ	Max	Min	Max	Min	Max	
t_{PLH}	Propagation Delay	3.5	7.0	9.5	3.5	14.0	3.5	11.0	ns
t_{PHL}	C_0 to S_n	3.0	7.0	9.5	3.0	14.0	3.0	11.0	
t_{PLH}	Propagation Delay	3.0	7.0	9.5	3.0	17.0	3.0	13.0	ns
t_{PHL}	A_n or B_n to S_n	3.0	7.0	9.5	3.0	14.0	3.0	11.5	
t_{PLH}	Propagation Delay	3.0	5.7	7.5	3.0	10.5	3.0	8.5	ns
t_{PHL}	C_0 to C_4	3.0	5.4	7.0	2.5	10.0	3.0	8.0	
t_{PLH}	Propagation Delay	3.0	5.7	7.5	3.0	10.5	3.0	8.5	ns
t_{PHL}	A_n or B_n to C_4	2.5	5.3	7.0	2.5	10.0	2.5	8.0	

Unit Loading/Fan Out: See Section 2 for U.L. Definitions

Pin Names	Description	54F/74F U.L. HIGH/LOW	Input I_{IH}/I_{IL} Output I_{OH}/I_{OL}
A_0–A_3	A Operand Inputs	1.0/2.0	20 μA/ −1.2 mA
B_0–B_3	B Operand Inputs	1.0/2.0	20 μA/ −1.2 mA
C_0	Carry Input	1.0/1.0	20 μA/ −0.6 mA
S_0–S_3	Sum Outputs	50/33.3	−1 mA/20 mA
C_4	Carry Output	50/33.3	−1 mA/20 mA

Figure 5.6 TTL 74F283A adder (courtesy of National Semiconductor Corporation).

The fan-out and dynamic characteristics of TTL 74283 are also shown in Figure 5.6. Note the various propagation delays introduced by the device. (What is the fan-out and how does the delay compare with the delays introduced by the PBA of Figure 5.1, built using TTL 7400 and 7410 NAND gates?)

Figure 5.7 shows the CMOS four-bit adder MC14008B. The typical propagation delay of this circuit is 160 ns, compared with 120 ns of TTL 74283.

The following examples illustrate the typical applications of adder circuits.

Example 5.1

Figure 5.8 shows the cascade of two 74283s to form an eight-bit adder. Note that the propagation delay of this circuit will be twice that of the 74283.

Example 5.2 BCD-to-Excess-3 code converter

Figure 5.9 shows the use of a four-bit PBA in designing the BCD-to-Excess-3 code converter. The BCD code word is connected to one set of the four input bits, and a constant of 3 (i.e., 0011) is connected to the other set. The output is the BCD input plus 3, which is the Excess-3 code.

Example 5.3 Adder/subtractor

Figure 5.10 shows the four-bit adder/subtractor circuit using a 74283. When the SUBTRACT signal is high, C_{in} will be 1 and the operand B will be complemented, thereby adding the twos complement of B to A. When the SUBTRACT signal is low, regular addition takes place.

5.2.4 Decimal Adder

For computational devices used in business environments (e.g., cash registers, calculators), the input and the output must be in decimal form. It is convenient to retain the input decimal numbers in the BCD form, perform computations on those numbers in BCD, and produce results in decimal form, thus avoiding the inaccuracies brought about by the decimal-to-binary-to-decimal conversion. An adder used in this environment should be able to add two decimal digits expressed in BCD, producing a sum digit and a carry digit. In order for this adder to be used at any stage in the decimal addition, a CARRY-IN input bit from the previous stage of the computation will also be needed. Thus, the adder will have eight inputs and five outputs.

Because a decimal digit has a maximum value of 9, the result of the addition of any two digits cannot exceed 18. If there is a carry from the

4-BIT FULL ADDER

The MC14008B 4-bit full adder is constructed with MOS P-channel and N-channel enhancement mode devices in a single monolithic structure. This device consists of four full adders with fast internal look-ahead carry output. It is useful in binary addition and other arithmetic applications. The fast parallel carry output bit allows high-speed operation when used with other adders in a system.

- Look-Ahead Carry Output
- High-Speed Operation – 160 ns typical from Sum_{in} to Sum_{out}
- Quiescent Current – 5.0 nA/package typical @ 5 Vdc
- Diode Protection on All Inputs
- All Outputs Buffered
- Supply Voltage Range = 3.0 Vdc to 18 Vdc
- Capable of Driving Two Low-power TTL Loads, One Low-power Schottky TTL Load or Two HTL Loads Over the Rated Temperature Range
- Pin-for-Pin Replacement for CD4008B

MAXIMUM RATINGS (Voltages referenced to V_{SS})

Rating	Symbol	Value	Unit
DC Supply Voltage	V_{DD}	-0.5 to +18	Vdc
Input Voltage All Inputs	V_{in}	-0.5 to V_{DD} +0.5	Vdc
DC Current Drain per Pin	I	10	mAdc
Operating Temperature Range AL Device	T_A	-55 to +125	°C
CL CP Device		-40 to +85	
Storage Temperature Range	T_{stg}	-65 to +150	°C

BLOCK DIAGRAM

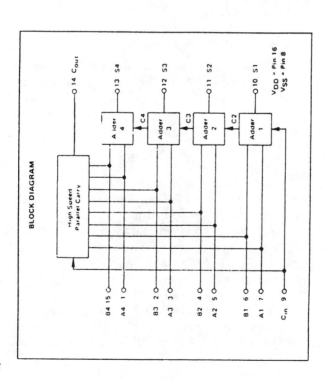

V_{DD} = Pin 16
V_{SS} = Pin 8

Figure 5.7 MC14008B adder (courtesy of Motorola Inc.).

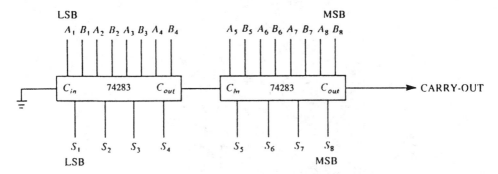

Figure 5.8   Eight-bit adder using two 7483s.

previous stage in the addition, the sum would be at most 19. The first column in Table 5.1 shows the results of the addition. In BCD coding, only digits 0 through 9 are allowed. Hence, the 19 sum values in the first column of the table must be converted into the values shown in the second column, where a carry to the next more significant stage is produced when the result exceeds 9. This process requires the Add-6 correction (discussed in Chapter 1) to the binary sum of column 1 produced by the adder. Figure 5.11 shows some examples.

Figure 5.12 shows the BCD adder circuit using two four-bit PBAs. The first PBA adds the two BCD digits and produces the five-bit binary SUM. The second PBA adds 6 to the binary SUM to produce the correct

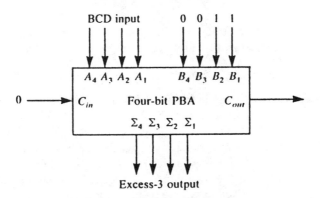

Figure 5.9   BCD-to-Excess-3 converter.

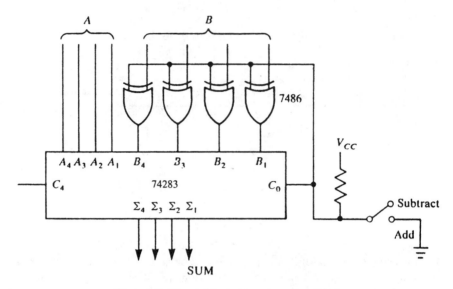

Figure 5.10 Adder/subtractor using 7483.

BCD digit and a carry. From Table 5.1, note that the correction is needed only when

(a) S_4 is 1, or
(b)   the binary sum is in the range of 1010 to 1111.

By plotting the six terms corresponding to (b) and grouping them on a K-map, we can see that (b) corresponds to

$$S_3 S_2 + S_3 S_1$$

Since the carry C is also produced when the Add-6 correction is performed, the condition for the correction and carry can be expressed as

$$C = S_4 + S_3 S_2 + S_3 S_1$$

The Add-6 correction circuit in Figure 5.12 implements this condition.

Note that the circuit in Figure 5.12 uses two MSI chips (PBAs) and an SSI chip if the AND-OR circuit is implemented with NAND gates. A TTL MSI circuit that implements the four-bit BCD adder is available (82S83). This MSI circuit achieves shorter propagation delays than the circuit in Figure 5.12 by using lookahead carry generation circuits. Also note that the circuit corresponding to the second PBA in Figure 5.12 can be simplified considerably, since only a constant value of 0110 is added to the

TABLE 5.1 BCD Addition

Binary sum					BCD sum					Decimal sum
S_4	S_3	S_2	S_1	S_0	C	Z_3	Z_2	Z_1	Z_0	
0	0	0	0	0	0	0	0	0	0	0
0	0	0	0	1	0	0	0	0	1	1
0	0	0	1	0	0	0	0	1	0	2
0	0	0	1	1	0	0	0	1	1	3
0	0	1	0	0	0	0	1	0	0	4
0	0	1	0	1	0	0	1	0	1	5
0	0	1	1	0	0	0	1	1	0	6
0	0	1	1	1	0	0	1	1	1	7
0	1	0	0	0	0	1	0	0	0	8
0	1	0	0	1	0	1	0	0	1	9
0	1	0	1	0	1	0	0	0	0	10
0	1	0	1	1	1	0	0	0	1	11
0	1	1	0	0	1	0	0	1	0	12
0	1	1	0	1	1	0	0	1	1	13
0	1	1	1	0	1	0	1	0	0	14
0	1	1	1	1	1	0	1	0	1	15
1	0	0	0	0	1	0	1	1	0	16
1	0	0	0	1	1	0	1	1	1	17
1	0	0	1	0	1	1	0	0	0	18
1	0	0	1	1	1	1	0	0	1	19

other operand, and hence a complete PBA circuit is not necessary. These enhancements have been used in the 82S83.

5.3 Magnitude Comparators

A magnitude comparator compares two n-bit numbers (A and B) and produces three outputs corresponding to $A = B$, $A > B$, and $A < B$. Figure 5.13 shows the truth table for a two-bit magnitude comparator.

Figure 5.14 shows the TTL 7485, a four-bit magnitude comparator. This device can be cascaded to form a comparator of almost any length. Figure 5.15 shows the use of 7485s to form a 24-bit comparator.

5.4 Shifters

As we have seen before, shifting a binary number left one bit is equivalent to multiplying it by 2, and shifting it one bit to the right is equivalent to

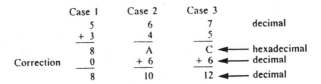

Case 1	Case 2	Case 3	
5	6	7	decimal
+ 3	4	5	
8	A	C ◄——	hexadecimal
Correction 0	+ 6	+ 6 ◄——	decimal
8	10	12 ◄——	decimal

Figure 5.11   Examples of BCD addition.

dividing it by 2. In addition to this *arithmetic shifting, shifters* are also useful in performing *logical shifts* on binary data.

Figure 5.16(a) shows an *n*-bit *simple shifter*, which shifts the input data (X) one bit to the left or to the right based on the control input D (Direction), to produce the *n*-bit output Y. Input S controls the shift/no shift operation. The operation of the circuit can be described as the following:

If $S = 0$, no shift (i.e., $Y_i = X_i$)

If $S = 1$ and $D = 0$, right shift (i.e., $Y_i = X_{i+1}$)

If $S = 1$ and $D = 1$, left shift (i.e., $Y_i = X_{i-1}$)

where $0 \leq i \leq (n - 1)$. The AND-OR circuit for this shifter is shown in Figure 5.16(b).

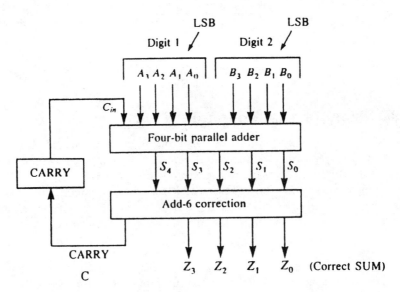

Figure 5.12   BCD adder.

A	B			
$a_1\,a_0$	$b_1\,b_0$	$A = B$	$A > B$	$A < B$
0 0	0 0	1	0	0
0 0	0 1	0	0	1
0 0	1 0	0	0	1
0 0	1 1	0	0	1
0 1	0 0	0	1	0
0 1	0 1	1	0	0
0 1	1 0	0	0	1
0 1	1 1	0	0	1
1 0	0 0	0	1	0
1 0	0 1	0	1	0
1 0	1 0	1	0	0
1 0	1 1	0	0	1
1 1	0 0	0	1	0
1 1	0 1	0	1	0
1 1	1 0	0	1	0
1 1	1 1	1	0	0

Figure 5.13 Truth table for a magnitude comparator.

The simple shifter can be generalized to the *p-bit shifter* shown in Figure 5.17. The shift direction is controlled by D. The amount of shift can be 0, 1, 2, ..., p bits to the left or right and is controlled by the set of control signals $S_1, \ldots, S_q$, where $2^q \geq p$.

Figure 5.18 shows the TTL 74S350, a four-bit shifter with tristate outputs. It can shift the four-bit data by zero, one, two, or three bits under the control of the two control signals S_0 and S_1. OE is the output enable signal. The operation of the circuit is represented by the truth table. It can perform the logical shifting (where logic-0s are inserted at either end of the shifting field), arithmetic shifting (where the sign bit is extended during the right shift), and end-around shifting (where the data word forms a continuous loop), by connecting I_{-1}, I_{-2}, and I_{-3} to I_0 through I_3 appropriately. (The determination of these connections is left as an exercise.)

5.5 Arithmetic/Logic Units (ALUs)

Several arithmetic/logic units are available that are capable of performing addition and subtraction and various logic operations on two operands.

Figure 5.19 shows a popular four-bit ALU (TTL 74181). There are two four-bit inputs (A and B) and the CARRY-IN C_n. The output consists of the four-bit F and the CARRY-OUT bit C_{n+4}. In addition to the Generate (G) and Propagate (P) outputs that are useful in interconnecting the device

PIN CONFIGURATION

LOGIC SYMBOL

LOGIC SYMBOL (IEEE/IEC)

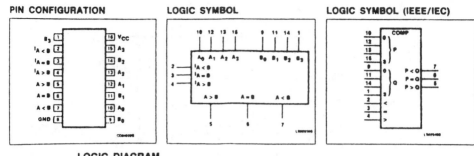

LOGIC DIAGRAM

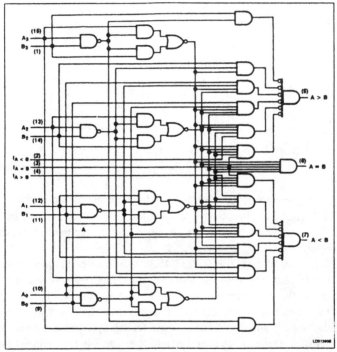

FUNCTION TABLE

COMPARING INPUTS				CASCADING INPUTS			OUTPUTS		
A_3, B_3	A_2, B_2	A_1, B_1	A_0, B_0	$I_{A>B}$	$I_{A<B}$	$I_{A=B}$	$A>B$	$A<B$	$A=B$
$A_3 > B_3$	X	X	X	X	X	X	H	L	L
$A_3 < B_3$	X	X	X	X	X	X	L	H	L
$A_3 = B_3$	$A_2 > B_2$	X	X	X	X	X	H	L	L
$A_3 = B_3$	$A_2 < B_2$	X	X	X	X	X	L	H	L
$A_3 = B_3$	$A_2 = B_2$	$A_1 > B_1$	X	X	X	X	H	L	L
$A_3 = B_3$	$A_2 = B_2$	$A_1 < B_1$	X	X	X	X	L	H	L
$A_3 = B_3$	$A_2 = B_2$	$A_1 = B_1$	$A_0 > B_0$	X	X	X	H	L	L
$A_3 = B_3$	$A_2 = B_2$	$A_1 = B_1$	$A_0 < B_0$	X	X	X	L	H	L
$A_3 = B_3$	$A_2 = B_2$	$A_1 = B_1$	$A_0 = B_0$	H	L	L	H	L	L
$A_3 = B_3$	$A_2 = B_2$	$A_1 = B_1$	$A_0 = B_0$	L	H	L	L	H	L
$A_3 = B_3$	$A_2 = B_2$	$A_1 = B_1$	$A_0 = B_0$	L	L	H	L	L	H
$A_3 = B_3$	$A_2 = B_2$	$A_1 = B_1$	$A_0 = B_0$	X	X	H	L	L	H
$A_3 = B_3$	$A_2 = B_2$	$A_1 = B_1$	$A_0 = B_0$	H	H	L	L	L	L
$A_3 = B_3$	$A_2 = B_2$	$A_1 = B_1$	$A_0 = B_0$	L	L	L	H	H	L

H = HIGH voltage level
L = LOW voltage level
X = Don't care

Figure 5.14 Four-bit magnitude comparator (TTL 7485) (courtesy of Signetics Corporation).

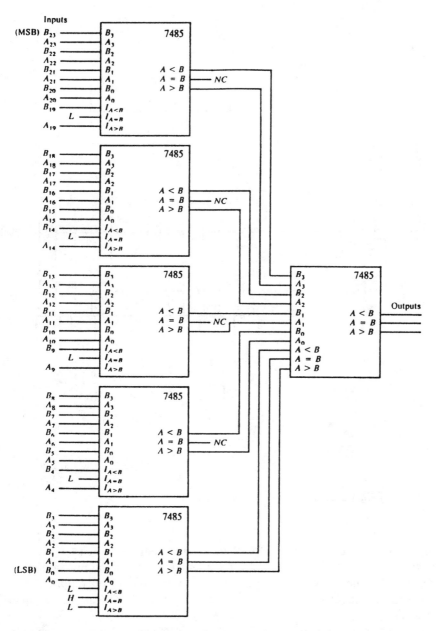

Figure 5.15 A 24-bit comparator using 7485s (courtesy of Signetics Corporation).

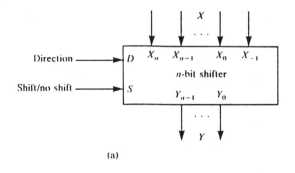

(a)

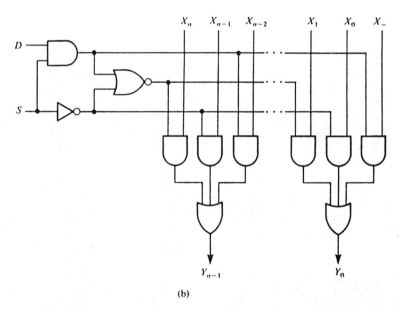

(b)

Figure 5.16 Simple shifter. (a) Block diagram; (b) circuit.

to a lookahead carry unit 74182, there is an equality output $(A = B)$. The function of the MODE and SELECT inputs are shown in the function table. The MODE input selects the arithmetic or logic mode of operation. The 4-bit SELECT input determines one of the 16 possible operations in either mode.

This device can be used with both active-low and active-high operands, as shown by the function tables and the corresponding pinouts. When the mode control (M) is high, the device performs one of the 16 logic operations on operands A and B, and all the carries are inhibited. When the mode

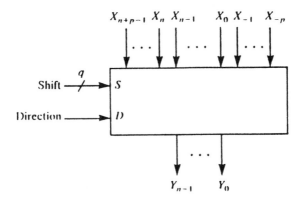

Figure 5.17 *p*-Bit shifter.

control is low, all the carries are enabled and the device performs one of the 16 arithmetic operations. An internal carry lookahead circuit is used to generate the CARRY-OUT (C_{n+4}). Several 74181s can be cascaded to form ripple-carry mode ALUs for operands with larger numbers of bits. When a high-speed operation is needed, the P and G outputs are used to interface with the lookahead carry generator 74182.

Figure 5.20 shows the lookahead carry generator 74182, which provides the carry lookahead across a group of four 74181s.

Figure 5.21 shows a 32-bit ALU using 74181s and 74182s.

5.6 Multipliers

The shift-and-add algorithm for multiplication that was described in Chapter 1 can be used to implement a multiplier using shifter and adder circuits as the components. A better implementation of this circuit is obtained by using *shift registers*. (Chapter 7 provides the details of the shift register operation and shows the multipler implementation.) Another implementation of a multiplier is as a fully combinational device (e.g., a two-level circuit starting from the truth table). These implementations form the two ends of a range of possible multiplier configurations. The fully combinational implementation is the fastest but requires extensive hardware as the number of bits in the operands increases. The shift-and-add implementation, though slower, is the simplest. In addition, the shift-and-add implementation enables the cascading of multiplier circuits designed for *n*-bit multiplication to form multipliers for numbers with multiples of *n*-bits.

Here, we will examine another implementation scheme for the multiplier. Figure 5.22 shows a four-bit multiplier circuit. The multiplicand

LOGIC SYMBOL

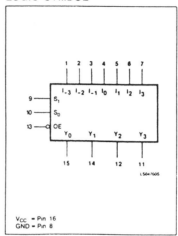

Vcc = Pin 16
GND = Pin 8

FUNCTION TABLE

OE	S₁	S₀	I₃	I₂	I₁	I₀	I₋₁	I₋₂	I₋₃	Y₃	Y₂	Y₁	Y₀
H	X	X	X	X	X	X	X	X	X	Z	Z	Z	Z
L	L	L	D₃	D₂	S₁	D₀	X	X	X	D₃	D₂	D₁	D₀
L	L	H	H	X	D₂	D₁	D₀	D₋₁	X	X	D₂	D₁	D₀
L	H	L	X	X	D₁	D₀	D₋₁	D₋₂	X	D₁	D₀	D₋₁	D₋₂
L	H	H	X	X	X	D₀	D₋₁	D₋₂	D₋₃	D₀	D₋₁	D₋₂	D₋₃

H = HIGH voltage level
L = LOW voltage level
X = Don't care
(Z) = HIGH impedance (off) state
D_n = HIGH or LOW state of referenced I_n input

LOGIC EQUATIONS

$$Y_0 = \bar{S}_0 \cdot \bar{S}_1 \cdot I_0 + S_0 \cdot \bar{S}_1 \cdot I_{-1} + \bar{S}_0 \cdot S_1 \cdot I_{-2} + S_0 \cdot S_1 \cdot I_{-3}$$

$$Y_1 = \bar{S}_0 \cdot \bar{S}_1 \cdot I_1 + S_0 \cdot \bar{S}_1 \cdot I_0 + \bar{S}_0 \cdot S_1 \cdot I_{-1} + S_0 \cdot S_1 \cdot I_{-2}$$

$$Y_2 = \bar{S}_0 \cdot \bar{S}_1 \cdot I_2 + S_0 \cdot \bar{S}_1 \cdot I_1 + \bar{S}_0 \cdot S_1 \cdot I_0 + S_0 \cdot S_1 \cdot I_{-1}$$

$$Y_3 = \bar{S}_0 \cdot \bar{S}_1 \cdot I_3 + S_0 \cdot \bar{S}_1 \cdot I_2 + \bar{S}_0 \cdot S_1 \cdot I_1 + S_0 \cdot S_1 \cdot I_0$$

Figure 5.18 Four-bit shifter (TTL 74S350) (courtesy of Signetics Corporation).

A and the multiplier *B* are each four bits long. The implementation parallels the longhand binary multiplication procedure. Each partial product is obtained by ANDing the corresponding bit of *B* with each bit of *A*. The partial products are shifted and accumulated using the four-bit adders.

The middle two stages of the circuit can be converted into a module, as shown in Figure 5.23. This module multiplies the four-bit number *A* by the two-bit number $(b_1 b_0)$ and adds the result to the four-bit number P_i, to produce P_{i+1}. That is,

Logic Symbols

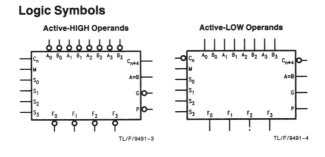

Active-HIGH Operands   Active-LOW Operands

TL/F/9491-3   TL/F/9491-4

**Pin Assignment
for DIP, SOIC and Flatpak**

Figure 5.19 Four-bit ALU (74181) (courtesy of National Semiconductor).

$$P_{i+1} = A \times b_1 \times 2 + A \times b_0 + P_i$$

This is a general 4×2 multiplier block; two such blocks can be used to build a 4×4 multiplier, as shown in Figure 5.24.

In fact, this type of multiplier module is available as an MSI module. The TTL 74LS261, which is shown in Figure 5.25, is one such module. It multiplies the four-bit number B by the two-bit number M to produce the product Q. B_0 and M_0 are sign bits. This module uses twos complement arithmetic. Several fast multiplication algorithms and the corresponding hardware modules are now available. (For further details, see the books by Ercegovac and Lang listed in the references at the end of this chapter.)

'F181 Operation Table

	S_0	S_1	S_2	S_3	Logic (M = H)	Arithmetic (M = L, C_0 = Inactive)	Arithmetic (M = L, C_0 = Active)
a. All Input Data Inverted	L	L	L	L	$\overline{A}$	A minus 1	A
	H	L	L	L	$\overline{A} \cdot \overline{B}$	A • B minus 1	A • B
	L	H	L	L	$\overline{A} + \overline{B}$	A • $\overline{B}$ minus 1	A • $\overline{B}$
	H	H	L	L	Logic "1"	minus 1 (2s comp.)	Zero
	L	L	H	L	$\overline{A} + B$	A plus (A + $\overline{B}$)	A plus (A + $\overline{B}$) plus 1
	H	L	H	L	$\overline{B}$	A • B plus (A + $\overline{B}$)	A • B plus (A + $\overline{B}$) plus 1
	L	H	H	L	$A \oplus \overline{B}$	A minus B minus 1	A minus B
	H	H	H	L	$A + \overline{B}$	A + $\overline{B}$	A + $\overline{B}$ plus 1
	L	L	L	H	$\overline{A} \cdot B$	A plus (A + B)	A plus (A + B plus 1)
	H	L	L	H	$A \oplus B$	A plus B	A plus B plus 1
	L	H	L	H	B	A • $\overline{B}$ plus (A + B)	A • $\overline{B}$ plus (A + B) plus 1
	H	H	L	H	$A + B$	A + B	A + B plus 1
	L	L	H	H	Logic "0"	A plus A (2 × A)	A plus A (2 × A) plus 1
	H	L	H	H	$A \cdot \overline{B}$	A plus A • B	A plus A • B plus 1
	L	H	H	H	$A \cdot B$	A plus A • $\overline{B}$	A plus A • $\overline{B}$ plus 1
	H	H	H	H	A	A	A plus 1
b. All Input Data True	L	L	L	L	$\overline{A}$	A	A plus 1
	H	L	L	L	$\overline{A} + \overline{B}$	A + B	A + B plus 1
	L	H	L	L	$\overline{A} \cdot \overline{B}$	A + $\overline{B}$	A + $\overline{B}$ plus 1
	H	H	L	L	Logic "0"	minus 1 (2s comp.)	Zero
	L	L	H	L	$\overline{A} \cdot \overline{B}$	A plus A • $\overline{B}$	A plus A • $\overline{B}$ plus 1
	H	L	H	L	$\overline{B}$	A • $\overline{B}$ plus (A + B)	A • B plus (A + B) plus 1
	L	H	H	L	$A \oplus B$	A minus B minus 1	A minus B
	H	H	H	L	$A \cdot B$	A • $\overline{B}$ minus 1	A • $\overline{B}$
	L	L	L	H	$\overline{A} + B$	A plus A • B	A plus A • B plus 1
	H	L	L	H	$\overline{A \oplus B}$	A plus B	A plus B plus 1
	L	H	L	H	B	A • B plus (A + $\overline{B}$)	A • B plus (A + B) plus 1
	H	H	L	H	$A \cdot B$	A • B minus 1	A • B
	L	L	H	H	Logic "1"	A plus A (2 × A)	A plus A (2 × A) plus 1
	H	L	H	H	$A + \overline{B}$	A plus (A + B)	A plus (A + B) plus 1
	L	H	H	H	$A + B$	A plus (A + $\overline{B}$)	A plus (A + $\overline{B}$) plus 1
	H	H	H	H	A	A minus 1	A
c. A Input Data Inverted; B Input Data True	L	L	L	L	$\overline{A}$	A minus 1	A
	H	L	L	L	$\overline{A} + B$	A • $\overline{B}$ minus 1	A • B
	L	H	L	L	$\overline{A} \cdot \overline{B}$	A • B minus 1	A • B
	H	H	L	L	Logic "1"	minus 1 (2s comp.)	Zero
	L	L	H	L	$\overline{A} \cdot B$	A plus (A + B)	A plus (A + B) plus 1
	H	L	H	L	B	A • $\overline{B}$ plus (A + B)	A • $\overline{B}$ plus (A + B) plus 1
	L	H	H	L	$A \oplus B$	A plus B	A plus B plus 1
	H	H	H	L	$A + B$	A + B	A + B plus 1
	L	L	L	H	$\overline{A} + B$	A plus (A + B)	A plus (A + B) plus 1
	H	L	L	H	$\overline{A \oplus B}$	A minus B minus 1	A minus B
	L	H	L	H	$\overline{B}$	A • B plus (A + $\overline{B}$)	A • B plus (A + $\overline{B}$) plus 1
	H	H	L	H	$A + \overline{B}$	A + $\overline{B}$	A + $\overline{B}$ plus 1
	L	L	H	H	Logic "0"	A plus A (2 × A)	A plus A (2 × A) plus 1
	H	L	H	H	$A \cdot B$	A plus A • $\overline{B}$	A plus A • $\overline{B}$ plus 1
	L	H	H	H	$A \cdot B$	A plus A • B	A plus A • B plus 1
	H	H	H	H	A	A	A plus 1
d. A Input Data True; B Input Data Inverted	L	L	L	L	$\overline{A}$	A	A plus 1
	H	L	L	L	$\overline{A} \cdot B$	A + $\overline{B}$	A + $\overline{B}$ plus 1
	L	H	L	L	$\overline{A} + B$	A + B	A + B plus 1
	H	H	L	L	Logic "0"	minus 1 (2s comp.)	Zero
	L	L	H	L	$\overline{A} + B$	A • B plus (A + $\overline{B}$)	A • $\overline{B}$ plus (A + B) plus 1
	H	L	H	L	B	A • B plus (A + $\overline{B}$)	A • $\overline{B}$ plus (A + B) plus 1
	L	H	H	L	$A \oplus \overline{B}$	A plus B	A minus B
	H	H	H	L	$A \cdot B$	A • B minus 1	A • B
	L	L	L	H	$\overline{A} \cdot B$	A plus A • B	A plus A • B plus 1
	H	L	L	H	$A \oplus B$	A minus B minus 1	A minus B
	L	H	L	H	$\overline{B}$	A • $\overline{B}$ plus (A + B)	A • $\overline{B}$
	H	H	L	H	$A \cdot \overline{B}$	A • $\overline{B}$ minus 1	A • $\overline{B}$
	L	L	H	H	Logic "1"	A plus A (2 × A)	A plus A (2 × A) plus 1
	H	L	H	H	$A + B$	A plus (A + $\overline{B}$)	A plus (A + $\overline{B}$) plus 1
	L	H	H	H	$A + \overline{B}$	A plus (A + B)	A plus (A + B) plus 1
	H	H	H	H	A	A minus 1	A

Pinout (each section): C_n, A_0 B_0 A_1 B_1 A_2 B_2 A_3 B_3, C_{n+4}, M, S_0, S_1, S_2, S_3, A=B, G, P, F_0 F_1 F_2 F_3 — 'F181

Figure 5.19 *Continued*

LOGIC DIAGRAM

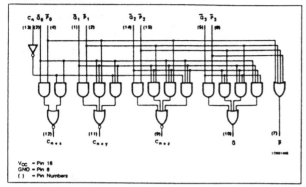

Vcc = Pin 16
GND = Pin 8
( ) = Pin Numbers

The logic equations provided at the outputs are:

$$C_{n+x} = G_0 + P_0 C_n$$
$$C_{n+y} = G_1 + P_1 G_0 = P_1 P_0 C_n$$
$$C_{n+z} = G_2 + P_2 G_1 + P_2 P_1 G_0$$
$$\overline{G} = \overline{G_3 + P_3 G_2 + P_3 P_2 G_1 + P_3 P_2 P_1 G_0}$$
$$\overline{P} = \overline{P_3 P_2 P_1 P_0}$$

The '182 can also be used with binary ALU's in an active LOW or active HIGH input operand mode. The connections to and from the ALU to the carry lookahead generator are identical in both cases.

FUNCTION TABLE

			INPUTS								OUTPUTS		
C_n	$\overline{G}_0$	$\overline{P}_0$	$\overline{G}_1$	$\overline{P}_1$	$\overline{G}_2$	$\overline{P}_2$	$\overline{G}_3$	$\overline{P}_3$	C_{n+x}	C_{n+y}	C_{n+z}	$\overline{G}$	$\overline{P}$
X	H	H							L				
L	H	X							L				
X	L	X							H				
H	X	L							H				
X	X	X	H	H						L			
X	H	H	H	X						L			
L	H	X	H	X						L			
X	X	X	L	X						H			
X	L	X	X	L						H			
H	X	L	X	L						H			
X	X	X	X	X	H	H					L		
X	X	X	H	H	H	X					L		
X	H	H	H	X	H	X					L		
L	H	X	H	X	H	X					L		
X	X	X	X	X	L	X					H		
X	X	X	L	X	X	L					H		
X	L	X	X	L	X	L					H		
H	X	L	X	L	X	L					H		
	X		X	X	X	X	H	H				H	
	X		X	X	H	H	H	X				H	
	X		H	H	H	X	H	X				H	
	H		H	X	H	X	H	X				H	
	X		X	X	X	X	L	X				L	
	X		X	X	L	X	X	L				L	
	X		L	X	X	L	X	L				L	
	L		X	L	X	L	X	L				L	
		H		X		X		X					H
		X		H		X		X					H
		X		X		H		X					H
		X		X		X		H					H
		L		L		L		L					L

H = HIGH voltage level
L = LOW voltage level
X = Don't care

LOGIC SYMBOL

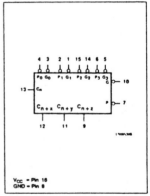

Vcc = Pin 16
GND = Pin 8

Figure 5.20 Lookahead carry generator (74182) (courtesy of Signetics Corporation).

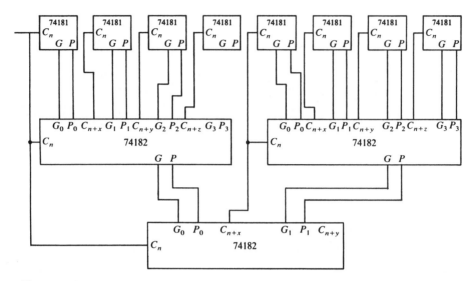

Figure 5.21 A 32-bit ALU. (Note: A and B inputs and F outputs of 74181s are not shown.)

5.7 Decoders

A *code* is a string of several bits. With an n-bit code, it is possible to represent 2^n unique values. In digital system design, it is often necessary to invoke a set of operations based on the bit pattern (or the code) in a certain field. For instance, consider the *operation code*, which is part of every instruction in a digital computer. The operation code specifies the operation to be performed. This code must be decoded by the computer's control unit to determine which operation to perform.

The devices that translate the n-bit pattern into one of the 2^n possible values are called *decoders*. Thus, a decoder generally translates a binary value into a nonbinary one. An *encoder*, however, converts a nonbinary value into a binary one. A device that transforms a nonbinary value into another nonbinary value is called a *code converter*. However, there are exceptions to this general classification. For example, the TTL 7447—a device that transforms the BCD input into a seven-segment output to drive a seven-segment display device—is traditionally called a BCD-to-seven-segment decoder rather than a code converter.

An *n-input decoder* (*a binary decoder*, to be more specific) is a combinational circuit that has n binary inputs and 2^n outputs. The input values can be considered to represent integers from 0 to $(2^n - 1)$. When the input

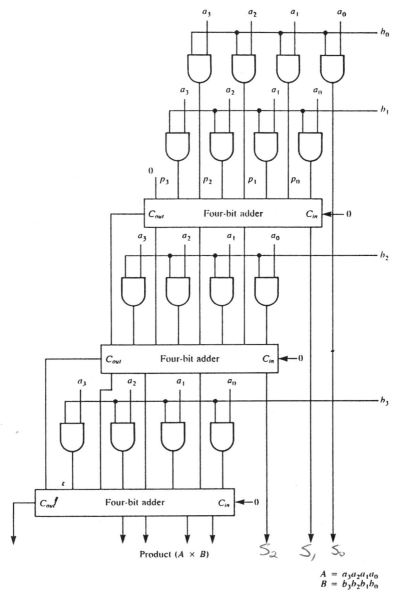

Figure 5.22 A 4 × 4 multiplier.

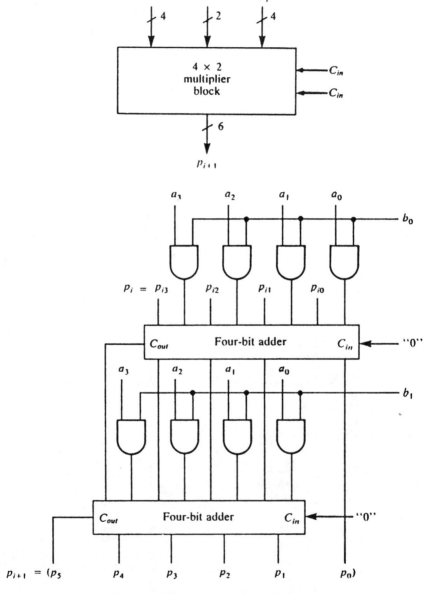

Figure 5.23 A 4 × 2 multiplier block.

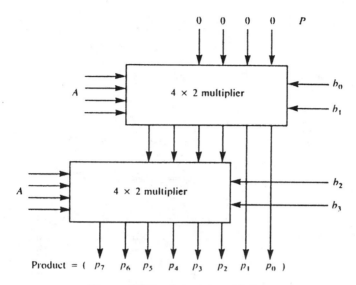

Figure 5.24 A 4 × 4 multiplier.

represents the integer i, the ith output will be high and all other outputs will be low. The outputs are numbered from 0 to $(2^n - 1)$. If the input code does not use all the 2^n combinations, the decoder need not have all the 2^n outputs. For example, a BCD-to-decimal decoder will have four inputs and only ten outputs. A general decoder symbol is shown below:

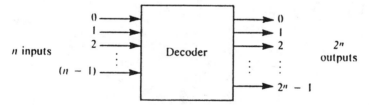

Figure 5.26(a) shows the truth table for a 3-to-8 decoder. The three inputs are designated A, B, and C, with C as the LSB. The outputs are numbered from 0 through 7. The circuit is shown in (b). Note that it is not always necessary to draw a truth table for a decoder. Since there will be only one 1 in any output column of the truth table for a decoder, a circuit can easily be realized by just using an AND gate with the appropriate number of inputs for each output of the decoder.

Figure 5.27 shows a 3-to-8 decoder IC (TTL 74138). This IC is also designated as a 1-of-8-decoder to emphasize the selection of one of the eight outputs. Also note that this IC can be used as a *demultiplexer*. (Demulti-

LOGIC SYMBOL

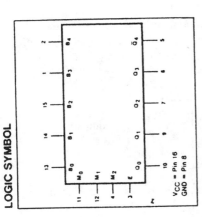

Pins: B₄ = 2, B₃ = 1, B₂ = 15, B₁ = 14, B₀ = 13, M₀ = 11, M₁ = 12, M₂ = 4, E = 3, Q̄₄ = 5, Q₃ = 6, Q₂ = 7, Q₁ = 9, Q₀ = 10

V_{CC} = Pin 16
GND = Pin 8

FUNCTION TABLE

INPUTS				OUTPUTS				
E	M_2	M_1	M_0	$\bar{Q}_4$	Q_3	Q_2	Q_1	Q_0
L	X	X	X	$\bar{q}_4$	q_3	q_2	q_1	q_0
H	L	L	L	H	L	L	L	L
H	L	L	H	$\bar{B}_4$	B_4	B_3	B_2	B_1
H	L	H	L	$\bar{B}_4$	B_4	B_3	B_2	B_1
H	L	H	H	$\bar{B}_4$	$\bar{B}_3$	$\bar{B}_2$	$\bar{B}_1$	$\bar{B}_0$
H	H	L	L	B_4	$\bar{B}_4$	$\bar{B}_3$	$\bar{B}_2$	$\bar{B}_1$
H	H	L	H	B_4	$\bar{B}_4$	$\bar{B}_3$	$\bar{B}_2$	$\bar{B}_1$
H	H	H	L	B_4	B_4	B_3	B_2	B_1
H	H	H	H	H	L	L	L	L

H = HIGH voltage level
L = LOW voltage level
q = Lower case letters indicate the state of the referenced output one setup time prior to the HIGH-to-LOW Enable transition.
B_n = The logic level on the referenced multiplicand input.

FUNCTIONAL DESCRIPTION

The "261" is designed to perform binary multiplication in two's complement form, two bits at a time.

The M inputs are for the multiplier bits and the B inputs are for the multiplicand. The Q outputs represent the partial product as a recoded base-4 number. This recoding effectively reduces the Wallace-Tree (summing) hardware requirements by a factor of two.

The outputs represent partial products in one's-complement notation generated as a result of multiplication. A simple rounding scheme using two additional gates is needed for each partial product to generate two's complement.

The leading (most significant) bit of the product is inverted for ease in extending the sign to left justify the partial-product bits.

Figure 5.25 Multiplier module (TTL 74LS261) (courtesy of Signetics Corporation).

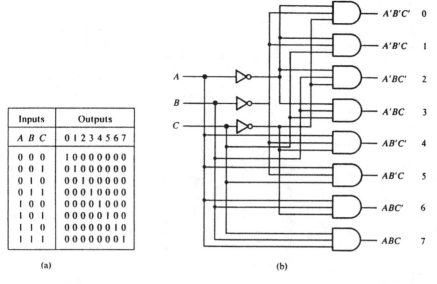

Inputs	Outputs
A B C	0 1 2 3 4 5 6 7
0 0 0	1 0 0 0 0 0 0 0
0 0 1	0 1 0 0 0 0 0 0
0 1 0	0 0 1 0 0 0 0 0
0 1 1	0 0 0 1 0 0 0 0
1 0 0	0 0 0 0 1 0 0 0
1 0 1	0 0 0 0 0 1 0 0
1 1 0	0 0 0 0 0 0 1 0
1 1 1	0 0 0 0 0 0 0 1

(a)   (b)

Figure 5.26   3-to-8 decoder. (a) Truth table; (b) circuit.

plexing is described later in this chapter.) Note that the outputs are active-low. That is, the selected output will be low and all other outputs will be high. The IC is enabled only when E_1 and E_2 are low and E_3 is high; otherwise, none of the outputs is selected (i.e., all the outputs will be high). The multiple ENABLE inputs of the IC are useful in expanding the function of the IC. (See the IC manuals listed in the reference section for further details.)

An n-to-$2''$ decoder can be viewed as a device that produces all the $2''$ minterms of an n-variable function. Thus, to build a combinational circuit with n inputs and m outputs, an n-to-$2''$ decoder can be used. Corresponding to each output of the circuit, an OR gate with the appropriate number of inputs is needed. The outputs of the decoder corresponding to the minterms that produce the output form the inputs to the OR gate.

Example 5.4

Implement the following function using decoders:

$$F(A, B, C) = \Sigma m\ (1, 3, 7)$$
$$G(A, B, C = \Sigma m\ (2, 3, 6)$$

DESCRIPTION

The '138 decoder accepts three binary weighted inputs (A_0, A_1, A_2) and when enabled, provides eight mutually exclusive, active LOW outputs ($\overline{0}$ – $\overline{7}$). The device features three Enable Inputs: two active LOW ($\overline{E}_1$, $\overline{E}_2$) and one active HIGH (E_3). Every output will be HIGH unless $\overline{E}_1$ and $\overline{E}_2$ are LOW and E_3 is HIGH. This multiple enable function allows easy parallel expansion of the device to a 1-of-32 (5 lines to 32 lines)

decoder with just four '138s and one inverter.

The device can be used as an eight output demultiplexer by using one of the active LOW Enable inputs as the Data input and the remaining Enable inputs as strobes. Enable inputs not used must be permanently tied to their appropriate active HIGH or active LOW state.

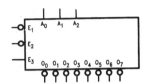

FUNCTION TABLE

INPUTS						OUTPUTS							
$\overline{E}_1$	$\overline{E}_2$	E_3	A_0	A_1	A_2	$\overline{0}$	$\overline{1}$	$\overline{2}$	$\overline{3}$	$\overline{4}$	$\overline{5}$	$\overline{6}$	$\overline{7}$
H	X	X	X	X	X	H	H	H	H	H	H	H	H
X	H	X	X	X	X	H	H	H	H	H	H	H	H
X	X	L	X	X	X	H	H	H	H	H	H	H	H
L	L	H	L	L	L	L	H	H	H	H	H	H	H
L	L	H	H	L	L	H	L	H	H	H	H	H	H
L	L	H	L	H	L	H	H	L	H	H	H	H	H
L	L	H	H	H	L	H	H	H	L	H	H	H	H
L	L	H	L	L	H	H	H	H	H	L	H	H	H
L	L	H	H	L	H	H	H	H	H	H	L	H	H
L	L	H	L	H	H	H	H	H	H	H	H	L	H
L	L	H	H	H	H	H	H	H	H	H	H	H	L

H = HIGH voltage level
L = LOW voltage level
X = Don't care

LOGIC DIAGRAM

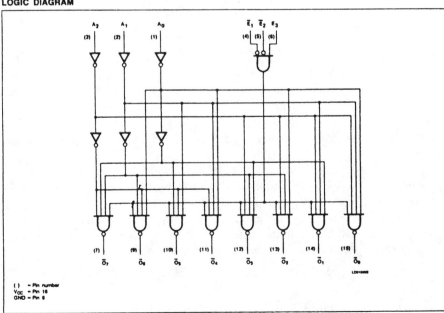

Figure 5.27   TTL 3-to-8 decoder (courtesy of Signetics Corporation).

The circuit is shown below:

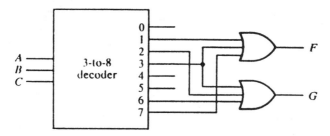

This is not an efficient implementation of the circuit, since the decoder is more complex than the two-level circuit required to implement these functions. Nevertheless, this is an example of an untraditional design using MSI components. A more efficient realization of these functions—through the use of multiplexer ICs—is described later in this chapter.

5.8 Code Converters

As mentioned earlier, a code converter is a combinational circuit that translates the input code word into the corresponding new code word. The BCD-to-7-segment decoder and the BCD-to-Excess-3 conversion circuit of Chapter 3 are examples of code converters. Several code converters are available as MSI components. A partial list follows:

	TTL	CMOS
BCD-to-7 segment decoders	7446	4495
(designed to drive various display types)	7447	4511
	7448	4513
Excess-3-to-decimal (one-of-ten) decoder	7443	
Excess-3-gray-to-decimal (one-of-ten) decoder	7444	

Figure 5.28 shows the TTL 7446, a BCD-to-7 segment decoder. The four-bit BCD input is translated into a seven-bit code. Each output drives a segment of a 7-segment display device. The segment will be on when the output of the decoder connected to it is low. Figure 5.28 shows the 7-segment display device, the display patterns for 0 through 9, and the truth table for the code converter. Note that the outputs of the converter are active-low and are compatible with the display device inputs.

The active-low lamp test (LT) input overrides all other inputs and enables a check to be made on possible display malfunctions.

LOGIC SYMBOL

PIN CONFIGURATION

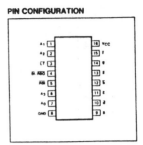

V_{CC} = Pin 16
GND = Pin 8

FUNCTIONAL DESCRIPTION

The "46A" and "47A" 7-segment decoders accept a 4-bit BCD code input and produce the appropriate outputs for selection of segments in a 7-segment matrix display used for representing the decimal numbers "0-9." The seven outputs ($\bar{a}$, $\bar{b}$, $\bar{c}$, $\bar{d}$, $\bar{e}$, $\bar{f}$, $\bar{g}$) of the decoder select the corresponding segments in the matrix shown in Figure A. The numeric designations chosen to represent the decimal numbers are shown in Figure B, together with the resulting displays for input code configurations in excess of binary "9."

The "46A" and "47A" have provisions for automatic blanking of the leading and/or trailing edge zeroes in a multidigit decimal number, resulting in an easily readable decimal display conforming to normal writing practice. In an 8-digit mixed integer fraction decimal representation, using the automatic blanking capability, 0070.0500 would be displayed as 70.05. Leading edge zero suppression is obtained by connecting the Ripple Blanking Output ($\overline{BI}/\overline{RBO}$) of a decoder to the Ripple Blanking Input ($\overline{RBI}$) of the next lower stage device. The most significant decoder stage should have the $\overline{RBI}$ input grounded; and, since suppression of the least significant integer zero in a number is not usually desired, the $\overline{RBI}$ input of this decoder stage should be left open. A similar procedure for the fractional part of a display will provide automatic suppression of trailing edge zeroes.

The decoder has an active LOW input Lamp Test which overrides all other input combinations and enables a check to be made on possible display malfunctions. The $\overline{BI}/\overline{RBO}$ terminal of the decoder can be OR-tied with a modulating signal via an isolating buffer to achieve pulse duration intensity modulation. A suitable signal can be generated for this purpose by forming a variable frequency multivibrator with a cross coupled pair of open collector gates.

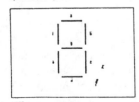

NOTE
b $\overline{BI}/\overline{RBO}$ is wire-AND logic serving as blanking input (BI) and/or ripple-blanking output (RBO)
    The blanking out (BI) must be open or held at a HIGH level when output functions 0 through 15 are desired, and, ripple-blanking input ($\overline{RBI}$) must be open or at a HIGH level if blanking of a decimal 0 is not desired

TRUTH TABLE

DECIMAL OR FUNCTION	INPUTS						OUTPUTS							
	$\overline{LT}$	$\overline{RBI}$	A_3	A_2	A_1	A_0	$\overline{BI}/\overline{RBO}$(b)	$\bar{a}$	$\bar{b}$	$\bar{c}$	$\bar{d}$	$\bar{e}$	$\bar{f}$	$\bar{g}$
0	H	H	L	L	L	L	H	L	L	L	L	L	L	H
1	H	X	L	L	L	H	H	H	L	L	H	H	H	H
2	H	X	L	L	H	L	H	L	L	H	L	L	H	L
3	H	X	L	L	H	H	H	L	L	L	L	H	H	L
4	H	X	L	H	L	L	H	H	L	L	H	H	L	L
5	H	X	L	H	L	H	H	L	H	L	L	H	L	L
6	H	X	L	H	H	L	H	H	H	L	L	L	L	L
7	H	X	L	H	H	H	H	L	L	L	H	H	H	H
8	H	X	H	L	L	L	H	L	L	L	L	L	L	L
9	H	X	H	L	L	H	H	L	L	L	H	H	L	L
10	H	X	H	L	H	L	H	H	H	H	L	L	H	L
11	H	X	H	L	H	H	H	H	H	L	L	H	H	L
12	H	X	H	H	L	L	H	H	L	H	H	L	L	L
13	H	X	H	H	L	H	H	L	H	H	L	H	L	L
14	H	X	H	H	H	L	H	H	H	H	L	L	L	L
15	H	X	H	H	H	H	H	H	H	H	H	H	H	H
$\overline{BI}$(b)	X	X	X	X	X	X	L	H	H	H	H	H	H	H
$\overline{RBI}$(b)	H	L	L	L	L	L	L	H	H	H	H	H	H	H
$\overline{LT}$	L	X	X	X	X	X	H	L	L	L	L	L	L	L

H = HIGH voltage level
L = LOW voltage level
X = Don't care

NUMERICAL DESIGNATIONS—RESULTANT DISPLAYS

LOGIC DIAGRAM

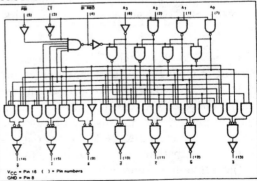

V_{CC} = Pin 16   ( ) = Pin numbers
GND = Pin 8

Figure 5.28   BCD-to-7-segment decoder (courtesy of Signetics Corporation).

When multiple-digit decimal numbers are to be displayed, each digit uses a 7-segment display device. The leading and trailing 0s typically are not displayed. For example, 0074.5600 is displayed as 74.56. Leading 0 suppression is obtained by connecting the Ripple Blanking Output (RBO) of a decoder to the ripple blanking input (RBI) of the device at the next lower stage. The RBI of the decoder at the most significant stage is connected to the Ground. The RBI inputs of the other stages in the integer portion are left open, since the trailing 0s in the integer portions are not suppressed. The trailing 0s in the fracation portion are suppressed by similar connections. (Signetics has replaced TTL 7446 with 4511 and 4543.)

5.9 Encoders

An encoder performs the reverse function of a decoder. The 2^n-to-n decoder shown below generates an n-bit code word as a function of the combination of values on its 2^n inputs.

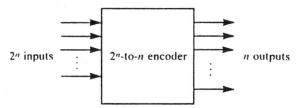

Usually only one of the inputs is 1 and all the others are 0. The output is a binary code word corresponding to the input that is at 1.

The classical design procedure can be used to derive the circuit for an encoder. However, since only one input is 1 at any time, only n rows are needed in the truth table rather than 2^n rows. The remaining rows contribute don't-care conditions. If more than one input is 1 at any time, the encoder generally will not produce a valid output. Nevertheless, there are exceptions. The so-called *priority encoders* (TTL 74148 is an example) allow more than one of their inputs to be active at any time. Each input has a priority assinged to it. The code word produced as the output corresponds to that of the highest-priority input among all the inputs that are active (see problem 5.11).

The encoder design procedure is demonstrated in Example 5.5.

Example 5.5

A truth table for a 4-to-2 encoder is shown in Figure 5.29(a). W, X, Y, and Z are the inputs and D_0 and D_1 are the outputs. Corresponding to each input at 1 is an output bit pattern (code word). The output functions are

Inputs				Outputs	
W	X	Y	Z	D_0	D_1
1	0	0	0	0	0
0	1	0	0	0	1
0	0	1	0	1	0
0	0	0	1	1	1

(a)

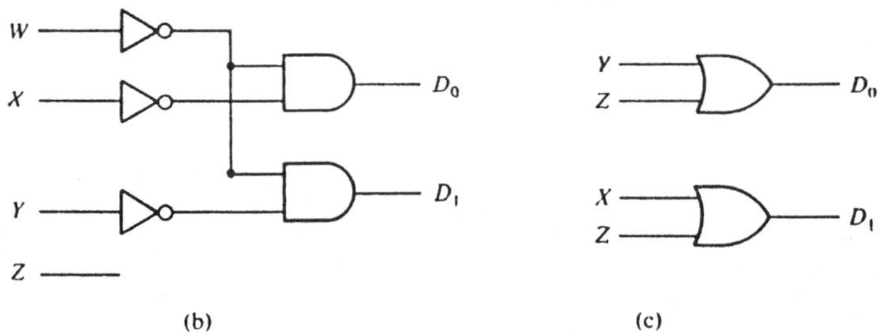

(b)

(c)

Figure 5.29   A 4-to-3 encoder. (a) Truth table; (b) circuit; (c) alternate circuit.

$$D_0 = W'X'$$
$$D_1 = W'Y' \quad (5.9)$$

Note that these functions are already simplified: it is sufficient to have $W = 0$ and $X = 0$ for D_0 to be 1, no matter what the values of Y and Z are. Similarly, the fact that $W = 0$ and $Y = 0$ is sufficient for D_1 to be 1. Such observations may not always be possible. In such cases, it will be necessary to complete the truth table to include all the *don't-care* conditions, and one of the minimization techniques can be used to simplify the circuit. The 4-to-2 encoder circuit is shown in Figure 5.26(b).

Note that the outputs of this encoder can also be expressed as

$$D_0 = Y + Z$$
$$D_1 = X + Z \quad (5.10)$$

Figure 5.29(c) illustrates the implementation.

Because encoders can be built easily using OR gates, as shown above, encoder ICs are not available as off-the-shelf components. However, several priority encoders (TTL 74147, 74148, 74278, and 74LS348 along with CMOS 4532) are available.

5.10 Multiplexers

Multiplexing is the process of channeling information from one of several sources to a single destination. For example, in a computer system with several input devices, all the devices are connected through a multiplexer to the common channel carrying data to the processing unit. Only one of the (selected) devices will transmit data at any time. Because all the devices share a common channel, multiplexing results in less expensive interconnection hardware.

Information is commonly multiplexed from many sources onto a single transmission line through *Frequency Division Multiplexing* (FDM) and *Time Division Multiplexing* (TDM). In FDM, each signal is *modulated* up to a specific frequency band and then transmitted. For example, in telephone conversations in which 3 kilohertz (KHz) is the highest frequency to each signal, one signal may be modulated to 3 to 6 KHz and another from 7 to 10 KHz for transmission on the same line. In TDM, a common multiplexing mode in digital circuits, each signal source is allocated a time slot in which it transmits on the common transmission line. Thus, if there are 10 devices, each might transmit for 1 μs during each 10-μs transmission cycle.

A *multiplexer* (*selector*) is thus a switch connecting one of its several inputs to the output. A set of n control inputs is needed to select one of the 2^n inputs. A four-input multiplexer is shown below:

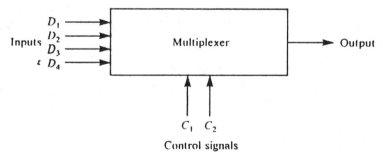

The operation of this multiplexer can be described by the following table:

C_1	C_2	Output
0	0	D_1
0	1	D_2
1	0	D_3
1	1	D_4

Although there are six inputs, a complete truth table with 2^6 rows is not required to design the circuit, since the output simply assumes the value of one of the four inputs, depending on the control signals C_1 and C_2. That is,

$$\text{Output} = D_1 \cdot C_1' \cdot C_2' + D_2 \cdot C_1' \cdot C_2 + D_3 \cdot C_1 C_2' + D_4 \cdot C_1 C_2 \quad (5.11)$$

The circuit for realizing this multiplexer is shown in Figure 5.30.

Each of the inputs D_1, D_2, D_3, and D_4 and the output of this multiplexer circuit is a single line. If the application requires that the data lines to be multiplexed have more than one bit each, the circuit shown in Figure 5.30 must be duplicated once for each bit of the data.

Figure 5.31 shows the functional details of the quad two-input multiplexer IC (74158). It is a 2-to-1 multiplexer, except that each input is four bits (a quad) wide. I_{0a}, I_{0b}, I_{0c}, and I_{0d} designate the four bits of the input I_0, and I_{1a}, I_{1b}, I_{1c}, and I_{1b} designate the other input I_1. The output is the complement of the selected input. S selects one of the inputs to be connected to the output. E is the enable signal and must be low for the circuit to be enabled. The truth table in the figure shows the operational details.

Although the primary use for a multiplexer is to switch several data channels onto one data channel when data are transmitted from several sources to a designation, it can be used to design combinational circuits as well.

Example 5.6

Figure 5.32 shows a circuit that connects one of the four inputs A, B, C, or D to the output. Each input is four bits wide. Three 74158s are used.

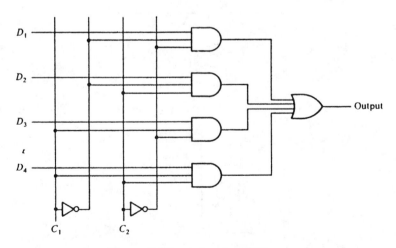

Figure 5.30 A 4-to-1 multiplexer.

FUNCTIONAL DESCRIPTION

The "158" is a Quad 2-Input Multiplexer which selects four bits of data from two sources under the control of a common Select input (S), presenting the data in inverted form at the four outputs ($\overline{Y}$). The Enable input ($\overline{E}$) is active LOW. When $\overline{E}$ is HIGH, all of the outputs ($\overline{Y}$) are forced HIGH regardless of all other input conditions.

Moving data from two groups of registers to four common output busses is a common use of the "158." The state of the Select input determines the particular register from which the data comes. It can also be used as a function generator. The device is useful for implementing gating functions by generating any four functions of two variables with one variable common.

The device is the logic implementation of a 4-pole, 2-position switch where the position of the switch is determined by the logic levels supplied to the Select Input. Logic equations are shown below:

$$\overline{Y}_a = \overline{E} \cdot (I_{1a} \cdot S + I_{0a} \cdot \overline{S})$$

$$\overline{Y}_b = \overline{E} \cdot (I_{1b} \cdot S + I_{0b} \cdot \overline{S})$$

$$\overline{Y}_c = \overline{E} \cdot (I_{1c} \cdot S + I_{0c} \cdot \overline{S})$$

$$\overline{Y}_d = \overline{E} \cdot (I_{1d} \cdot S + I_{0d} \cdot \overline{S})$$

LOGIC SYMBOL

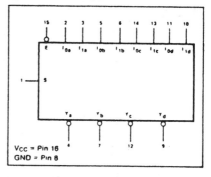

FUNCTION TABLE, '158

ENABLE	SELECT INPUT	DATA INPUTS		OUTPUT
$\overline{E}$	S	I_0	I_1	$\overline{Y}$
H	X	X	X	H
L	L	L	X	H
L	L	H	X	L
L	H	X	L	H
L	H	X	H	L

H = HIGH voltage level
L = LOW voltage level
X = Don't care

LOGIC DIAGRAM, '158

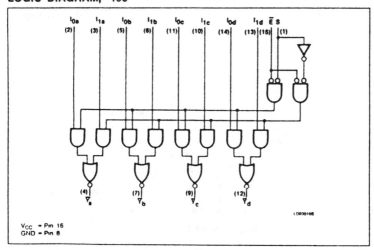

Figure 5.31 TTL multiplexer (courtesy of Signetics Corporation).

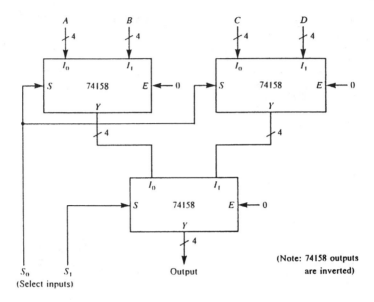

Figure 5.32   A data selector. (Note: 74158 outputs are inverted.)

(TTL 74153 is a dual 4-to-1 multiplexer IC. The circuit in this example can also be realized by using two 74153s.)

Example 5.7

Implement $F(A, B, C)$, $= \Sigma m$ (2, 3, 5, 7) using a multiplexer. Figure 5.33 shows such an implementation. The inputs A, B, and C are con-

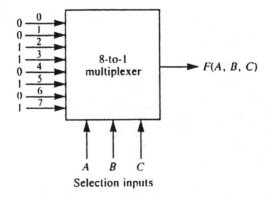

Figure 5.33   Logic realization using a multiplexer.

nected to the three selection inputs of the multiplexer. The other eight inputs of the multiplexer are set to the output values that correspond to the eight input conditions. Note that in this implementation no extra gates are required, though they would be if we were designing with decoders. It is possible to reduce the number of inputs on the multiplexer to implement this function.

Consider the truth table for the function in Example 5.7:

Minterm	ABC	F
0	000	0
1	001	0
2	010	1
3	011	1
4	100	0
5	101	1
6	110	0
7	111	1

We can partition the truth table based on the values of B and C, as shown below (this is often called the *folded-table* method):

BC	A	Minterm	F	
00	0	0	0	$F = 0$
	1	4	0	
01	0	1	0	$F = A$
	1	5	1	
10	0	2	1	$F = A'$
	1	6	0	
11	0	3	1	$F = 1$
	1	7	1	

From the partitioned truth table we can see that as long as $B = 0$ and $C = 0$, F is 0 regardless of what A is; $F = A$ when $B = 0$ and $C = 1$; $F = A'$ when $B = 1$ and $C = 0$; and $F = 1$ when $B = 1$ and $C = 1$. If B and C are used to select one of the four inputs of a multiplexer and the four inputs are connected to 0, A, A', and 1 respectively, the function F in

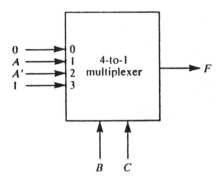

Figure 5.34   Implementation using a multiplexer.

Figure 5.34 is realized using a multiplexer with fewer inputs than the one in Figure 5.33.

The above procedure can be generalized to a function of n variables:

1.   Partition the truth table of the n-variable function into $2^{(n-1)}$ partitions, each partition corresponding to one combination of the values of the $n - 1$ least significant variables. (If the variable order in the truth table of the function from left to right is *ABCDE*, then *BCDE* are the least significant variables and *A* is the most significant.)

2.   For each partition, observe the value of the output (F) as related to the value of the most significant variable (A). F should be equal to one of the following values: 0, 1, A, or A'.

3.   Use a multiplexer with $n - 1$ selection inputs and $2^{(n-1)}$ data inputs. Partition the data inputs into four groups as determined in step 2, and connect them to 0, 1, A, and A'.

Multipliers are commonly available as IC packages in the following configurations: quad 2-to-1, dual 4-to-1, 8-to-1 and 16-to-1. These can be used to realize 2, 3, 4, and 5 variable functions, respectively. As seen above, a quad IC can be used to implement four different circuits, and a dual IC can implement two different circuits of the same number of variables.

5.11   Demultiplexers

Demultiplexing is the reverse of multiplexing. That is, a *demultiplexer* (or a *distributor*) distributes its input signal to one of its several outputs. A demultiplexer with n control signals and 2^n outputs is shown below:

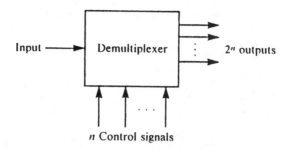

n Control signals

A typical application for a demultiplexer is switching the output data from a digital computer to one of several output devices.

In fact, a decoder with an *Enable* input can be used as a demultiplexer. Consider the 2-to-4 decoder shown in Figure 5.35(a). As shown in the truth table in (b), the outputs of the decoder are valid when the ENABLE signal is 1; otherwise, all the outputs will be at 0. If we treat the ENABLE signal as the input and the data inputs A and B as select control signals, the decoder circuit will perform demultiplexing. As can be seen from (b), the selected output will have the same values as that of the ENABLE input signal.

Figure 5.36 shows a demultiplexer/decoder IC (74139). This IC has two devices. Each can be used as either a 2-to-4 decoder or a 1-to-4 demultiplexer. When used as a demultiplexer, the ENABLE acts as the data input. As seen from the truth table, when the ENABLE is high, all the outputs are deselected. When the ENABLE is low, the output selected by the select signals A_0 and B_0 is low; thus, it acts as a 1-to-4 demultiplexer.

Example 5.8

Figure 5.37 shows the realization of a one-of-eight demultiplexer using the two parts of a 74139. Two of the three select signals needed (Q and R) are directly connected to the select signals of the top and bottom

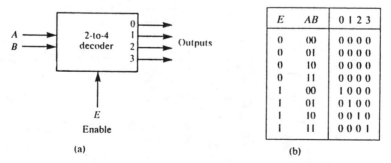

E	AB	0 1 2 3
0	00	0 0 0 0
0	01	0 0 0 0
0	10	0 0 0 0
0	11	0 0 0 0
1	00	1 0 0 0
1	01	0 1 0 0
1	10	0 0 1 0
1	11	0 0 0 1

(a)   (b)

Figure 5.35 2-to-4 decoder/1-to-4 demultiplexer. (a) Circuit; (b) truth table.

DESCRIPTION

The '139 is a high-speed, dual 1-of-4 decoder/demultiplexer. This device has two independent decoders, each accepting two binary weighted inputs (A_0, A_1) and providing four mutually exclusive active LOW outputs ($\bar{0} - \bar{3}$). Each decoder has an active LOW Enable ($\bar{E}$). When $\bar{E}$ is HIGH, every output is forced HIGH. The Enable can be used as the Data input for a 1-of-4 demultiplexer application.

PIN CONFIGURATION

LOGIC SYMBOL

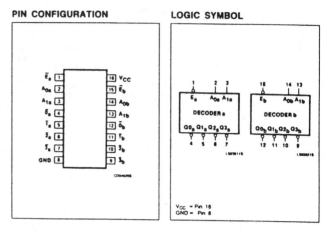

LOGIC DIAGRAM

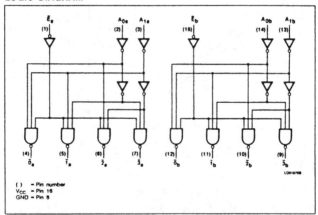

FUNCTION TABLE

INPUTS			OUTPUTS			
$\bar{E}$	A_0	A_1	$\bar{0}$	$\bar{1}$	$\bar{2}$	$\bar{3}$
H	X	X	H	H	H	H
L	L	L	L	H	H	H
L	H	L	H	L	H	H
L	L	H	H	H	L	H
L	H	H	H	H	H	L

H = HIGH voltage level
L = LOW voltage level

Figure 5.36 TTL decoder/demultiplexer (courtesy of Signetics Corporation).

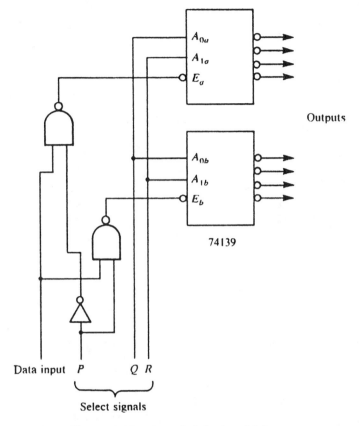

Figure 5.37 One-of-eight demultiplexer.

parts of the 74139. Select signal P is combined with the data input D to enable either the top or the bottom demultiplexer, depending on whether the value of P is 0 or 1, respectively. Two NAND gates are used to complete the circuit.

The following examples further illustrate the use of multiplexers and demultiplexers.

Example 5.9

Figure 5.38 shows a TDM scheme. The 8 data lines I_0 through I_7 are multiplexed onto the transmission line by an 8-to-1 multiplexer (TTL 74151). The three select signals (S_0, S_1, and S_2) are driven by the three outputs of a *counter* that counts from 000 to 111 and returns to 000

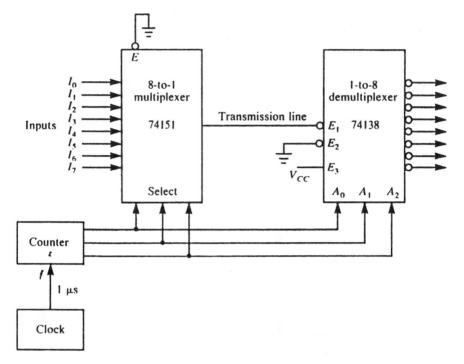

Figure 5.38   A TDM scheme.

to repeat the cycle. The counter is driven by a *clock* signal. (Clocks are described in Chapter 6 and counters are described in Chapter 7.) For the purposes of this example, assume that the counter increments every μs, thereby providing an 8 μs cycle for the multiplexer. That is, each input is connected to the transmission line for 1 μs every 8 μs. The transmission line is connected to the input of a 1-to-8 demultiplexer (TTL 74138) that distributes the incoming signal to one of the 8 outputs selected by the control signals A_0, A_1, and A_2, which are again driven by the clock and counter circuit.

This circuit is suitable when the transmission line is not too long. When the data are transmitted over longer distances (say, over telephone lines), the same clock and counter circuitry is not used for synchronizing the operation of the multiplexer and the demultiplexer. Rather, the destination circuit waits for a synchronizing signal from the source to activate itself. From that point on, the two remain synchronized. For example, since the multiplexer output is low when it is disabled, the destination can monitor when the transmission line changes to high as the indication of a synchronizing signal and start a local clock and counter circuit.

Our purpose in this example is to illustrate the use of a multiplexer and a demultiplexer, rather than provide the details of data communication through telephone lines. Additional hardware is needed to implement the above scheme using telephone lines as the transmission media. A modulator is needed to modulate the input signal and a demodulator is needed to demodulate the signal at the destination. These functions are performed by devices known as *modems* (MOdulator/DEModulator).

The circuit discussed above transmits one bit at a time from each input to the output. If each data unit to be transmitted is n bits long, then either n identical copies of this circuit can be used to transmit data in *parallel* or the single circuit is used n times to transmit data in *series*. The former mode is more expensive and the latter one is slower.

Figure 5.39 shows the parallel transmission circuit, assuming that each data unit is four bits wide. Here, four copies of the circuit in Figure 5.38 are used. The clock/counter circuit is common to all four copies.

Example 5.10

Figure 5.40 shows a circuit that multiplexes 16 data lines onto a bus that drives a 7-segment display through the BCD-to-7-segment decoder (TTL 7448). Two TTL 74251 multiplexers are combined to obtain 16-to-1 multiplexing. Each input line is 4 bits wide (BCD). The four-bit input multiplexed over the bus is transmitted to the 7448 to form the inputs A_0 through A_3.

Example 5.11

It is required to display the data on eight BCD input lines on eight 7-segment displays (one for each digit). Assume that the data arrive on input lines 1 through 8 in 8 consecutive time slots of t μs each. The data input cycle of $8t$ μs repeats with new data each cycle.

Figure 5.41 shows the circuit. The 8-to-1 multiplexer channels one of the inputs to the BCD-to-7-segment decoder, which in turn drives the digit selected by the 3-to-8 decoder. Since the multiplexer and the 3-to-8 decoder are driven by the same clock and counter circuit, the ith input is associated with the ith digit of the display ($1 \leq i \leq 8$). The segment and digit drivers are the interface blocks that might be needed to drive the display unit. The structure of these blocks depends on the capabilities of the other ICs selected for the design. If

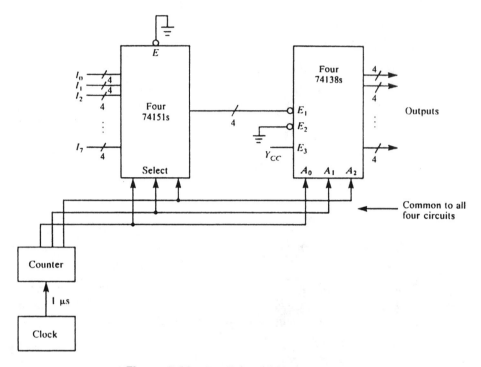

Figure 5.39 Parallel transmission circuit.

the time period t that is selected is short enough, the display appears continuous even though it is being driven one digit at a time.

The above example illustrates the use of different types of ICs in one design and circuit minimization using the multiplexer scheme. The selection of the appropriate ICs and the corresponding detailed design are left as an exercise.

5.12 Majority Circuits

A common method used in building fault-tolerant systems is to apply the concept of triple modular redundancy (TMR). Here, three identical copies of the system are built, each performing the transformation of the data inputs into corresponding outputs. Then the outputs of the three systems are compared and the correct output is the one produced by at least two among the three. That is, the three copies vote and the majority wins. With this scheme, even when one of the copies is not operating properly (i.e., faulted), the

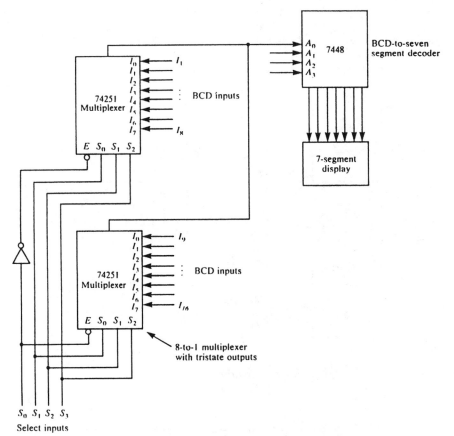

Note: Only one set of 74251s is shown; three more sets driven by the same select inputs and driving A_1, A_2, and A_3 inputs of 7448 are not shown.

Figure 5.40 A sample multiplexer circuit. (Note: Only one set of 74251s is shown; three more sets driven by the same select inputs and driving A_1, A_2, and A_3 inputs of 7448 are not shown.)

system produces the correct output. Majority circuits are useful in implementing such fault tolerant schemes.

Figure 5.42 shows a three input majority circuit. The output of this circuit is 1 when two of its inputs are 1; otherwise it is 0. Thus the function realized by this circuit is

$$Z = AB + AC + BC$$

Note that it is not necessary in this case to develop the complete truth table

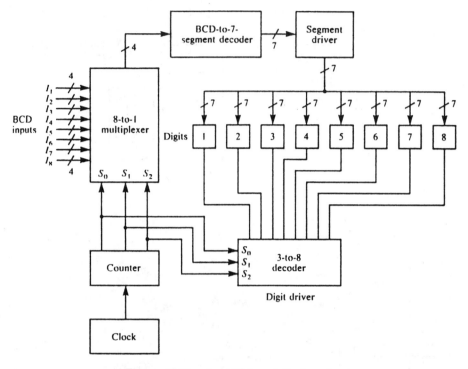

Figure 5.41 A multiplexed display circuit.

and derive the function from it. The above function can be derived just by examining the input conditions to be satisfied for the circuit to provide an output of 1. This is another example of special situations where the classical function derivation procedures of Chapters 2 and 3 need not be followed.

In the majority circuit of Figure 5.42 all inputs have equal weights (i.e., each has one vote). In a generalized majority gate shown in Figure

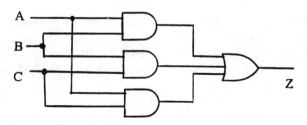

Figure 5.42 Three-input majority circuit.

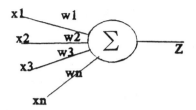

Figure 5.43 The generalized majority gate.

5.43, each input has a weight w associated with it. Then the output is 1 when the weighted sum of all the inputs is equal to or exceeds a certain threshold value T. That is,

$$Z = 1 \text{ if } \sum_{i=1}^{n} w_i x_i \geq T$$

$$= 0 \text{ Otherwise.}$$

In fact, the AND, OR and NOT gates are special cases of the generalized majority gate. In the case of the n-input AND, all weights are set to 1 and the threshold T is n. For an n-input OR, the weights are each 1 and T is 1. What are the values for w and T for a NOT gate?

The generalized forms of majority gates are known as threshold logic gates. Refer to the book by Muroga for a detailed treatment of threshold logic.

5.13 Modular Design Methodology

The classical design procedure of Chapter 4 is oriented toward implementing a two-level gate circuit. Although this procedure always yields a minimum circuit, it becomes impractical as the circuit being designed becomes increasingly complex. The most typical way to manage the circuit's complexity is to "divide and conquer." That is, partition the overall design problem into simpler subproblems until the complexity of the resultant subsystem can be managed easily. In logic circuit design, the subsystem can be realized by MSI- and LSI-level ICs. When ICs that suit the subsystem are not available, an SSI-level implementation is used. A thorough familiarity with available off-the-shelf ICs is thus necessary to cost-effectively partition and realize the digital system.

Another advantage of this modular design methodology is that system performance can be enhanced by improving subsystem performance. In addition, it is easier to replace one of the modules by a higher-performance equivalent without having to redesign the entire system.

Example 5.12 illustrates the modular design methodology.

Example 5.12

It is required to design a circuit that adds two floating-point numbers.

Figure 5.44(a) shows the top-level block diagram of the circuit. $F1$ and $F2$ are the two floating-point numbers. (See Chapter 1 for details on floating-point representations.) Assuming that each number is n bits long (the exponent, sign, and mantissa combined), a truth table representation of this circuit will need to handle $2n$ variables, which comes unwieldly even for small n.

In the addition of two floating-point numbers, the mantissa and exponent computations are handled separately. As such, the floating-point adder can be partitioned into two subsystems, as shown in Figure 5.44(b).

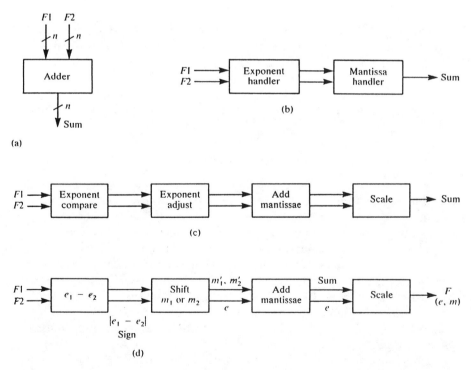

Figure 5.44 Top-down design of a floating-point adder. (a) Top-level diagram; (b) first partition; (c) refined partitions; (d) partitions with hardware structure definition.

Let

$$F1 = m_1 \times b^{e_1}$$

and

$$F2 = m_2 \times b^{e_2}$$

where m_1 and m_2 are the mantissae and e_1 and e_2 are the base-b exponents. The sum of these two numbers is

$$(m_1 + m_2)b_e \qquad \text{if } e_1 = e_2 = e$$
$$(m_1 + m_2 b^{(e_2 - e_1)})b^{e_1} \qquad \text{if } e_1 > e_2$$
$$(m_1 b^{(e_2 - e_1)} + m_2)b^{e_2} \qquad \text{if } e_1 < e_2$$

That is, the mantissae can be added if the two exponents are equal. If not, the lower exponent is incremented (and the corresponding mantissa is shifted right) to make it equal to the higher mantissa, and then the mantissae are added. For example, if

$$F1 = 0.5 \times 10^{-3}$$

and

$$F2 = 0.75 \times 10^{-4}$$

then the exponent of $F2$ is first adjusted, resulting in

$$F2 = 0.75 \times 10^{-3}$$

and the sum is

$$(0.5 + 0.075) \times 10^{-3} = 0.575 \times 10^{-3}$$

If after the addition the sum of the mantissae is greater than 1, it is shifted right (and the exponent is incremented) to obtain a fraction as the mantissa. Such an adjustment is also needed if a normalized representation (see Chapter 1) is used.

The subsystems in Figure 5.44 (b) can be partitioned further into the four subsystems shown in (c). Each of these blocks can now be implemented independently.

The "exponent compare" can be implemented by using either a comparator or a subtractor. A subtractor is used in (d). The subtractor supplies the magnitude and sign of $e_1 - e_2$ to the "exponent adjust" block. A shifter and an incrementer (e.g., an add-1 circuit) are needed to implement this block. The "add mantissa" block receives the adjusted mantissae that are ready to be added, and can be implemented by an adder. The "normalize" unit shifts the sum and adjusts the exponent appropriately and is similar to the exponent-adjust unit in its hardware structure.

This top-down methodology, in which the complex system is partitioned into subsystems (and subsystems into less complex subsystems, and so on), is convenient for the description, documentation, and implementation of a complex digital system and is used extensively by modern-day designers. Note that in the top levels of this procedure, we are more interested in the *behavior* of the subsystems than in their hardware *structure*. As the design progresses to the lower levels, structural details are added to each partition.

5.14 Summary

This chapter introduced the most popular combinational circuits that are available as SSI, MSI, and LSI devices. When implementing logic circuits using these off-the-shelf components, identification of the usability of such functions for the design at hand is required. The modular (top-down) design methodology used to design complex circuits with standard modules was illustrated. This chapter concentrated on the random logic design using the SSI and MSI components. It should be noted that the random logic implementations using such components as decoders and multiplexers are slower than the corresponding circuits built with two levels of gates using the classical design procedure. But the number of ICs—and hence the attendant costs of the interconnections and circuit boards—is reduced by using ICs that implement the popular combinational circuits. The IC manufacturers continue to bring newer ICs into the market. The periodicals listed in the references below should be consulted for announcements of new ICs.

References

Computer, IEEE Computer Society, New York (Monthly).
Computer Design, PennWell, Littleton, MA (Semimonthly).
Ercegovac, M. D. and Lang, T. *Digital Systems and Hardware/Firmware Algorithms*, John Wiley, New York, 1985.
Greenfield, J.D. *Practical Digital Design Using ICs*, John Wiley, New York, 1983.
Muroga, S. *Threshold Logic and Its Applications*, John Wiley, New York, NY, 1971.
National Advanced Bipolar Logic Data Book, Santa Clara, CA, 1995.
The TTL Data Book Series, Texas Instruments, Dallas, TX, 1984.
TTL Manual, Signetics, Sunnyvale, CA, 1978.

Problems

5.1 (a) Design circuits for half- and full subtractors. The half-subtractor has two inputs and two outputs (DIFFERENCE and BORROW). The full subtractor has three inputs (two bits from the current stage and a BORROW-IN from the previous stage) and two outputs (DIFFERENCE and BORROW).

(b) Connect a half-subtractor and three full subtractors to realize the subtraction circuit for four-bit numbers.

5.2 Draw the block diagram of a 16-bit adder/subtractor using the four-bit adder/subtractor circuit of Figure 5.10.

5.3 Implement the full adder using a 3-to-8 decoder.

5.4 Use two 4-to-1 multiplexers to implement a full adder.

5.5 Design the circuit for the addition of two Excess-3 digits along the lines of the decimal adder in Section 5.2. Derive the correction circuit needed and simplify it as much as possible.

5.6 Design a circuit whose inputs correspond to a BCD digit and the output is the 9s complement of the input.

5.7 Draw a block diagram for the BCD adder of Figure 5.12. Use four of these BCD adders and four 9s complementers of Problem 5.6 to design a four digit BCD adder/subtractor (applying the method of Figure 5.10).

5.8 (a) Design a 16-bit adder using TTL 74283s.

(b) Use 74182 as the carry lookahead circuit for the circuit in (a).

(c) How fast is the circuit in (b) compared with that in (a)?

5.9 Consider two binary numbers:

$$A = a_0 a_1 a_2 \cdots a_n$$

and

$$B = b_0 b_1 b_2 \cdots b_n$$

$A = B$ if and only if $a_i = b_i$ ($0 \le i \le n$). $A > B$ if $a_0 > b_0$ and $A < B$ if $a_0 < b_0$. If for any $j \ge 1$, $a_i = b_i$ for all i ($0 \le i \le j - 1$), then $A > B$ if $a_u > b_j$ and $A < B$ if $a_j < b_j$. Use these conditions to design a four-bit magnitude comparator.

5.10 Extend the 4×4 multiplier of this chapter to an 8-by-8 multiplier. Compare the speed of this multiplier built of SSI ICs of your choice, with that of any 8×8 multiplier available off-the-shelf.

5.11 Design an 8-to-3 priority encoder. Inputs are numbered from 0 through 7, and if two or more inputs are 1 simultaneously, the output code corresponds to the lowest-numbered input among such inputs; that is, the lowest-numbered input has priority.

5.12 Show how two 2-to-1 multiplexers can be connected to form a 3-to-1 multiplexer without any additional gates.

5.13 Realize a 32-to-1 multiplexer using two 16-to-1 multiplexers and a 2-to-1 multiplexer.

5.14 Implement the simple shifter using a cascade of 4-to-1 multiplexers, with no additional gates.

5.15 Implement the generalized 3-bit shifter using a cascade of 8-to-1 multiplexers.

5.16 Use a 4-to-10 decoder to realize the BCD-to-Excess-3 converter. Use a minimum number of additional gates.

5.17 Obtain the operational details of TTL 74185 (code converter) from the IC manual and design a binary-to-BCD converter for (a) a six-bit and (b) an eight-bit binary input.

5.18 It is required to generate the following functions of two four-bit operands, A and B:

(a) A plus B
(b) A minus B
(c) $A \cdot B$ (logical AND)
(d) $A + B$ (logical OR)

Design a circuit using a TTL 74181.
   The four functions above must be generated in four consecutive time slots of 1 μs each. Extend the circuit using a clock and counter type circuit and a decoder.

5.19 Implement the multiple-output functions of Problem 3.16 using appropriate decoders.

5.20 Implement the functions of Problem 3.16 using 8-to-1 multiplexers. Implement each output independently.

6

Synchronous Sequential Circuits

6.1 Introduction

The digital circuits we have examined so far do not possess any memory. That is, the output of the combinational circuit at any time is a function of the inputs at that time. In practice, most digital systems contain memory elements in addition to the combinational logic portion, thus making them *sequential* circuits. The output of a sequential circuit at any time is a function of its external inputs and the internal *state* at that time. The state of the circuit is defined by the contents of the memory elements in the circuit and is a function of the previous states and the inputs to the circuit.

Consider, for example, the *decimal counter* circuit used earlier in this book. This counter has 10 internal states, one corresponding to each count. The counter moves from one count to the next in response to the external increment input. Thus, the counter has one external *input*. The current count is the *present state* of the counter. The *next state* is the new count (which is one more than its present count), achieved in response to the increment signal. At any time, the count (or the present state) is used by other circuits in the system. Thus, the *output* of the counter is its present state. In general, the output and the next state are both functions of the present state and the external input.

Figure 6.1 shows a block diagram of a sequential circuit with m inputs, n outputs, and p internal memory elements. The output of the p memory elements combined constitutes the state of the circuit at time t (i.e., the *present state*). The combinational logic determines the output of the circuit at time t and provides the next-state information to the memory elements based on the external inputs and the present state. Based on the next-state information at t, the contents of all the memory elements change to the next state, which is the state at time $(t + \Delta t)$, where Δt is a time increment sufficient for the memory elements to make the transition. We will denote $(t + \Delta t)$ as $(t + 1)$ in this chapter.

There are two types of sequential circuits: *synchronous* and *asynchronous*. The behavior of a synchronous circuit depends on the signal values at discrete points of time. The behavior of an asynchronous circuit depends on the order in which the input signals change, and these changes can occur at any time.

The discrete time instants in a synchronous circuit are determined by a controlling signal, usually called a *clock*. A clock signal makes 0 to 1 and 1 to 0 transitions at regular intervals. Figure 6.2 shows two clock signals (one is the complement of the other), along with the various terms used to describe the clock. A pair of 0-to-1 and 1-to-0 transitions constitutes a *pulse*. That is, a pulse consists of a *rising edge* and a *falling edge*. The time between these transitions (edges) is the *pulse width*. The *period* (T) of the clock is the time between two corresponding edges of the clock, and the *clock frequency* is the reciprocal of its period. Although the clock in Figure 6.2 is shown with a regular period T, the intervals between the two pulses do not need to be equal.

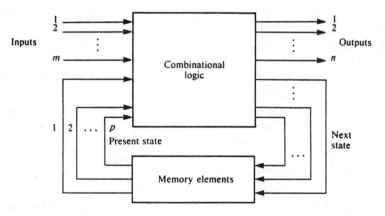

Figure 6.1 Block diagram of a sequential circuit.

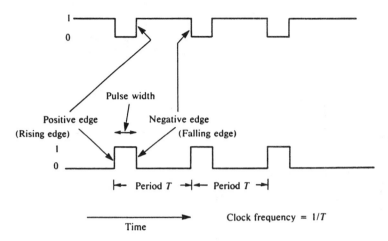

Figure 6.2 Clock.

Synchronous sequential circuits use flip-flops as memory elements. As discussed earlier, a flip-flop is an electronic device that can store either a 0 or a 1. That is, a flip-flop can stay in one of the two logic states, and a change in the inputs to the flip-flop is needed to bring about a change of state. Typically, there will be two outputs from a flip-flop: one corresponds to the normal state (Q) and the other corresponds to the complement state (Q'). We will examine four popular flip-flops in this chapter.

Asynchronous circuits use time-delay elements (*delay lines*) as memory elements. The delay line shown in Figure 6.3(a) introduces a propagation delay (Δt) into its input signal. As shown in (b), the output signal is the same as the input signal, except that it is delayed by Δt. For instance, the 0-to-1 transition of the input at t_1 occurs on the output at t_2, Δt later. Thus, if delay lines are used as memory elements, the next-state information at time t forms their input and the next state is achieved at ($t + \Delta t$). In practice, the propagation delays introduced by the combinational circuit's logic gates may be sufficient to produce the needed delay, thereby not necessitating a physical time-delay element. In such cases, the model of Figure 6.1 reduces to a combinational circuit with *feedback* (i.e., a circuit whose outputs are fed back as inputs). Thus, an asychronous circuit may be treated as a combinational circuit with *feedback*. Because of the feedback, the changes occurring in the output as a result of input changes may in turn contribute to further changes in inputs—and the cycle of changes may continue to make the circuit unstable if the circuit is not properly designed. In general, asynchronous circuits are difficult to analyze and design. If properly designed, however, they tend to be faster than synchronous circuits.

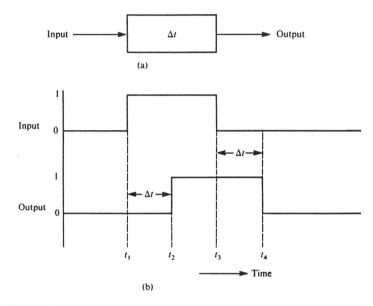

Figure 6.3 Delay element. (a) Block diagram; (b) I/O characteristics.

A synchronous sequential circuit generally is controlled by pulses from a *master clock*. The flip-flops in the circuit make a transition to the new state only when a clock pulse is present at their inputs. In the absence of a single master clock, the operation of the circuit becomes unreliable, since two clock pulses arriving from different sources at the inputs of the flip-flops cannot be guaranteed to arrive at the same time (because of unequal path delays). This phenomenon is called *clock skewing*. Clock skewing can be avoided by analyzing the delay in each path from the clock source and inserting additional gates in paths with shorter delays to make the delays of all paths equal.

In this chapter we will describe the analysis and design procedures for synchronous sequential circuits and defer the discussion of asynchronous circuits to Chapter 8.

6.2   Flip-Flops

As mentioned earlier, a flip-flop is a device that can store either a 0 or a 1. When the flip-flop contains a 1, it is said to be *set* (i.e., $Q = 1$, $Q' = 0$) and when it contains a 0 it is *reset* (i.e., $Q = 0$, $Q' = 1$). We will introduce the logic properties of four popular types of flip-flops in this section.

6.2.1 Set-Reset (*SR*) Flip-Flops

An *SR* flip-flop has two inputs: *S* for setting and *R* for resetting the flip-flop. An ideal *SR* flip-flop can be built using a cross-coupled NOR circuit, as shown in Figure 6.4(a). The operation of this circuit is illustrated in (b). When inputs $S = 1$ and $R = 0$ are applied at any time t, Q' assumes a value of 0 (one gate delay later). Since Q' and R are both at 0, Q assumes a value of 1 (another gate delay later). Thus, in two gate delay times the circuit settles at the set state. We will denote the two gate delay times as Δt. Hence, the state at time $(t + \Delta t)$ or $(t + 1)$, designated as $Q(t + 1)$ is 1. If *S* is changed to 0, as shown in the second row of (b), an analysis of the circuit indicates that the Q and Q' values do not change. If R is then changed to 1, the output values change to $Q = 0$ and $Q' = 1$. Changing R back to 0 does not alter the output values. When $S = 1$ and $R = 1$ are applied, both outputs assume a value of 0, regardless of the previous state of the circuit. This condition is not desirable, since the flip-flop operation requires that one output always be the complement of the other. Further, if now the input condition changes to $S = 0$ and $R = 0$, the state of the circuit depends on the order in which the inputs change from 1 to 0. If *S* changes faster than *R*, the circuit attains the reset state; otherwise, it attains the set state.

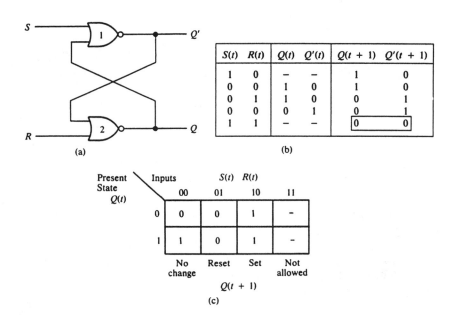

Figure 6.4 *SR* flip-flop. (a) Logic diagram; (b) partial truth table; (c) state table.

Thus, the cross-coupled NOR gate circuit forms an *SR* flip-flop. The input condition $S = 1$ and $R = 0$ *sets* the flip-flop; the condition $S = 0$ and $R = 1$ *resets* the flip-flop. $S = 0$ and $R = 0$ constitute a "no change" condition. (The input condition $S = 1$ and $R = 1$ is not permitted to occur on the inputs.)

The transitions of the flip-flop from the present state $Q(t)$ to the next state $Q(t + 1)$ for various input combinations are summarized in Figure 6.4(c). This table is called the *state table*. It has four columns (one corresponding to each input combination) and two rows (one corresponding to each state the flip-flop can be in).

Recall that the outputs of the cross-coupled NOR gate circuit in Figure 6.4 do not change instantaneously once there is a change in the input condition. The change occurs after a delay of Δt, which is the equivalent of at least two gate delays. This is an asynchronous circuit, since the outputs change as the inputs change. The circuit is also called the *SR latch*. As the name implies, such a device is used to *latch* (i.e., store) the data for later use. In the following circuit, the data (1 or a 0) on the INPUT line are latched by the flip-flop:

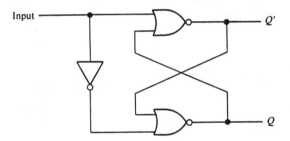

A clock input can be added to an asynchronous circuit to construct a *clocked SR flip-flop*, as shown in Figure 6.5(a). As long as the clock stays at 0 (i.e., in the absence of the clock pulse), the outputs of the two AND gates (S_1 and R_1) are 0, and hence the state of the flip-flop does not change. The S and R values are impressed on the flip-flop inputs (S_1 and R_1) only during the clock pulse. Thus, the clock controls all the transitions of this synchronous circuit. The graphic symbol for the clocked SR flip-flop is shown in (b).

Given the present state and the S and R input conditions, the next state of the flip-flop can be determined, as shown in the *characteristic table* in (c). This table is obtained by rearranging the state table in Figure 6.4(c) so that the next state can be determined easily once the present state and the input condition are known. The *characteristic equation* derived from the characteristic table shows the flip-flop's operation in the form of an equation.

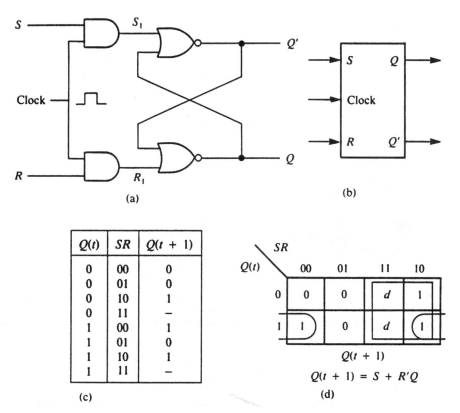

Figure 6.5 Clocked *SR* flip-flop. (a) Circuit; (b) graphic symbol; (c) characteristic table; (d) characteristic equation.

It is derived by representing and simplifying $Q(t + 1)$ in terms of S, R, and $Q(t)$ on a K-map, as shown in (d) where the "not allowed" input condition is treated as a don't-care.

An SR flip-flop can also be formed by cross-coupling two NAND gates. The analysis of such a circuit is left as an exercise.

As mentioned earlier, it takes at least two gate delay times for the state transition to occur after there has been a change in the input condition of the flip-flop. Thus the pulse width of the clock controlling the flip-flop must be at least equal to this delay, and the inputs should not change until the transition is complete. If the pulse width is longer than the delay, the state transitions resulting from the first input condition change are over-ridden by any subsequent changes in the inputs during the clock pulse. If it is necessary to recognize all the changes in the input conditions, however, the pulse width

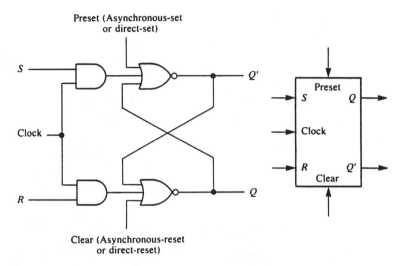

Figure 6.6 A clocked *SR* flip-flop with preset/clear.

must be short enough. The pulse width and the clock frequency thus must be adjusted to accommodate the flip-flop circuit transition time and the rate of input change. (The timing characteristics and requirements of flip-flops are further discussed later in this chapter.)

Figure 6.6 shows the graphic symbol for an *SR* flip-flop with both clocked (*S* and *R*) and asynchronous inputs (*preset* and *clear*). A clock is not required to activate the flip-flop through the asynchronous inputs. Asynchronous (or direct) inputs are not used during the regular operation of the flip-flop. They generally are used to initialize the flip-flop to either the set or the reset state. For instance, when the circuit power is turned on, the state of the flip-flops cannot be determined. The direct inputs are used to initialize the state, either manually through a ''master clear'' switch or through a power-up circuit that pulses the direct input of all the flip-flops in the circuit.

We will now examine three other commonly used flip-flops. The preset, clear, and clocked configurations discussed above apply to these flip-flops as well. In the remaining portions of this chapter, if a reference to a signal does not show a time associated with it, it is assumed to be the current time *t*.

6.2.2 *D* Flip-Flops

Figure 6.7 shows a *D* (delay or data) flip-flop and its state table. The *D* flip-flop assumes the state of the *D* input; that is, $Q(t + 1) = 1$ if $D(t) = 1$, and $Q(t + 1) = 0$ if $D(t) = 0$. The function of this flip-flop is to introduce a unit

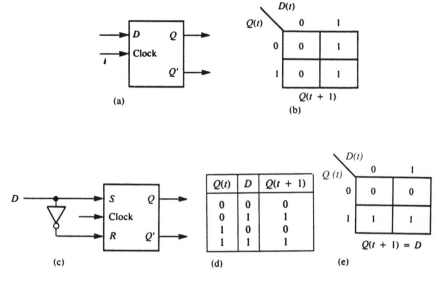

Figure 6.7 *D* flip-flop. (a) Graphic symbol; (b) state table; (c) flip-flop realized from an *SR* flip-flop; (d) characteristic table; (e) characteristic equation.

delay (Δt) in the signal input at *D*. Hence this flip-flop is known as a *Delay flip-flop*. It is also called a *data flip-flop*, since it stores the data on the *D* input line.

The *D* flip-flop is a modified *SR* flip-flop that is obtained by connecting *D* to an *S* input and *D'* to an *R* input, as shown in (c). A clocked *D* flip-flop is also called a *gated D-latch*, in which the clock signal gates the data into the latch.

The next state of the *D* flip-flop is the same as the data input at any time, regardless of the present state. This is illustrated by the characteristic table shown in (d) and the characteristic equation in (e).

6.2.3 *JK* Flip-Flops

The *JK* flip-flop is a modified *SR* flip-flop in that the *J* = 1 and *K* = 1 input combination is allowed to occur. When this combination occurs, the flip-flop complements its state. The *J* input corresponds to the *S* input, and the *K* input corresponds to the *R* input of an *SR* flip-flop. Figure 6.8 shows the graphic symbol, the state table, the characteristic table and equation, and the realization of a *JK* flip-flop using an *SR* flip-flop. (See problem 6.3 for a hint on how to convert one type of flip-flop into another.)

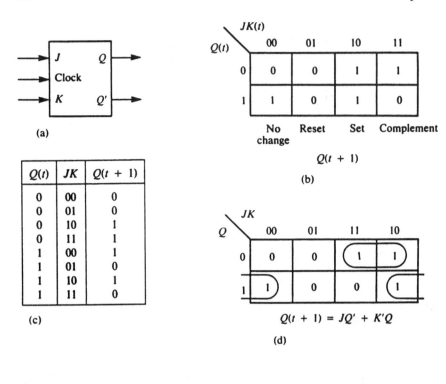

(a)

(b)

$Q(t + 1)$

(c)

(d)

$$Q(t + 1) = JQ' + K'Q$$

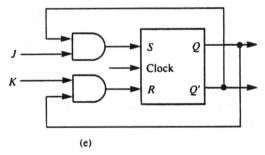

(e)

Figure 6.8 *JK* flip-flop. (a) Graphic symbol; (b) state table; (c) characteristic table; (d) characteristic equation; (e) realization of *JK* flip-flop using an *SR* flip-flop.

6.2.4 *T* Flip-Flops

Figure 6.9 shows the graphic symbol, state table, characteristic table, and equation for a *T* (toggle) flip-flop. This flip-flop complements its state when $T = 1$ and remains in the same state as it was when $T = 0$. A *T* flip-flop can

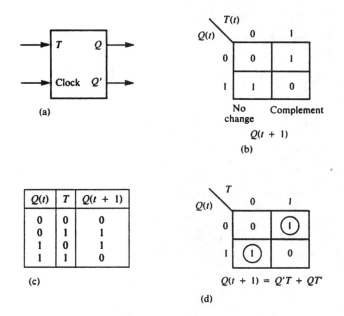

Figure 6.9 *T* flip-flop. (a) Graphic symbol; (b) state table; (c) characteristic table; (d) characteristic equation.

be realized by connecting the *J* and *K* inputs of a *JK* flip-flop, as shown in Figure 6.10.

6.2.5 Characteristic and Excitation Tables

The characteristic table of a flip-flop is useful in the analysis of sequential circuits, since it provides the next-state information as a function of the present state and the inputs. The characteristic tables of all the flip-flops are given in Figure 6.11 for ready reference.

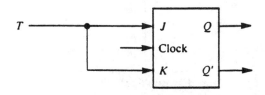

Figure 6.10 *T* flip-flop realized from a *JK* flip-flop.

Q(t)	SR	Q(t+1)
0	00	0
0	01	0
0	10	1
0	11	-
1	00	1
1	01	0
1	10	1
1	11	-

(a)

Q(t)	D	Q(t+1)
0	0	0
0	1	1
1	0	0
1	1	1

(b)

Q(t)	JK	Q(t+1)
0	00	0
0	01	0
0	10	1
0	11	1
1	00	1
1	01	0
1	10	1
1	11	0

(c)

Q(t)	T	Q(t+1)
0	0	0
0	1	1
1	0	1
1	1	0

(d)

Figure 6.11 Characteristic tables. (a) *SR* flip-flop; (b) *D* flip-flop; (c) *JK* flip-flop; (d) *T* flip-flop.

The *excitation tables* (or the *input tables*) shown in Figure 6.12 for each flip-flop are useful in designing sequential circuits, since they describe the excitation (or input condition) required to bring the state transition of the flip-flop from $Q(t)$ to $Q(t + 1)$. These tables are derived from the state tables of the corresponding flip-flops. Consider the state table for the *SR*

Q(t)	Q(t+1)	SR	D	JK	T
0	0	0d	0	0d	0
0	1	10	1	1d	1
1	0	01	0	d1	1
1	1	d0	1	d0	0

Note: d = don't-care (0 or 1)

Figure 6.12 Excitation tables.

flip-flop shown in Figure 6.4. For a transition of the flip-flop from state 0 to 0 (as shown by the first row of the state table), the input can be either $SR = 00$ or 01. That is, an SR flip-flop makes a transition from 0 to 0 as long as S is 0 and R is either 1 or 0. This excitation requirement is shown as $SR = 0d$ in the first row of the excitation table. A transition from 0 to 1 requires an input of $SR = 10$; a transition from 1 to 0 requires $SR = 01$ and that from 1 to 1 requires $SR = d0$. Thus, the excitation table accounts for all possible transitions. The excitation tables for the other three flip-flops are similarly derived.

6.3 Timing Characteristics of Flip-Flops

Consider the cross-coupled NOR circuit forming an SR flip-flop. Figure 6.13(a) shows a timing diagram, assuming that the flip-flop is at state 0 to begin with. At t_1, input S changes from 0 to 1. In response to this, Q changes to 1 at t_2, a delay of Δt after t_1. Δt is the time required for the circuit to settle to the new state. At t_3, S goes to 0, with no change in Q. At t_4, R changes to 1, and hence Q changes to 0, Δt time later at t_5. At t_6, R changes to 0, with no effect on Q.

Note that the S and R inputs should each remain at their new data value at least for time Δt for the flip-flop to recognize the change in the input condition (i.e., to the make the state transition). This time is called the *hold time*. Now consider the timing diagram for the clocked SR flip-flop shown in Figure 6.13(b). The clock pulse width $w = t_8 - t_1$. S changes to

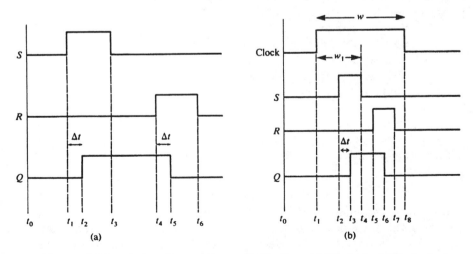

Figure 6.13   Timing diagram for an SR flip-flop. (a) Unclocked; (b) clocked.

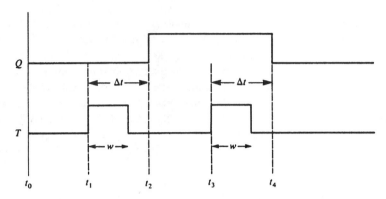

Figure 6.14 Timing diagram for a T flip-flop.

1 at t_2, and in response to it Q changes to 1 at t_3, Δt later. Since the clock pulse is still at 1 when R changes to 1 at t_5, Q changes to 0 at t_6. If the pulse width were to be $w_1 = t_4 - t_1$, only the change in S would have been recognized. Thus, in the case of a clocked SR flip-flop, the clock pulse width should at least equal Δt for the flip-flop to change its state in response to a change in the input. If the pulse width is greater than Δt, the S and R values should change no more than once during the clock pulse, since the flip-flop circuit will keep changing states as a result of each input change and registers only the last input change. As such, the clock pulse width is a critical parameter for the proper operation of the flip-flop.

Consider the timing diagram of Figure 6.14 for a T flip-flop. When T changes to 1 at t_1, the flip-flop changes from its original state of 0 at t_2, Δt time later. Since the T flip-flop circuit contains a feedback path from its outputs to its input, if the T input stays at 1 longer (i.e., beyond t_2), the output would be fed back to the input and the flip-flop changes state again. To avoid this oscillation, w must always be less than Δt.

In order to avoid such problems resulting from clock pulse width, flip-flops in practice are designed either as *master-slave flip-flops* or as *edge-triggered flip-flops*, which are described next.

6.3.1 Master-Slave Flip-Flops

The master-slave configuration is shown in Figure 6.15(a). Here, two flip-flops are used. The clock controls the separation and connection of the circuit inputs from the inputs of the master, and the inverted clock controls the separation and connection of slave inputs from the master outputs. In prac-

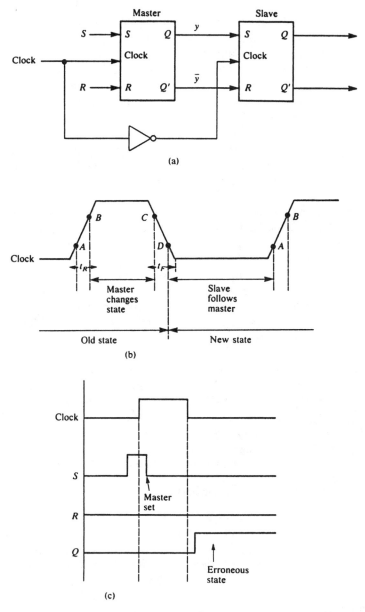

Figure 6.15 Master-slave flip-flop. (a) Circuit; (b) timing diagram; (c) one's catching problem.

tice, the clock signal takes a certain amount of time to make the transition from 0 to 1 and 1 to 0, as shown by t_R and t_F, respectively, in the timing diagram (b). As the clock changes from 0 to 1, at point A the slave stage is disconnected from the master stage; at point B, the master is connected to the circuit inputs and changes its state based on the inputs. At point C, as the clock makes its transition from 1 to 0, the master stage is isolated from the inputs, and at D, the slave inputs are connected to the outputs of the master stage. The slave flip-flop changes its state based on its inputs, and the slave stage is isolated from the master stage at A again. Thus, the master-slave configuration results in at most one state change during each clock period, thereby avoiding the race conditions resulting from clock pulse width.

Note that the inputs to the master stage can change after the clock pulse while the slave stage is changing its state without affecting the operation of the master-slave flip-flop, since these changes are not recognized by the master until the next clock pulse. Master-slave flip-flops are especially useful when the input of a flip-flop is a function of its own output.

Consider the timing diagram of Figure 6.15(c) for a master-slave flip-flop. Here, S and R initially are both 0. The flip-flop should not change its state during the clock pulse. However, a glitch in the S line while clock is high sets the master stage, which in turn is transferred to the slave stage, resulting in an erroneous state. This is called a *one's catching* problem and can be avoided by ensuring that all the input changes are complete and the inputs stable well before the leading edge of the clock. This timing requirement is known as the *setup* time (t_{setup}). That is, $t_{setup} > w$, the clock pulse width. This can be achieved either by a narrow clock pulse width (which is difficult to guarantee) or by a large setup time (which reduces the flip-flop's operating speed). Edge-triggered flip-flops are preferred over master-slave flip-flops because of the one's catching problem associated with the latter.

6.3.2 Edge-Triggered Flip-Flops

Edge-triggered flip-flops are designed so that they change their state based on input conditions at either the rising or the falling edge of the clock. The rising edge of the clock triggers a positive edge-triggered flip-flop (as shown in Figure 6.13), and the falling edge of the clock triggers a negative edge-triggered flip-flop. Any change in input values after the occurrence of the triggering edge will not bring about a state transition in these flip-flops until the next triggering edge.

The triggering edge can be formed by inserting a resistor-capacitor circuit at the clock input of the flip-flop. The resistor-capacitor circuit recognizes the change in its input by producing a spike, as shown at top of next page:

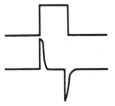

The flip-flop circuit can be designed to recognize either the positive or the negative spike.

Figure 6.16 shows another common circuit that recognizes the change in its input by producing a short pulse. Assume that Z is at 0 at a steady state. When X changes from 0 to 1, P also changes instantaneously, while Y changes after a delay introduced by the inverter. For the duration of the delay, the AND gate has both its inputs at 1 producing a short pulse in Z, as shown by the timing diagram. The pulse width depends on the propagation delay through the inverter.

Figure 6.17(a) shows the most common trailing edge-triggered flip-flop circuit, built out of three cross-coupled NOR flip-flops. Flip-flops 1 and 2 serve to set the inputs to the third flip-flop at appropriate values based on the clock and D inputs. Consider the clock and D input transitions shown in (b). Flip-flop 3 is reset initially (i.e., $Q = 0$). When the clock goes to 1 at t_0, point W goes to 0 (after one gate delay). Since Z remains at 0, flip-flop 3 does not change its state. While the clock pulse is at 1, X and Y follow D (i.e., $X = D'$ and $Y = D$), as at t_1. When the clock changes to 0 at t_2, Z changes to 1 (after a delay) at t_3, but W remains 0. Consequently, flip-flop 3 changes its state to 1 (after a delay). Thus the state change is brought about by the trailing edge of the clock.

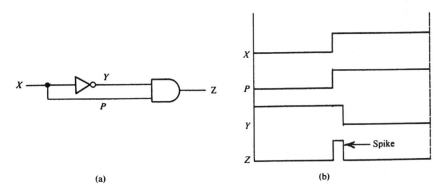

Figure 6.16 Spike generating circuit. (a) Circuit; (b) timing.

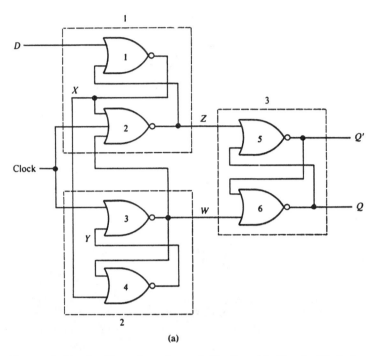

Figure 6.17   Trailing edge-triggered flip-flop. (a) Circuit; (b) timing.

While the clock is at 0, change in the D input does not change either Z or W, as shown at t_4 and t_5.

Z is 1 and W is 0 at t_6 when the clock changes to 1 and Z goes to 0. At t_7, the D input changes to 0. Since the clock is at 1, X and Y change accordingly (after a delay). These changes result in changing W to 1 at the trailing edge of the clock at t_8. Since $Z = 0$ and $W = 1$, flip-flop 3 changes to 0.

As can be seen by the timing diagram shown in (b), after the trailing edge of the clock pulse, either W or Z becomes 1. When Z is 1, D is blocked at gate 1. When W is 1, D is blocked at gates 2 and 4. This blocking requires one gate delay after the trailing edge of the clock, and hence D should not change until this blocking occurs. Thus, the *hold* time is one gate delay.

Note that the total time required for the flip-flop transition is three gate delays after the trailing edge—one gate delay for W and Z to change and two gate delays after that for Q and Q' to change. Thus, if we add t_{setup} of two gate delays to the transition time of three gate delays, the minimum clock period is five gate delays if the output of the flip-flop is fed directly

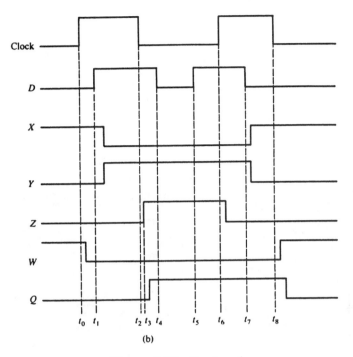

(b)

Figure 6.17 *Continued*

back to its input. If additional circuitry is in the feedback path, as is usually the case with most sequential circuits, the minimum clock period increases correspondingly.

A leading edge-triggered flip-flop can be designed using cross-coupled NAND circuits along the lines of the circuit shown above.

6.4 Flip-Flop ICs

Figure 6.18 shows the TTL 7474, a dual positive edge-triggered D flip-flop IC. The triangle at the clock input in the graphic symbol indicates positive edge triggering. (Negative edge triggering is indicated by a triangle along with a bubble at the input, as shown in *the case of 74LS73, in* Figure 6.19.) S_D and R_D are active-low asynchronous set and reset inputs, respectively, and operate independently of the clock. The data on the D input are transferred to the Q output at the positive clock edge. The D input must be stable one setup time (20 ns) prior to the positive clock edge. The positive transition time of the clock (i.e., from 0.8 V to 2.0 V) should be equal to or

LOGIC SYMBOL

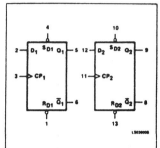

LOGIC DIAGRAM

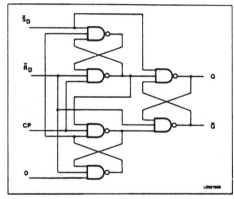

MODE SELECT — FUNCTION TABLE

OPERATING MODE	INPUTS				OUTPUTS	
	$\overline{S}_D$	$\overline{R}_D$	CP	D	Q	$\overline{Q}$
Asynchronous Set	L	H	X	X	H	L
Asynchronous Reset (Clear)	H	L	X	X	L	H
Undetermined[1]	L	L	X	X	H	H
Load "1" (Set)	H	H	↑	h	H	L
Load "0" (Reset)	H	H	↑	l	L	H

H = HIGH voltage level steady state.
h = HIGH voltage level one set-up time prior to the LOW-to-HIGH clock transition.
L = LOW voltage level steady state.
l = LOW voltage level one set-up time prior to the LOW-to-HIGH clock transition.
X = Don't care.
↑ = LOW-to-HIGH clock transition.
NOTE:
(1) Both outputs will be HIGH while both $\overline{S}_D$ and $\overline{R}_D$ are LOW, but the output states are unpredictable if $\overline{S}_D$ and $\overline{R}_D$ go HIGH simultaneously.

Figure 6.18 *D* flip-flop IC (7474) (courtesy of Signetics Corporation).

less than the clock-to-output delay time for the reliable operation of the flip-flop.

The 7473 and 74LS73 shown in Figure 6.19 are dual master-slave *JK* flip-flop ICs. The 7473 is positive pulse-triggered (note the absence of the triangle on the clock input in a graphic symbol). *JK* information is loaded into the master while the clock is high and transferred to the slave during the high-to-low transition. For the conventional operation of this flip-flop, the *JK* inputs must be stable while the clock is high. The flip-flop also has direct set and reset inputs.

The 74LS73 is a negative edge-triggered flip-flop. The *JK* inputs must be stable one setup time (20 ns) prior to the high-to-low transition of the clock. This flip-flop has an active-low direct reset input.

The 7475 shown in Figure 6.20 has four bistable latches. Each two-bit latch is controlled by an active-high enable input (*E*). When enabled, the

AC ELECTRICAL CHARACTERISTICS $T_A = 25°C$, $V_{CC} = 5.0V$

	PARAMETER	TEST CONDITIONS	74 $C_L = 15pF$, $R_L = 400\Omega$		74LS $C_L = 15pF$, $R_L = 2k\Omega$		74S $C_L = 15pF$, $R_L = 280\Omega$		UNIT
			Min	Max	Min	Max	Min	Max	
f_{MAX}	Maximum clock frequency	Waveform 1	15		25		75		MHz
t_{PLH} t_{PHL}	Propagation delay Clock to output	Waveform 1		25 40		25 40		9 9	ns
t_{PLH} t_{PHL}	Propagation delay Set or Reset to output	Waveform 2		25		25		6	ns
		Waveform 2 CP = HIGH		40		40		13.5	
t_{PHL}	Set or Reset to output	Waveform 2 CP = LOW		40		40		8	ns

NOTE:
Per industry convention, f_{MAX} is the worst case value of the maximum device operating frequency with no constraints on t_r, t_f, pulse width or duty cycle.

AC SET-UP REQUIREMENTS $T_A = 25°C$, $V_{CC} = 5.0V$

	PARAMETER	TEST CONDITIONS	74		74LS		74S		UNIT
			Min	Max	Min	Max	Min	Max	
$t_W(H)$	Clock pulse width (HIGH)	Waveform 1	30		25		6		ns
$t_W(L)$	Clock pulse width (LOW)	Waveform 1	37				7.3		ns
$t_W(L)$	Set or reset pulse width (LOW)	Waveform 2	30		25		7		ns
$t_s(H)$	Set-up time (HIGH) data to clock	Waveform 1	20		20		3		ns
$t_s(L)$	Set-up time (LOW) data to clock	Waveform 1	20		20		3		ns
t_h	Hold time data to clock	Waveform 1	5		5		2		ns

AC WAVEFORMS

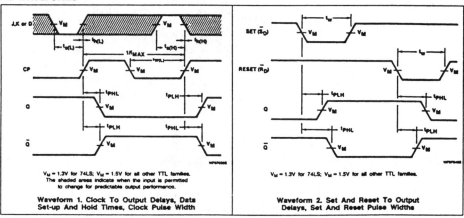

V_M = 1.3V for 74LS; V_M = 1.5V for all other TTL families.
The shaded areas indicate when the input is permitted to change for predictable output performance.

Waveform 1. Clock To Output Delays, Data Set-up And Hold Times, Clock Pulse Width

V_M = 1.3V for 74LS; V_M = 1.5V for all other TTL families.

Waveform 2. Set And Reset To Output Delays, Set And Reset Pulse Widths

Figure 6.18 *Continued*

LOGIC SYMBOL

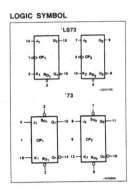

Figure 6.19 A dual *JK* flip-flop IC (74F109 and 74F112)) (courtesy of National Semiconductors Corporation).

data enter the latch and appear at the *Q* outputs. The *Q* outputs follow the data inputs as long as the enable is high. The latched outputs remain stable as long as the enable input stays low. The data inputs must be stable one setup time (20 ns) prior to the high-to-low transition of the enable for the data to be latched.

6.5 Analysis of Synchronous Sequential Circuits

The analysis of a synchronous sequential circuit is the process of determining the functional relation that exists between its outputs, its inputs, and its internal states. The contents of all the flip-flops in the circuit combined determine the internal state of the circuit. Thus, if the circuit contains *n* flip-flops, it can be in one of the 2^n states. Knowing the present state of the circuit and the input values at any time *t*, we should be able to derive its next state (i.e., the state at time $t + 1$) and the output produced by the circuit at *t*.

A sequential circuit can be described completely by a state table that is very similar to the ones shown for flip-flops in Figures 6.4 through 6.9. For a circuit with *n* flip-flops, there will be 2^n rows in the state table. If there are *m* inputs to the circuit, there will be 2^m columns in the state table. At the intersection of each row and column, the next-state and the output information are recorded. A *state diagram* is a graphical representation of the state table, in which each state is represented by a circle and the state transitions are represented by arrows between the circles. The input combination that brings about the transition and the corresponding output information are shown on the arrow. Analyzing a sequential circuit thus corre-

Logic Symbol

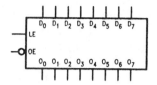

Unit Loading/Fan Out:

Pin Names	Description	54F/74F	
		U.L. HIGH/LOW	Input I_{IH}/I_{IL} Output I_{OH}/I_{OL}
D_0-D_7	Data Inputs	1.0/1.0	20 μA/−0.6 mA
LE	Latch Enable Input (Active HIGH)	1.0/1.0	20 μA/−0.6 mA
$\overline{OE}$	Output Enable Input (Active LOW)	1.0/1.0	20 μA/−0.6 mA
O_0-O_7	TRI-STATE Latch Outputs	150/40 (33.3)	−3 mA/24 mA (20 mA)

Truth Table

Inputs			Output
LE	$\overline{OE}$	D_n	O_n
H	L	H	H
H	L	L	L
L	L	X	O_n (no change)
X	H	X	Z

H = HIGH Voltage Level
L = LOW Voltage Level
X = Immaterial
Z = High Impedance State

Logic Diagram

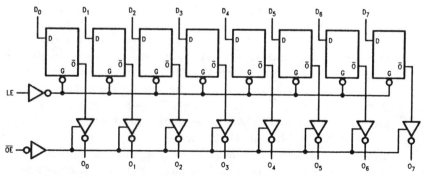

Figure 6.20 A latch IC (7475) (courtesy of Signetics Corporation).

sponds to generating the state table and the state diagram for the circuit. The state table or state diagram can be used to determine the output sequence generated by the circuit for a given input sequence if the *initial state* is known. It is important to note that for proper operation, a sequential circuit must be in its initial state before the inputs to it can be applied. Usually the power-up circuits are used to initialize the circuit to the appropriate state when the power is turned on. The following examples will illustrate the analysis procedure.

Example 6.1

Consider the sequential circuit shown in Figure 6.21(a). There is one circuit input X and one output Z, and the circuit contains one D flip-flop. To analyze the operation of this circuit, we can trace through the circuit for various input values and states of the flip-flop to derive the corresponding output and next-state values. Since the circuit has one flip-flop, it has two states (corresponding to $Q = 0$ and $Q = 1$). The present state is designated as Y (in this circuit, $Y = Q$). The output Z is a function of the state of the circuit Y and the input X at any time t. The next state of the circuit $Y(t + 1)$ is determined by the value of the D input at time t. Since the memory element is a D flip-flop, $Y(t + 1) = D(t)$.

Assume that $Y(t) = 0$ and $X(t) = 0$. Tracing through the circuit, we can see that $Z = 0$ and $D = 0$. Hence, $Z(t) = 0$ and $Y(t + 1) = 0$. The above state transition and output are shown in the top left blocks of the next-state and output tables in (b). Similarly, when $X(t) = 0$ and $Q(t) = 1$, $Z(t) = 1$ and $D(t) = 1$, making $Y(t + 1) = 1$, as shown in the bottom left blocks of these tables. The other two entries in these tables are similarly derived by tracing through the circuit. The two tables are merged into one, entry by entry, as shown in (c) to form the so-called *transition table* for the circuit. Each block of this table corresponds to a present state and an input combination. Corresponding next-state and output information is entered in each block, separated by a slash mark. From the table in (c), we can see that if the state of the circuit is 0, it produces an output of 0 and stays in the state 0 as long as the input values are 0s. The first 1 input condition sends the circuit to a 1 state with an output of 1. Once it is in the 1 state, the circuit remains in that state regardless of what the input is, but the output is the complement of the input X. The state diagram in (d) illustrates the operation of the circuit graphically. Here, each circle represents a state, and an arrow represents the state transition. The input value corresponding to that transition and the output of the circuit at that time are represented

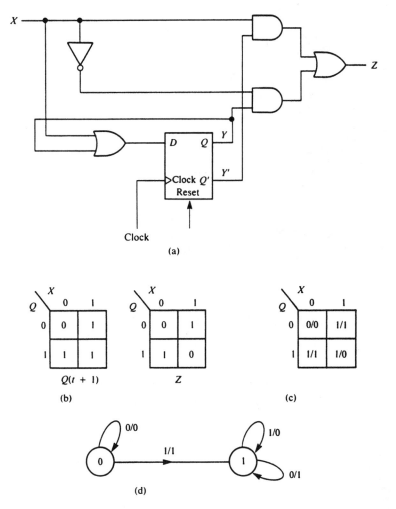

Figure 6.21 Sequential circuit analysis. (a) Circuit; (b) next-state and output tables; (c) transition table; (d) state diagram; (e) timing diagram for level input; (f) timing diagram for synchronous pulse input.

on each arrow, separated by a slash. We will generalize the state diagram and state table representations in the next example.

Since the flip-flop is a positive edge-triggered flip-flop, the state transition takes place only when the rising edge of the clock occurs. However, this fact cannot be explicitly shown in the above tables. A timing diagram can be used to illustrate these timing characteristics.

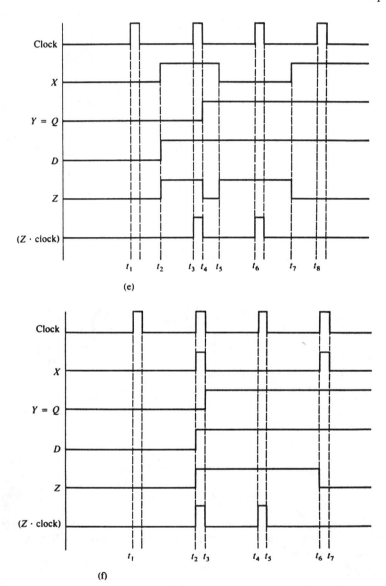

Figure 6.21 *Continued*

The operation of the circuit for a four–clock pulse period is shown in (e). The input X is assumed to make the transitions shown. The flip-flop is positive edge-triggered. The state change occurs as a result of the rising edge but takes a certain amount of time after the edge occurs. It is assumed that the new state is attained by the falling edge of the clock. Assuming an initial state of 0, at t_1, D is 0 and Z is 0, thus not affecting Y. At t_2, X changes to 1 and hence D and Z change to 1, neglecting the gate delays. At t_3, the positive edge of the clock starts the state transition, making Q reach 1 by t_4. Z goes to 0, since Q goes to 1 at t_4. X changes to 0 at t_5, thereby bringing Z to 1 but not changing D. Hence Y remains 1 through t_8, as does D. Corresponding transitions of Z are also shown. The last part of the timing diagram shows Z ANDed with the clock to illustrate the fact that output Z is valid only during the clock pulse. If X is assumed to be valid only during clock pulses, the timing diagram represents an input sequence of $X = 0101$ and the corresponding output sequence of $Z = 0110$. Note that the output sequence is the twos complement of the input sequence with the LSB occurring first and the MSB last.

The transitions of input X shown in (e) above make it an asynchronous-level input, since it is formed of high and low levels of arbitrary length and the level changes are not synchronized with the clock pulses. Figure 6.22 shows the four variations of inputs that are generally possible for a sequential circuit. If the level changes coincide with the clock edges, as in (c), the input is a synchronous-level input; otherwise, it is asynchronous-level input,

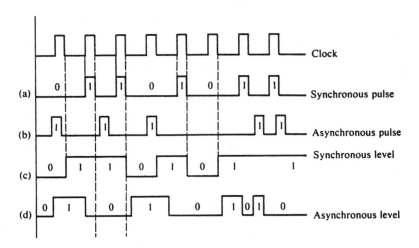

Figure 6.22 Types of input.

as in (d). If the input is in the form of pulses, it can be either synchronized to the clock (synchronous pulse) or not (asynchronous pulse). The asynchronous-pulse−type of input is not useful in synchronous circuits. The synchronous pulse input is realized by ANDing the input with the clock. When the input is synchronized with the clock (either the level or the pulse), we will assume that, for the purposes of analysis here, its value during the clock pulse determines the circuit transition and outputs. For instance, if the D input of a positive edge-triggered D flip-flop goes from 0 to 1 along with the rising edge of the clock, the flip-flop sees that the D input is 1 for its transition.

Figure 6.21(f) shows the timing diagram for the sequential circuit of Example 6.1 for a synchronous pulse input.

The circuit tracing procedure discussed here can be adopted for the analysis of simple circuits. As the circuit becomes more complex, the tracing becomes cumbersome. We will now illustrate a more systematic procedure for the derivation of the state table (and hence the state diagram).

From the analysis of the combinational circuit portion of Figure 6.21(a), the flip-flop input equation is

$$D = X + Y$$

and the circuit output equation is

$$Z = XY' + X'Y$$

Figure 6.23(a) shows the K-map for D, derived from the flip-flop input equation above. This table shows the value of the D input for each of the combinations of circuit input and present states. If we know the value of D, we can determine the corresponding next state of the flip-flop $Y(t + 1)$ in Figure 6.23(b) by using the flip-flop excitation table. In the case of the D flip-flop, the next state is the same as D, and hence the next-state table shown in (b) is identical to (a). The K-map for Z is shown in (c). Tables (b) and (c) are merged to form the transition table shown in (d). The transition table is the state table in which the states are represented in binary form. In general, each state in the circuit is designated by an alphabetic character. By assigning A to 0 and B to 1, we can obtain the state table shown in (e).

Example 6.2 illustrates the analysis procedure for a more complex circuit.

Example 6.2

Consider the sequential circuit shown in Figure 6.24(a). There are two clocked flip-flops, one input line X, and one output line Z. The Q outputs of the flip-flops (Y_1, Y_2) constitute the present state of the circuit at any time t. The signal values J and K determine the next

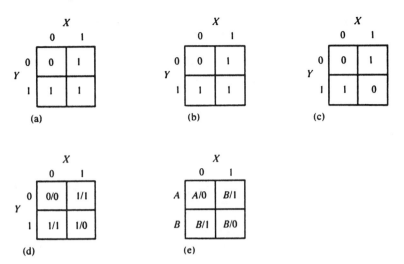

Figure 6.23 Sequential circuit analysis. (a) $D = X + Y$; (b) $Y(t + 1)$; (c) $Z = XY'$ $+ X'Y$; (d) transition table; (e) state table.

state $Y_1(t + 1)$ of the *JK* flip-flop. The value of D determines the next state $Y_2(t + 1)$ of the D flip-flop. Since both flip-flops are triggered by the same clock, their transitions occur simultaneously.

In practice, only one type of flip-flop is used in a circuit. Since an IC generally contains more than one flip-flop, using a single type of flip-flop in the circuit reduces the component count and hence the cost of the circuit. Different types of flip-flops have been used in the examples in this chapter for illustration purposes only.

By analyzing the combinational portion of the circuit, we can derive the flip-flop input (or excitation) equations:

$$J = XY_2 \quad \text{and} \quad K = X + Y_2'$$

and

$$D = Y_1'Y_2 + X'Y_2'$$

The circuit output equation is

$$Z = XY_1Y_2'$$

Because there are two flip-flops, there will be four states, and hence the state table shown in (b) will have four rows. The rows are identified with the *state vectors* $Y_1Y_2 = 00$, 01, 10, and 11. Input X can be 0 or 1, and hence the state table will have two columns.

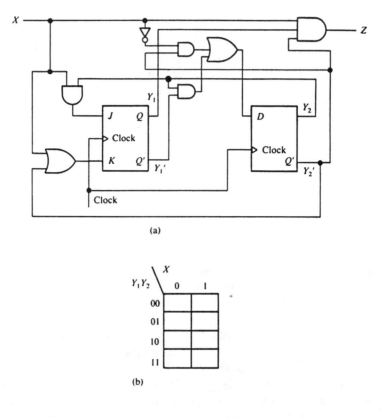

(a)

(b)

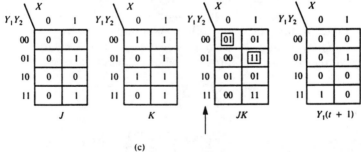

(c)

Figure 6.24 A sequential circuit. (a) Circuit diagram; (b) state table format; (c) JK flip-flop transitions; (d) D flip-flop transitions; (e) transition table; (f) next state table; (g) output; (h) state table; (i) state diagram; (j) an I/O sequence.

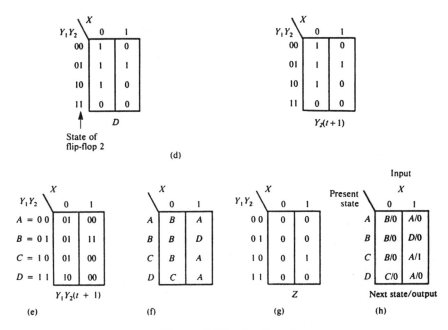

Figure 6.24 *Continued*

The next-state transitions of the JK flip-flop are shown in (c). Note that the maps for J and K are essentially K-maps derived from the input equations above, except that the last two rows must be interchanged to bring these tables into the K-map form. Tables for J and K are then merged, entry by entry, to derive the composite JK table —which makes it easier to derive the state transition of the JK flip-flop. Although both $Y_1(t)$ and $Y_2(t)$ values are shown in this table, only the $Y_1(t)$ value is required to determine $Y_1(t + 1)$ once the J and K values are known. For example, in the boxed entry at the top left of the table, $J = 0$ and $K = 1$; hence, from the characteristic table for the JK flip-flop (Figure 6.11), the flip-flop will reset, and $Y_1(t + 1)$ will equal 0. Similarly, in the boxed entry in the second row, $J = 1$ and $K = 1$. Hence, the flip-flop complements its state. Since the $Y_1(t)$ value corresponding to this entry is 0, $Y_1(t + 1) = 1$. This process is repeated six more times to complete the $Y_1(t + 1)$ table.

The analysis of D flip-flop transitions is shown in (d); $Y_2(t + 1)$ is derived from these transitions.

The transition tables for the individual flip-flops are then merged, column by column, to form the transition table in (e) for the entire

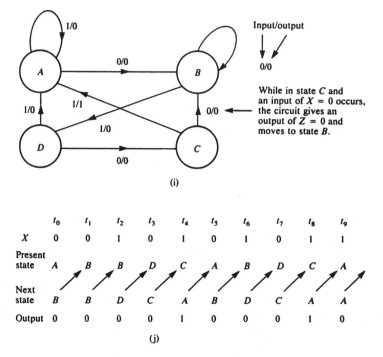

(i)

	t_0	t_1	t_2	t_3	t_4	t_5	t_6	t_7	t_8	t_9
X	0	0	1	0	1	0	1	0	1	1
Present state	A	B	B	D	C	A	B	D	C	A
Next state	B	B	D	C	A	B	D	C	A	A
Output	0	0	0	0	1	0	0	0	1	0

(j)

Figure 6.24 *Continued*

circuit. Instead of denoting the states by binary state vectors, letter designations can be used for each state, as shown in (e), and the next-state table shown in (f) is derived. The output table in (g) is derived from the circuit output equation shown above. The output and next-state tables are then merged to form the state table (h) for the circuit. The state table thoroughly depicts the behavior of the sequential circuit. The state diagram for the circuit derived from the state table is shown in (i).

Assuming a starting (or initial) state of A, and input sequence and the corresponding next-state and output sequences are shown in (j). Note that the output sequence indicates that the output is 1 only when the circuit input sequence is 0101. Thus, this is a 0101 sequence detector.

(Note that once a sequence is detected, the circuit goes into the starting state A and another complete 0101 sequence is required for the circuit to produce an output of 1. Can the state diagram in the above example be rearranged to make the circuit detect *overlapping* sequences? That is, the circuit should produce an output if a 01 occurs directly after the detection of a 0101 sequence. For example,

$$X = 000101010100101$$
$$Z = 000001010100001.)$$

If the *starting* or the initial state of the circuit and the input sequence are known, the state table and the state diagram for a sequential circuit permit a functional analysis whereby the circuit's behavior can be determined. Timing analysis is required when a more detailed analysis of the circuit parameters is needed. Figure 6.25 shows the timing diagram for the first five-clock pulses in Example 6.2.

This heuristic analysis can be formalized into the following step-by-step procedure for the analysis of synchronous sequential circuits:

1. Analyze the combinational part of the circuit to derive excitation equations for each flip-flop and the circuit output equations.
2. Note the number of flip-flops (p) and determine the number of states (2^p). Express each flip-flop input equation as a function of

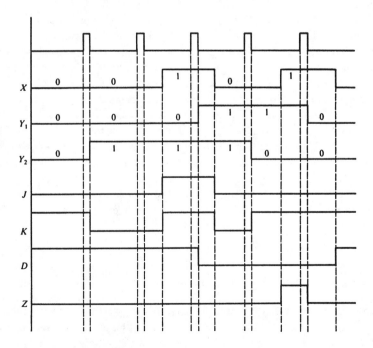

Figure 6.25 Timing diagram for Example 6.2.

circuit inputs and the present state and derive the transition table for each flip-flop, using the characteristic table for the flip-flop.

3. Derive the next-state table for each flip-flop and merge them into one, thus forming the transition table for the entire circuit.

4. Assign names to state vectors in the transition table to derive the next-state table.

5. Using the output equations, draw a truth table for each output of the circuit, and rearrange these tables into the state table form. If there is more than one output, merge the output tables column by column to form the circuit output table.

6. Merge the next-state and output tables into one to form the state table for the entire circuit.

7. Draw the state diagram.

It is not always necessary to follow this analysis procedure. Some circuits yield a more direct analysis, as shown in Example 6.3.

Example 6.3

Figure 6.26(a) is a sequential circuit made up of three T flip-flops. Recall that the T flip-flop complements its state when the T input is 1. Hence, if input X is held at 1, the flip-flop at the least significant end (*FF3*) complements at each clock pulse, while *FF2* complements only when Q_3 is 1, and *FF1* complements only when Q_2 is 1. If all the flip-flops are reset initially, the circuit starts with the state $Q_1Q_2Q_3 = 000$. The first clock pulse will take it to 001; the next clock pulse will take it to 010 and on to 111 at the next pulse. The circuit returns to 000 at the fourth clock pulse. If the circuit starts with the initial state of 100, one other four-state sequence is possible. Both state transition sequences are shown in (b). Thus, this circuit is a modulo-4 counter. If the desired counting sequence is other than what this circuit provides (say, 00-01-10-11), additional circuitry is needed to transform the state sequence to the required output sequence. Designing such output circuitry is left as an exercise.

6.6 Mealy and Moore Models

The state table and state diagram of Examples 6.1 and 6.2 correspond to the so-called *Mealy model* for the sequential circuit. The Mealy model represents a transition-assigned circuit, in which the circuit's output is associated with the state transition.

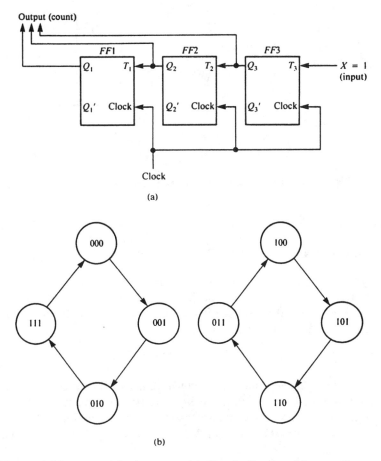

Figure 6.26 A modulo-4 counter. (a) Circuit diagram; (b) state diagram.

The state diagram of Example 6.3 corresponds to the *Moore model* for a sequential circuit. This model associates the output of the circuit with the present state of the circuit. That means that each state has a specified output associated with it and the outputs are not shown on the arcs of the state diagram. Figure 6.27 shows another example. States *A* and *B* are associated with an ouput of 0 and state *C* with an output of 1, as shown within the circles in the state diagram. Only one output column is needed in the state diagram, unlike with the Mealy model. A sequence of inputs to this circuit starting in state *A* and the corresponding next-state and output sequence are also shown in the figure. It should be noted that the output at any time is a function of the present state only and not of the next state.

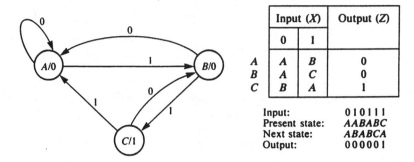

	Input (X)		Output (Z)
	0	**1**	
A	*A*	*B*	0
B	*A*	*C*	0
C	*B*	*A*	1

Input: 0 1 0 1 1 1
Present state: *A A B A B C*
Next state: *A B A B C A*
Output: 0 0 0 0 0 1

Figure 6.27 A Moore model circuit.

Example 6.4

Figure 6.28 shows another state diagram. Here, the circuit advances from state A to B, C, D, and back to A if input X is held at 0 and from D to C, B, A, and back to D if the input is held at 1. If the following outputs are assigned to the state:

$$A = 00, \quad B = 01, \quad C = 10, \quad D = 11,$$

the circuit behaves like an *up-down counter*, counting up when the input is 1 and down when the input is 0. Note that the outputs are strictly functions of the present state of the circuit. In fact, the flip-flop outputs form the output of the circuit, and no other output circuitry is needed.

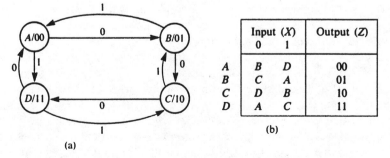

	Input (X)		Output (Z)
	0	**1**	
A	*B*	*D*	00
B	*C*	*A*	01
C	*D*	*B*	10
D	*A*	*C*	11

(b)

(a)

Figure 6.28 An up-down counter. (a) State diagram; (b) state table.

6.7 Design of Synchronous Sequential Circuits

The design of a sequential circuit is the process of deriving a logic diagram from the specification of the circuit's required behavior. The circuit's behavior is often expressed in words. The first step in the design is then to derive an exact specification of the required behavior in terms of either a state diagram or a state table. This is probably the most difficult step in the design, since no definite rules can be established to derive the state diagram or a state table. The designer's intuition and experience are the only guides. Once the description is converted into the state diagram or a state table, the remaining steps become mechanical. We will examine the classical design procedure through the examples in this section. It is not always necessary to follow this classical procedure, as some designs lend themselves to more direct and intuitive design methods. (The design of shift registers, described in Chapter 7, is one such example.) The classical design procedure consists of the following steps:

1. Deriving the state diagram (and state table) for the circuit from the problem statement.
2. Deriving the number of flip-flops (p) needed for the design from the number of states in the state diagram, by the formula

$$2^{p-1} < n \leq 2^p$$

   where n = number of states.
3. Deciding on the types of flip-flops to be used. (This often simply depends on the type of flip-flops available for the particular design.)
4. Assigning a unique p-bit pattern (state vector) to each state.
5. Deriving the state transition table and the output table.
6. Separating the state transition table into p tables, one for each flip-flop.
7. Deriving an input table for each flip-flop input using the excitation tables (Figure 6.12).
8. Deriving input equations for each flip-flop input and the circuit output equations.
9. Drawing the circuit diagram.

This design procedure is illustrated by the following examples.

Example 6.5

Design a sequential circuit that detects an input sequence of 1011. The sequences may overlap. A 1011 sequence detector gives an output of 1 when the input completes a sequence of 1011. Because overlap is allowed, the last 1 in the 1011 sequence could be the first bit of the next 1011 sequence, and hence a further input of 011 is enough to produce an output of 1. That is, the input sequence 1011011 consists of two overlapping sequences.

Figure 6.29(a) shows a state diagram. The sequence starts with a 1. Assuming a starting state of A, the circuit stays in A as long as the input is 0, producing an output of 0 waiting for an input of 1 to occur. The first 1 input takes the circuit to a new state B. So long as the inputs continue to be 1, the circuit has to stay in B waiting for a 0 to occur to continue the sequence and hence to move to a new state C. While in C, if a 0 is received, the sequence of inputs is 100, and the current sequence cannot possibly lead to 1011. Hence, the circuit returns to state A. But, if a 1 is received while in C, the circuit moves to a new state D, continuing the sequence. While in D, a 1 input completes the 1011 sequence. The circuit gives a 1 output and goes to B in preparation for a 011 for a new sequence. A 0 input while at B creates the possibility of an overlap, and hence the circuit returns to C so that it can detect the 11 subsequence required to complete the sequence.

Drawing a state diagram is purely a process of trial and error. In general, we start with an initial state. At each state, we move either to a new state or to one of the already-reached states, depending on the input values. The state diagram is complete when all the input combinations are tested and accounted for at each state. Note that the number of states in the diagram cannot be predetermined and various diagrams typically are possible for a given program statement. The amount of hardware needed to synthesize the circuit increases with the number of states. Thus, it is desirable to reduce the number of states if possible. State reduction techniques are described later in this chapter.

The state table for the example is shown in Figure 6.29(b). Since there are four states, we need two flip-flops. The four two-bit patterns are arbitrarily assigned to the states, and the transition table in (c) and the output table in (d) are drawn. From the output table, we can see that

$$Z = XY_1Y_2.$$

We will use an SR flip-flop and a T flip-flop. It is common practice to use one kind of flip-flop in a circuit. Different kinds are used here for illustration purposes only. The transitions of flip-flop 1 (SR) extracted from (c) are shown in the first table $Y_1(t + 1)$ of (e). From these transitions and using the excitation tables for the SR flip-flop

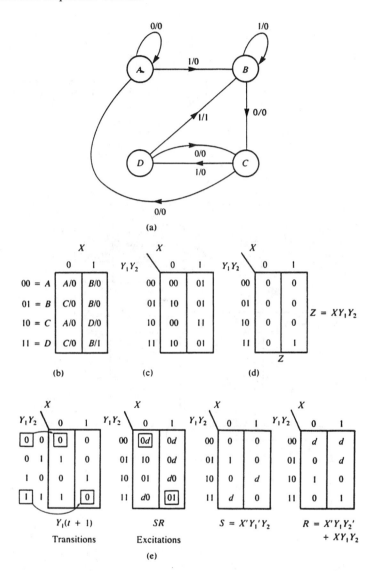

Figure 6.29 1011 sequence detector. (a) State diagram; (b) state table; (c) transition table; (d) output table; (e) flip-flop 1 (*SR*); (f) flip-flop 2 (*T*); (g) circuit diagram.

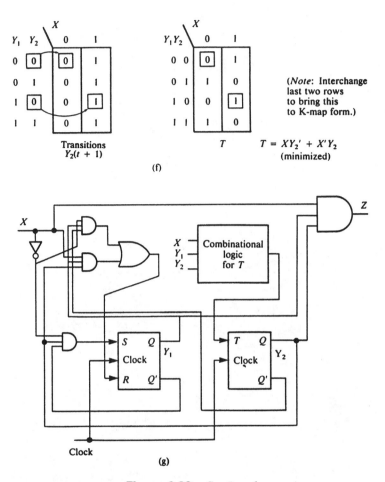

Figure 6.29 *Continued*

(Figure 6.12), the S and R excitations are derived. (For example, the 0-to-0 transition of the flip-flop requires that $S = 0$ and $R = d$, and a 1-to-0 transition requires that $S = 0$ and $R = 1$.) The S and R excitations (which are functions of X, Y_1, and Y_2) are separated into individual tables, and the excitation equations are derived. These equations are shown in (e). The input equation for the second flip-flop (T) is similarly derived and is shown in (f). The circuit diagram is shown in (g).

The complexity of a sequential circuit can be reduced by simplifying the input and output equations. In addition, a judicious allocation of state vectors to the states, also reduces circuit complexity. (State assignment procedures are described later in this chapter.)

Example 6.6

We will now derive the circuit for the modulo-4 up-down counter of Figure 6.28. The input X to the counter controls the direction of the count: up if $X = 0$ and down if $X = 1$. The state of the circuit (i.e., the count) itself is the circuit output. The state diagram is shown below.

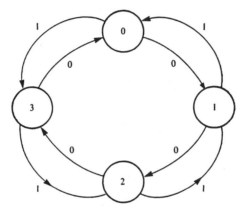

Derivation of a state diagram for this counter is straightforward, since the number of states and the transitions are completely defined by the problem statement. Note that only input values are shown on the arcs, since the output of the circuit is the state of the circuit itself. Since there are four states, we will need two flip-flops. The state assignment is also defined to be 00, 01, 10, and 11 to correspond to counts 0, 1, 2, and 3, respectively. The state table and transition table are shown below:

		X	
	Y_1Y_2	0	1
0	00	1	3
1	01	2	0
2	10	3	1
3	11	0	2

	X	
Y_1Y_2	0	1
00	01	11
01	10	00
10	11	01
11	00	10

The control signal is X. $X = 0$ implies count up; $X = 1$ implies count down.

We will use a JK and a D flip-flop. The input equations are derived below.

Y_1Y_2	X 0	1
00	0	1
01	1	0
10	1	0
11	0	1

Transitions of
flip-flop 1 (JK)

Y_1Y_2	X 0	1
00	0d	1d
01	1d	0d
10	d0	d1
11	d1	d0

JK
Inputs

Y_1Y_2	X 0	1
00	0	1
01	1	0
10	d	d
11	d	d

J
$J = XY_2' + X'Y_2$

Y_1Y_2	X 0	1
00	d	d
01	d	d
10	0	1
11	1	0

K
$K = XY_2' + X'Y_2$

Y_1Y_2	X 0	1
00	1	1
01	0	0
10	1	1
11	0	0

Transitions of
flip-flop 2 (D)

Y_1Y_2	X 0	1
00	1	1
01	0	0
10	1	1
11	0	0

Inputs D
$D = Y_2'$
(D is simplified,
by observation)

The circuit is shown below:

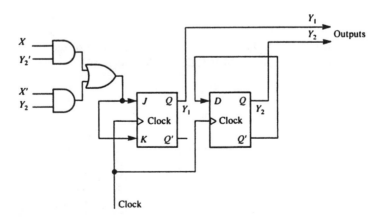

It is important to note that for the circuit to operate properly, a sequential circuit must be initialized to its starting state before the input sequence is applied. It is common practice to include a *master-clear*

circuit that can be used to initialize each flip-flop in the circuit once the power is turned on. Some circuits include a power-on reset circuit that performs the initialization as soon as the power is turned on. Chapter 8 provides the details of one such circuit.

6.7.1 Circuit Minimization

The implementation cost of the sequential circuit generally can be minimized by (a) reducing the number of memory elements in the circuit, (b) minimizing the combinational logic portion that produces the next-state information (e.g., the next-state logic defined by flip-flop input functions), and (c) minimizing the combinational logic portion that produces the outputs (i.e., the output logic). Note that in implementations using programmable devices minimization may not be necessary if the selected device provides sufficient AND and OR terms and flip-flops.

Reducing the number of memory elements is accomplished by reducing the number of states in the circuit if possible. The fewer states, the fewer flip-flops needed, and the associated circuitry is less complex. The state reduction procedures of Section 6.8 can be used to reduce the number of states once the state table is derived from the word description of the circuit behavior.

If the number of states (*N*) in the minimized state table is not a power of 2, some of the states of the circuit will be unused, since the total number of possible states is always a power of 2. These unused states can be used as don't-care conditions to minimize the circuitry.

Reducing the next state and the output logic is also accomplished through a judicious allocation of binary patterns (or state vectors) to the states in the circuit. The state vector allocation process is called the *state assignment*. We will examine the guidelines for optimal state assignment in Section 6.9

In practice, it is possible for some of the combinations of input values not to occur. In such cases, the nonoccurring input combinations contribute don't-care conditions that can be utilized to minimize the logic circuitry. Such *incompletely specified sequential circuits* are discussed in Section 6.10.

Some of the above minimization characteristics are illustrated informally in Example 6.7.

Example 6.7

Design a 00-01-10 sequence detector. That is, the circuit will have two inputs—say, *P* and *Q*—and one output, *Z*, which will be 1 when the above sequence of inputs occurs as shown on the next page:

	t_1	t_2	t_3	t_4	t_5	t_6	t_7
P	0	0	1	1	0	0	1
Q	0	1	0	1	0	1	0
Z	0	0	1	0	0	0	1

The state diagram and state table are shown in Figure 6.30. A is the initial state. There are four possible input combinations occurring at each state. The circuit waits in state A until a $PQ = 00$ input occurs to take it to state B, indicating that the first combination of the required sequence has occurred. The circuit moves to state C if the next required input combination, 01, occurs; the circuit returns to state A if a subsequent input combination of 10 occurs, producing $Z = 1$. The state diagram is completed by accounting for each of the four input combinations occurring at each state.

Let us now examine the state table to see whether the number of states can be reduced. First we compare states A and B. If the circuit is in either of these states, and if an input is applied and the output is observed, we can see that the circuit produces the same output for all input combinations regardless of whether the circuit was in state A or B. Now let us look at the next states. The circuit reaches the same next state from either state A or B under input combinations 00, 10, and 11. But under input combination 01, A goes to A, while B goes to C. Thus, if we can determine that states A and C are identical in their function, we can say that A and B are also identical and hence can be combined into one state. That is, A and B are identical if and only if A and C are identical. Let us now compare A and C. Their outputs are

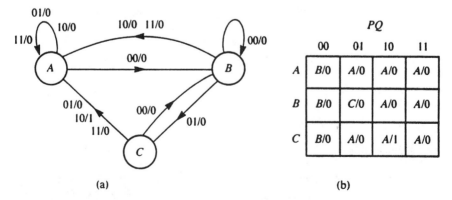

Figure 6.30 00-01-10 sequence detector. (a) State diagram; (b) state table.

identical for the three input combinations but differ for the input combination 10. Hence, these two states cannot be identical. Hence, A and B are also not identical. Similarly, B and C are also not identical, since their outputs differ. Thus, the state table cannot be reduced.

We need two flip-flops to implement this circuit. For simplicity, let us use D flip-flops. As mentioned earlier, we can reduce the complexity of the circuit by judiciously assigning state vectors to the states of the circuit. Figure 6.31 shows one such assignment: $A = 00$, $B = 01$, and $C = 11$. The vector 10 is not used. The transition table is

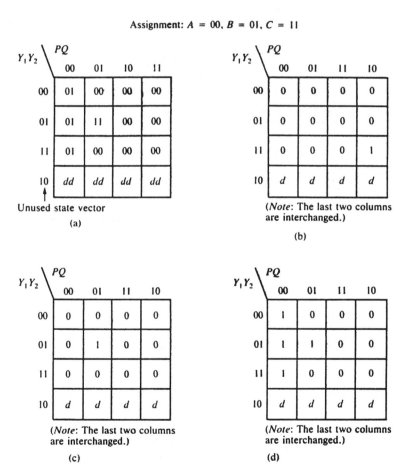

Figure 6.31 State assignment 1 for Example 6.7. (a) Transition table; (b) $Z = PQ'Y_1$; (c) $D_1 = P'QY_1'Y_2$; (d) $D_2 = P'Q' + P'Y_1'Y_2$.

obtained by substituting the state vectors for the states in the state table. The output table is also shown in the figure. Note that the unused state vector contributes don't-cares in both of these tables. Since we are using D flip-flops, D_1 and D_2 are the same as $Y_1(t + 1)$ and $Y_2(t + 1)$. The transition table is separated into individual tables for D_1 and D_2. Note that the column corresponding to combination 10 is interchanged with that corresponding to 11 to put the transition tables into the K-map form. From these tables, we can derive the flip-flop input equations and the circuit output equation. As can be seen from these equations, the circuit implementation requires 12 literals.

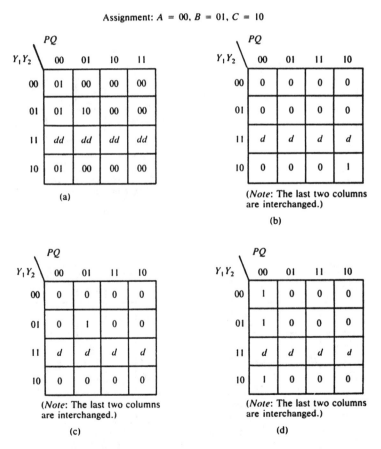

Assignment: $A = 00, B = 01, C = 10$

Figure 6.32 State assignment 2 for Example 6.7. (a) Transition table; (b) $Z = PQ'Y_1$; (c) $D_1 = P'QY_2$; (d) $D_2 = P'Q'$.

Figure 6.32 shows an alternate state assignment in which $A = 00$, $B = 01$, and $C = 10$. From the transition and output tables, we can derive the flip-flop input and circuit output equations; this implementation requires only 8 literals.

Thus, again, we can reduce the complexity of the circuit by judiciously allocating state vectors to the states. One procedure would be to try all possible combinations of state assignments to determine the one yielding the least complex circuit. But this procedure becomes unwieldy for circuits with large numbers of states. Section 6.9 describes some approaches to this state assignment problem.

It is interesting to observe the effect of using don't-cares for circuit minimization on the state table. Figure 6.33 shows the flip-flop input tables corresponding to the assignment in Figure 6.32. Here, each don't-care term used in the simplification has been changed to 1 and the other don't-cares are set to 0. The flip-flop input tables are merged to form the circuit tran-

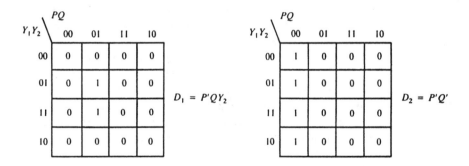

$$D_1 = P'QY_2$$

$$D_2 = P'Q'$$

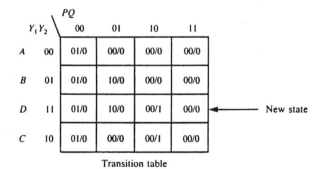

Transition table

Figure 6.33 Effect of utilizing don't-cares.

sition table. Here, D (corresponding to the unused state vector 11) is a new
state. Note that the circuit can never reach D from any other state. In addition, we can get out of state D to the initial state A by applying two of
the four possible input combinations to the circuit. Thus, the don't-cares did
not adversely affect the design.

6.8 State Reduction

During the generation of the state diagram from the word description of the
sequential circuit behavior, the diagram may contain some redundant states
(i.e., the states whose functions can be accomplished by other states). Because the number of flip-flops required to implement the circuit is directly
related to the number of states, eliminating redundant states may result in a
reduction of the number of flip-flops required. For example, if the state
diagram has 10 states, reducing the states by two would reduce the number
of flip-flops required to implement the circuit to three instead of four. However, if we reduce the number of states to only nine, we will still need four
flip-flops. Nonetheless, we will have an additional state vector that contributes don't-care conditions and hence this might result in better combinational
circuit minimization than the circuit with 10 states.

   Note that state reduction is possible only when (a) the circuit's behavior is not directly connected with the states of the circuit and (b) the reduced
circuit also produces the same output sequence as the original circuit for all
the input sequences and hence is equivalent to the original circuit (as far as
the external input/output behavior is concerned). If the outputs are directly
connected to states, as in the example of the counter described earlier, all
the states are required and hence state reduction is not possible.

   In this section, we will examine two state reduction procedures. The
first procedure uses *equivalence partitioning* of states, and the second uses
an *implication chart*. These procedures are based on the following
definitions.

***Definition*:** Two states A and B are said to be *equivalent* (i.e., $A = B$) if
and only if, for every possible input sequence, the sequential circuit produces
the same output sequence regardless of whether A or B is the initial state.

   This means that if A and B are equivalent, their corresponding X-
successors, for all X, are also equivalent. The X-successor of any state A is
the next state reached by the circuit in A when the input combination X is
applied.

***Definition*:** Two states A and B are said to be *k-equivalent*, if for each and
every input sequence of length k, the circuit produces an identical output

sequence, regardless of whether the starting state is A or B. Thus, two states are equivalent if they are k-equivalent for all k. If two states are k-equivalent, it can be shown that they are $(k - 1)$-equivalent, $(k - 2)$-equivalent, and so forth.

***Definition*:** Two states A and B are said to be *distinguishable* if at least one input sequence exists for which the output sequences are different depending on whether the starting state is A or B. The states are *k-distinguishable* if the length of the input sequence required to distinguish them is at least k.

The state reduction procedures use the above properties to find states that are 1-equivalent first, followed by 2-equivalent, 3-equivalent, and so on, until equivalent states are found.

6.8.1 Equivalence Partitioning

We will examine this procedure through Example 6.8.

Example 6.8

Consider the sequential circuit shown in Figure 6.34(a). There are eight states. The state table is separated into a next-state portion and an output portion for convenience. The equivalence partitioning process is shown in (b).

All the states in the table are 0-equivalent, since we cannot distinguish between the states with an input sequence of length 0. This property is shown by the partition P_0. This partition contains one *block* (or group) containing all the states in the circuit. This is always the starting partition for all circuits. Now we try to find the 1-equivalent states (i.e., those states that cannot be distinguished by applying an input sequence of length 1) by examining the output patterns of all the states. If two states have the same output pattern (i.e., similar output for each input combination), they are 1-equivalent, though the next states may not be equal. Note that both states A and C produce outputs of 0 and 0 under inputs of 0 and 1, respectively. Hence, they are 1-equivalent and are put in one block in partition P_1. Similarly, states B, F, and G all produce the same outputs under each input. So do states D, E, and H. Thus, this step yields three blocks in the partition P_1 corresponding to the 00, 01, and 10 output patterns. The states in each block in this partition are 1-equivalent to other states in the same block.

The next step is to examine the blocks in P_1 to obtain the partition P_2 of 2-equivalent states. Two states are 2-equivalent if they are 1-equivalent and their X-successors for all X are also 1-equivalent. This step is carried out by splitting blocks of P_1 whenever the X-successors of the states in each block are not in the same block of P_1. Let us examine the 0-successors for each block in P_1. For an input of 0, A and C reach G and G, respectively. That is, the 0-successor of (AC) is (GG) or simply (G). Similarly, the 0-successor of (BFG) is (EDH) and that of (DEH) is (ABC). Since (G) and (DEH) are each contained in one of the blocks in P_1, we retain the corresponding blocks (AC) and (BFG) as they are. But (ABC) is not contained by any block in P_1, although (AC) is. Hence, we split the corresponding block, (DEH), into (DH), whose 0-successors are (AC) and (E). Since (E) contains a single state, it no longer needs to be considered in the process. The 1-successors for the remaining blocks are (GF), (CEE), and (GF). (GF) is contained in a block of the previous partition, but (CEE) is not. Hence the corresponding block, (BFG), is split into (B) and (FG), since B reaches C and both F and G reach E. Since all the input combinations are now exhausted, this partition becomes the 2-equivalent partition P_2.

Partition P_3 is obtained similarly, by observing the 0- and 1-successors of the blocks in P_2. Note that P_2 and P_3 are the same, indicating that further partitioning is not possible. Hence, we stop the partitioning process, and $A = C$, $F = G$, and $D = H$.

The process of combining equivalent states is shown in Figure 6.34(c). Since $A = C$, either A or C can be eliminated from the state table. We eliminate C and replace each occurrence of C as the next state by A. Similarly, G and H are eliminated, replacing them with F and D, respectively. The reduced state table has five states and still needs three flip-flops for implementation.

The equivalence partitioning method is summarized below:

1. Start with P_0, which contains all the states of the circuit in one block.
2. Determine the blocks of P_1 by observing the output portion of the state table, grouping the states with the same output pattern into one block.
3. To determine the blocks of P_i from P_{i-1}, $i > 1$:
   (a) Determine the Xi-successors of each block in P_{i-1}; if the Xi-successors of each block are contained in a block of P_{i-1}, do not split the block. If they are not, split the block to obtain a new intermediate partition so that the Xi-successors are in the same blocks of P_{i-1}.
   (b) Repeat (a) for all Xi.
4. Repeat step (3) until no further partitioning is possible.

Present state	Next state X 0	1	Output X 0	1
A	G	F	0	0
B	E	C	0	1
C	G	G	0	0
D	A	G	1	0
E	B	A	1	0
F	D	E	0	1
G	H	E	0	1
H	C	F	1	0

(a)

P_0 = (ABCDEFGH)   0-equivalent
P_1 = (AC) (BFG) (DEH)   1-equivalent

↓   ↓   ↓

GG   EDH   ABC   0-successors
(AC) (BFG) (DH) (E)
GF   CEE   GF   1-successors

P_2 = (AC) (B) (FG) (DH) (E)   2-equivalent

↓   ↓   ↓

GG   DH   AC   0-successors
GF   EE   GF   1-successors

P_3 = (AC) (B) (FG) (DH) (E)   3-equivalent

(b)

Present state	Next state X 0	1	Output X 0	1	
A	~~G~~F	F	0	0	
B	E	~~C~~A	0	1	A = C
C	~~G~~F	~~G~~F	0	0	
D	A	~~G~~F	1	0	F = G
E	B	A	1	0	D = H
F	D	E	0	1	
G	~~H~~D	E	0	1	
H	~~C~~A	F	1	0	

(c)

Present state	Next state X 0	1	Output X 0	1
A	F	F	0	0
B	E	A	0	1
D	A	F	1	0
E	B	A	1	0
F	D	E	0	1

(d)

Figure 6.34 State reduction. (a) State table; (b) equivalence partitioning; (c) combining equivalent states; (d) reduced state table.

6.8.2 Implication Charts

The state reduction procedure using an *implication chart* is more systematic than the previous one and is suitable for implementation as a computer program. (For one such implementation, see the author's articles in *Electronic Design*, listed in the references at the end of this chapter. Several other implementations are available.)

Figure 6.35(a) shows the implication chart for the state table of Figure 6.34. The implication chart provides a mechanism for comparing each state with every other state in the state table to determine their equivalence. Along the horizontal axis, all states but the last in the state table are listed, and along the vertical axis all states but the first in the state table are listed. Thus, each block in the chart corresponds to a distinct state pair. We will designate each block by its coordinate state pair. For example, the topmost

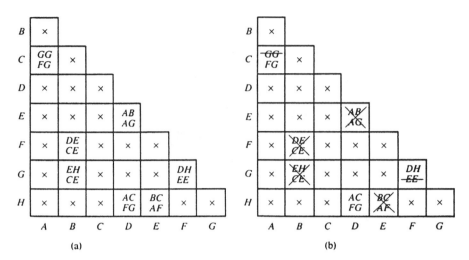

Figure 6.35 Implication chart method. (a) Implication graph (initial); (b) final graph.

block in the first column is block *A-B*, the one below it is *A-C*, and so on. The content of each block depends on the equivalency of its coordinate states. There can be types of entries in each block: (1) an ''×,'' indicating that the coordinate states of the block are not equivalent; (2) a ''√,'' indicating that the coordinate states are unconditionally equivalent; or (3) the next state pair corresponding to each input combination, indicating a conditional equivalence, wherein the coordinate states of the block are equivalent only if all the state pairs appearing as contents of the block are equivalent.

The implication chart starts off with all blocks empty. Corresponding to each block, we compare the coordinate states in the state table and enter the appropriate information into the block. For example, block *A-B* contains an × since states *A* and *B* have different output patterns and hence can never be equivalent. If we compare states *A* and *C*, we can see that since their output patterns are similar, they would be equivalent if all of their next-state-pairs were equivalent. We simply enter the next-state-pairs (*GG*) and (*FG*) into the block *A-C*. Since (*GG*) corresponds to a single state, it does not provide any restricting condition for equivalence and hence we can discard this entry. If two states are unconditionally equivalent (i.e., their output patterns are similar and all the next state pairs are equal), we would mark the corresponding block with a ''√.'' (We do not have such a condition in this example.) The first cycle through the implication chart is complete when all the blocks are filled.

During the second cycle, we examine each state-pair entry in conditional blocks to determine whether the state pair is equivalent. If it can be determined that a state pair is not equivalent, and $\times$ is entered into the corresponding block. Otherwise, the block is left unchanged. For example, the block A-C has the state pair (FG). Since the block F-G at this stage does not have an $\times$, we will leave the contents of A-C unchanged. Block B-F contains (DE) and (CE). Block D-E does not contain an $\times$, but C-E does. Thus, B and F are not equivalent. Hence, we mark B-F with an $\times$. To complete the cycle, this process is continued for each block that is not crossed. Figure 6.35(b) shows the implication chart at the end of the cycle.

We then repeat the above cycle on the implication chart until no more blocks can be crossed off. The coordinate states corresponding to uncrossed blocks are the equivalent state pairs. In this example, (AC), (FG), and (DH) are equivalent pairs.

It may be possible to merge the equivalent state pairs by using the transitivity condition: if $A = B$ and $B = C$, then $A = C$ and (ABC) is an equivalent set of states. No such merger is possible in the example above. Once the equivalent states are determined, the state table is reduced by elimination and substitution, as illustrated by the first method above.

The implication chart method of reduction is summarized below:

1. Draw the implication chart and mark each block with a "$\times$" or a "$\sqrt{}$" or next-state pairs, according to whether the coordinate states of the block are not equivalent, equivalent, or conditionally equivalent.
2. Scan the chart obtained in step 1 to eliminate entries with the same state as the components of the state pair.
3. Cross off the blocks in the implication chart if at least one of the entries in the block corresponds to a block that has already been crossed off.
4. Repet step 3 until no further blocks can be crossed off.
5. Note that the coordinate states corresponding to blocks that are not crossed off are equivalent.
6. Merge the equivalent state-pairs obtained in the preceding steps, if possible, to form larger equivalence partitions.
7. Eliminate and substitute equivalent states in the state table.

Example 6.9

Figure 6.36 shows another example of state reduction.

6.9 State Assignment

As we saw in Example 6.6, the choice of state assignments affects the complexity of the combinational logic portion of the sequential circuit. The num-

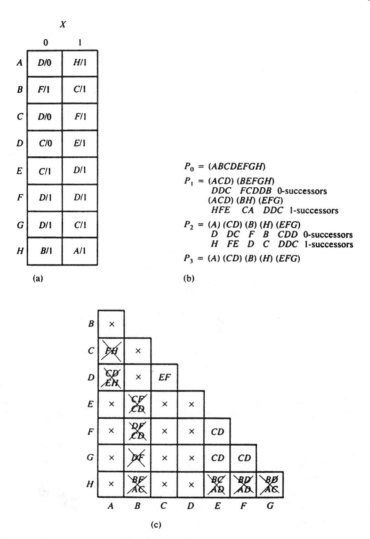

Figure 6.36 State reduction (Example 6.9). (a) State table; (b) equivalence partitioning; (c) implication chart.

ber of possible state assignments for a state table can be quite large. For example, to implement a sequential circuit with three states, we need two flip-flops; hence, there are four possible state vectors (00, 01, 10, and 11). If we assign one of these four for the first state, then we have three choices for the second state and two for the third state, thus making the total number of possible assignments equal to $4 \times 3 \times 2$, or 24. In general, if N_f is the number of flip-flops and N_s is the number of states, then

$$2^{N_f-1} < N_s \leq 2^{N_f}$$

and the number of state assignments possible is

$$(2^{N_f}!)/(2^{N_f} - N_s)!$$

But all these assignments are not *unique* (or *distinct*) with respect to their effect on the combinational logic. For example, if in a given assignment we complement a given bit position, we obtain a new assignment. However, the resulting logic equations will be of the same complexity. In fact, the next-state equations corresponding to the latter assignment can be obtained by complementing the state variable corresponding to the complemented bit position in the logic functions for the former assignment.

Similarly, swapping two columns in a state assignment will result in a new state assignment, which results in the swapping of two variables and does not affect the cost of the circuit.

Therefore, for sequential circuits with two states, there is actually just one unique assignment. For three- and four-state circuits, there are only three unique assignments out of the 24 possible assignments. The number of unique state assignments possible for a sequential circuit thus is derived by the following:

$$(2^{N_f} - 1)!/((2^{N_f} - N_s)!N_f!)$$

Table 6.1 shows the number of assignments possible and the unique assignments for various values of N_s. As can be seen, it is possible to try all the

TABLE 6.1   Number of State Assignments

		Number of:	
States	Flip-flops	State assignments possible	Unique assignments
2	1	2	1
3	2	24	3
4	2	24	3
5	3	6,720	140
6	3	20,160	420
7	3	40,320	840
8	3	40,320	840
9	4	415×10^7	10,810,800
10	4	219×10^8	75,675,600

possible unique assignments on circuits with up to four states to determine the optimal assignment. Beyond four states, this process becomes impractical.

The state assignment problem has been an active area of research and has resulted in various procedures to derive both good (i.e., adequate) and optimal assignments. The reader is referred to the works listed in the references at the end of this chapter for details. In this section, we will provide a set of rules to arrive at a 'good' state assignment. For simplicity, we will ignore the output logic minimization aspects and assume D flip-flop implementations. In general, a good state assignment for D flip-flop implementation would be a good assignment for implementations using other types of flip-flop as well.

Example 6.10

Consider the three-state circuit in Example 6.7. Figure 6.37 shows the third possible unique assignment and the corresponding next-state equations. The next-state equations for all the three assignments are listed below.

Assignment 1	Assignment 2	Assignment 3
$A = 00, B = 01,$ $C = 11$	$A = 00, B = 01,$ $C = 10$	$A = 00, B = 11,$ $C = 01$
$D_1 = P'QY_1'Y_2$ $D_2 = P'Q' + P'Y_1'Y_2$	$D_1 = P'QY_2$ $D_2 = P'Q'$	$D_1 = P'Q'$ $D_2 = P'Q' + P'Y_1$

Assignment 2 yields the least complex circuit of the three. This as-

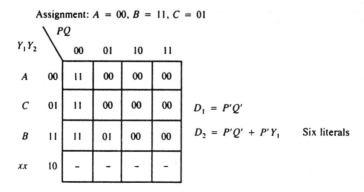

Figure 6.37 Assignment 3 (Example 6.10).

signment (a) minimizes the number of 1s in the next-state portion of the transition table, and (b) organizes the 1s in the transition table to allow better grouping for minimization (when the transition table is converted into a K-map for each D input in the implementation).

The following rules are based on these two observations and serve as guidelines for obtaining a good state assignment.

***Definition*:** Two state vectors are said to be *adjacent* if they differ in only one bit position. Two states, A and B, are said to be adjacent if their state vectors are adjacent. We will note this adjacency by (AB).

RULE 1:

Adjacent state vectors should be assigned to states that have the same next state for a given input condition.

The assignments made by the adjacency requirements of this rule cluster 1s in the column corresponding to the input combination. Since the present-state assignments are adjacent, the 1s in the column can be readily grouped.

RULE 2:

Adjacent state vectors should be assigned to states that are the next states of a single present state under logically adjacent input combinations.

The purpose of this rule is to cluster 1s along each row of the transition table under adjacent input combinations, thereby enabling better grouping.

Let us apply these rules to the previous example. Rule 1 implies (AB), (BC), and (AC); Rule 2 implies (AB), (BC), and (CA). Obviously, all these adjacency conditions cannot be met by any one assignment. Nevertheless, we can try to satisfy as many of them as possible. The assignment shown below satisfies (AB) and (AC), but not (BC). Here, $A = 00$, $B = 01$, and $C = 10$, which is the optimal assignment (i.e., assignment 2).

	0	1
0	A	C
1	B	

The next rule imposes a more severe constraint on the assignment. It requires that a pair of states that constitute the next state to a present state pair should also be adjacent. It tries to combine the premises of Rules 1 and 2 to bring together the 1s in the rows and columns to form larger groups of 1s. We will make the following definitions before providing Rule 3.

Definition: An *implication graph* for a sequential circuit is a flow graph whose nodes are pairs of states and the arcs between the nodes represent the state transitions.

Example 6.11

Consider the following state table. (The output portion is not shown for the sake of simplicity.)

	X	
	0	1
P	R	Q
Q	S	R
R	P	S
S	Q	R

The implication graph is shown below:

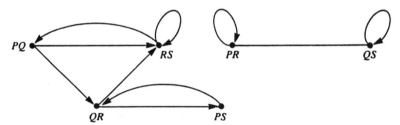

Starting from an arbitrary state pair PQ, the next-state-pairs are RS, corresponding to $X = 0$, and QR, corresponding to $X = 1$. The transitions from RS are to PQ and to RS itself. The transitions from QR are to PS and RS. The transitions from PS are both to QR. We cannot grow this implication graph any further. For completeness, we started another implication graph with the state pair PR, which is missing from the previous implication graph. The transitions from PR are to PR itself and to QS. The transitions from QS are to QS itself and to the single state R, which is not shown in the implication graph.

To summarize the implication graph construction process, we start with an arbitrary state pair and include a new node into the implication graph, depending on the transitions of the first pair under each input combination. We continue with this process with the new nodes until no new nodes are generated. Transitions to single states are not shown on the implication graph. Note that the implication graph provides essentially the same information as an implication chart, but in a graphical form.

Definition: An implication graph is said to be *complete* if it contains all possible pairs of states in the sequential circuit as its nodes. The two implication graphs above combined constitute the complete implication graph for the same circuit. Each implication graph is a *subgraph* of the complete implication graph.

Definition: An implication subgraph is said to be *closed* if all outgoing arcs in the subgraph terminate on the nodes within the subgraph and if every state in the circuit is represented by at least one node in the subgraph. (Note that arcs may enter the closed subgraph but cannot originate within the subgraph and then exit.)

The closed implication subgraph thus provides sufficient information to assign adjacent state vectors to states. The following rule is used.

RULE 3:

Using the adjacencies suggested by Rules 1 and 2, construct an implication graph. The implication graph need not be complete. Next, establish the chain of adjacencies using a closed implication subgraph (if possible). If a closed implication subgraph cannot be established, construct a contiguous implication graph with enough nodes to derive the adjacency chain.

Rules 1 and 3 are considered more important than Rule 2. Rule 4 provides the conditions for assigning state vectors so that the total number of 1s in the K-map is at a minimum.

RULE 4:

The all-zero state vector must be assigned to the "most-transferred-to" state.

In practice, this may not be desirable. The "starting state" of the circuit is usually assigned an all-zero state vector, so that a master-clear signal at the power-up can put the circuit in the appropriate stating state. If the starting state is not the most-transferred-to state, additional circuitry will be needed to initialize the circuit.

To summarize the guidelines for state assingment, we generate the adjacency requirements from Rules 1, 2, and 3, with Rules 1 and 3 taking precedence over Rule 2. Then, using an appropriate K-map, we enter the most-transferred-to state into the all-zero block (minterm 0). We then enter the other states into the blocks of the K-map to satisfy as many of the adjacencies derived above as possible, utilizing the blocks corresponding to the fewest 1s first. We will illustrate the assignment process through the following examples.

Example 6.12

Consider the state table of Figure 6.38(a). Rule 1 suggests (*BC*), (*AD*), and (*BD*). Rule 2 suggests (*BD*), (*AE*), (*BE*), (*AB*), (*CD*). Starting with the state pair (*BC*), an implication graph is constructed, as shown in (b). The closed subgraph is enclosed by dotted lines. Note that starting with (*AD*) would have yielded an implication graph with a single node,

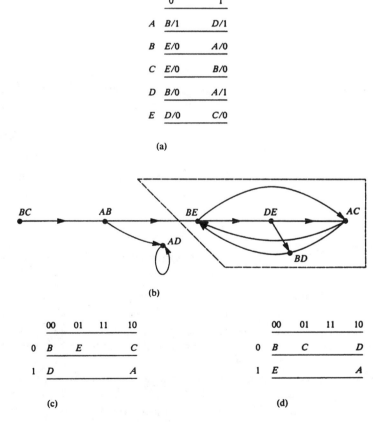

Figure 6.38 State assignment (Example 6.12). (a) State table; (b) transition diagram; (c) Assignment 1; (d) Assignment 2.

while starting with (*BD*) would have yielded the same closed subgraph as in (b).

The assignment is shown in (c). Starting with the blank three-variable K-map, *B*—the most-transferred-to state—is assigned to the 000 block. Then *D* and *E* are made adjacent to *B* (as suggested by the implication graph), and hence *D* and *E* cannot be made adjacent. *C* is then made adjacent to *B* in order to satisfy the adjacency conditions of Rule 1. Lastly, *A* is made adjacent to *D*. Note the order in which the assignments are made, so that the blocks corresponding to the fewest 1s in the state vector are used first.

An alternate assignment is shown in (d). This assignment satisfies all of the Rule 1 conditions and does not use the conditions from the implication graph. We will now compare the complexity of the next-state circuits obtained by these two state assignments.

Figure 6.39 shows the transition table corresponding to the first assignment. Because the next state is a function of four variables (X, Y_1, Y_2, and Y_3), three four-variable K-maps are needed, one for each *D* input. The six blocks corresponding to the unused state vectors ($Y_1Y_2Y_3 = 110, 011$, and 111 under $X = 0$ and 1) are marked as don't-cares. As can be seen from the transition table, D_1 is 1 at three combinations of input and the present state ($XY_1Y_2Y_3 = 1000, 1001$, and 1010). Corresponding blocks in the K-map for D_1 are marked with 1s. Similarly, the other two K-maps are completed. As seen by the minimized equations, the implementation requires 12 literals.

Figure 6.40 shows the derivation of flip-flop input equations corresponding to the alternate assignment shown in Figure 6.38(d). This implementation requires 17 literals. Thus, the first assignment is better than the second one.

Example 6.13

Figure 6.41 shows one more example of state assignment.

6.10 Incompletely Specified Sequential Circuits

A sequential circuit is said to be *incompletely specified* if some of the states or outputs are not specified for all the combinations of inputs and present states. That is, the state table contains don't-cares. As we have seen earlier in this chapter, don't-cares in the state table arise because of the unused state vectors, when the number of states in the reduced state table is less than 2^{N_f}, where N_f is the number of flip-flops required to implement the circuit. The other reason for the don't-cares in the state table is that some

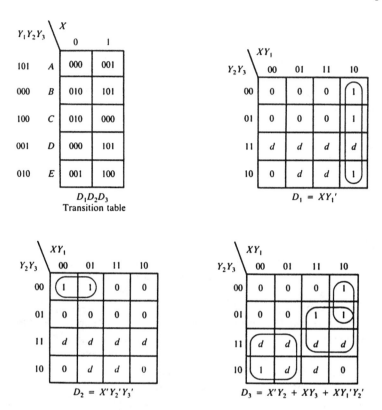

Figure 6.39 Assignment 1 (Example 6.12).

of the input combinations do not occur either on the circuit as a whole or when it is in certain states. The former case yields a state table with a column of don't-cares corresponding to each nonoccurring input combination. In the latter case, only certain entries will be don't-cares.

Concerning hardware realization, incompletely specified circuits have an advantage over completely specified ones since the unspecified states and outputs can be treated as don't-cares, thereby achieving better circuit minimization. In fact, the circuit design procedures of this chapter are applicable to these circuits as well, except for the state reduction techniques. Because some of the next states and outputs are not specified, the state equality considerations do not hold in the case of incompletely specified circuits; instead, we use the so-called *compatibility* considerations. We will describe

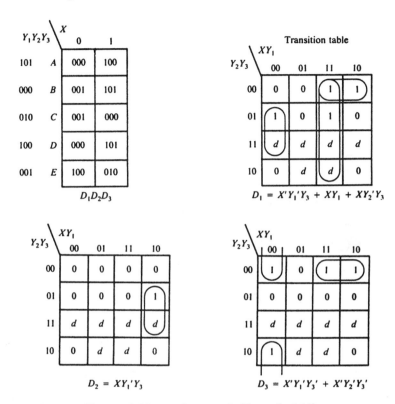

Figure 6.40 Assignment 2 (Example 6.12).

the state reduction procedure for incompletely specified circuits, after the following definitions.

Definition: An input sequence is said to be *applicable* to state S of a sequential circuit if and only if when the circuit is in state S and the input sequence is applied, all the next states are specified except possibly the last element of the next-state sequence.

For example, the input sequences 01111 and 01110 are applicable to state A of the circuit, described by Figure 6.42, while the input sequence 011111 is not.

Definition: Two states, A and B, of an incompletely specified circuit are said to be *compatible* if and only if for each input sequence applicable to A and B, the same output sequence will be produced when the outputs are

$$X$$

	0	1
P	*T*/0	*R*/1
Q	*T*/0	*P*/0
R	*Q*/0	*P*/1
S	*T*/1	*R*/0
T	*Q*/0	*U*/1
U	*S*/0	*P*/1

RULE 1:

(*PQS*), (*RT*), (*QRU*), (*PS*)

RULE 2:

(*RT*), (*PT*), (*PQ*), (*QU*), (*PS*)

RULE 3:

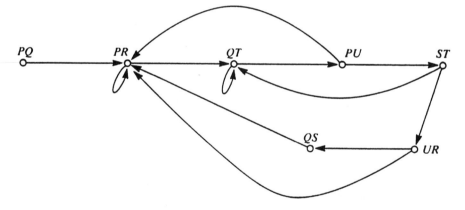

RULE 4:

The most-transferred-to states are *P* and *T*. *P* is assigned 000 since it is the starting state.

	00	01	11	10
0	*P*	*R*		*U*
1	*Q*		*S*	*T*

Figure 6.41 State assignment (Example 6.13).

	X	
	0	**1**
A	*B/-*	*D/0*
B	*B/1*	*C/0*
C	*E/0*	*A/1*
D	*E/0*	*-/-*
E	*B/1*	*C/0*

State table

State A:

Input sequence:	0 1 1 1 1	
Next-state sequence:	*B C A D* -	**Applicable**

Input sequence:	0 1 1 1 1	
Next-state sequence:	*B C A D* -	**Not applicable**

Input sequence:	0 1 1 1 0	
Next-state sequence:	*B C A D E*	**Applicable**

Figure 6.42   Input sequence applicability.

specified, whether *A* or *B* is the starting state. The compatibility of *A* and *B* is denoted by (*AB*).

For example, states *B* and *E* of the incompletely specified circuit in Figure 6.42 produce the output sequence 10001 for an input sequence of 01011. This can be shown to be true for each applicable input sequence. Thus, *B* and *E* are compatible.

The compatibility condition can be further defined as follows. Two states, *A* and *B*, of an incompletely specified circuit are compatible if and only if: (a) the outputs produced by *A* and *B* are the same when both outputs are specified for each possible input combination, and (b) the next states of *A* and *B* are compatible, when both next states are specified, for each possible input combination.

Note that the compatibility relation, unlike the equality relation, is not transitive. That is, if *A* and *B* are compatible and *B* and *C* are compatible, we cannot then say that *A* and *C* are also compatible. For *A*, *B*, and *C* to be compatible, it is required that (*AB*), (*BC*), and (*AC*) each be true.

Definition: A set of compatible states is called a *compatibility class*. A *maximal compatible* is a compatibility class that will not remain a compatible class if a state not presently in the compatibility class is added to it. In the above example, (*ABC*) is a maximal compatible, while (*AB*), (*AC*), and (*BC*) are not.

We can similarly define *incompatibility* of two states (i.e., states that are not compatible), *incompatible class*, and *maximal incompatible*.

We will use an implication chart to derive the compatible and incompatible classes for any incompletely specified circuit. Figure 6.43(a) shows the implication chart for the state table of Figure 6.42. As before, if the

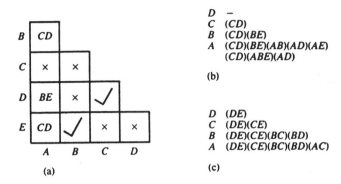

Figure 6.43 State reduction. (a) Implication chart; (b) compatible classes; (c) incompatible classes.

next-state-pair consists of a single state, it is not entered into the implication chart. Similarly, if one of the next states is unspecified, the state pair is not entered, since the unspecified state is compatible to all states. Two states are incompatible (shown with ×) if their outputs differ at least for one input combination. Again an unspecified output is equal to all the outputs. Compatible states are marked with a "√" in the corresponding block of the implication chart.

As before, we continue to scan the implication chart to mark any incompatibility until no further changes are made. From the implication chart, we can obtain the maximal compatibility and incompatibility classes.

In (b), the compatibility classes are collected from the implication chart, starting from the last column of the implication chart corresponding to state D. There are no compatible classes. From column C, we obtain one class (CD). At column B we add (BE). We cannot combine these two. At column A we obtain (AB), (AD), and (AE). Now, (AB), (AE), and (BE) can be combined to form (ABE). No other combinations are possible. Thus, (CD) (ABE) (AD) is the maximal compatible class.

To derive the maximal incompatible class, shown in (c), this process is repeated with the incompatible pairs in the implication chart. The maximal compatibility and incompatibility classes thus obtained are used to reduce the states in the incompletely specified circuit.

6.10.1 State Reduction Procedure

The state reduction of an incompletely specified circuit is performed by selecting a set of compatibility classes that meet the following conditions.

1. *Completeness*—All the states in the original incompletely specified circuit must be represented at least once in the set of compatibility classes selected.
2. *Consistency*—The chosen set of compatibility classes must be closed; that is, the implied next states of each compatibility class in the chosen set must be contained by some compatibility class within the set.
3. *Minimality*—The number of compatible classes must be a minimum.

Once the maximal compatibility classes are derived from the implication chart, these conditions are applied to select a set of compatible classes. Each class in this chosen set forms a state in the reduced state table.

In general, the number of maximal compatible classes can be larger than the number of states in the original circuit, and the selection of a minimal set to meet the conditions above is a trial-and-error process. However, the following conditions can be sued to derive the upper and lower bounds on the number of states in the reduced circuit.

The upper bound on the number of states in the reduced state table is the minimum of NS and NC, where NS is the number of states in the original circuit, and NC is the number of classes in the maximal compatible.

The lower bound is obtained by the maximum of NI_i, $i = 1$ to n, where NI_i is the number of states in the ith maximal incompatible class and there are n maximal incompatible classes. This condition is true since there must be one state in the reduced circuit corresponding to each state in an incompatible class. The largest of the maximal incompatibles thus determines the minimum number of states required.

To use the state reduction algorithm, follow these steps:

1. Construct the implication chart for the incompletely specified circuit.
2. Derive the maximal compatibles and the upper bound on the number of states required.
3. Derive the maximal incompatibles and the lower bound on the number of states required.
4. By trial and error, find a set of maximal compatibles that satisfy the completeness, consistency, and minimality conditions.
5. Construct the reduced state table.

The following examples illustrate this procedure.

Example 6.14

From Figure 6.43, we can see that we need a minimum of two and a maximum of three states in the reduced state table of the incompletely specified circuit of Figure 6.42. Furthermore, in order to satisfy the completeness condition, the two maximal compatibles (*ABE*) and (*CD*) are chosen. This selection also satisfies the minimality condition. We will draw a *closer table*, shown below, to test for the consistency condition. A closure table shows the next-state sets for each maximal compatible under each input condition. If the consistency condition is to be met, each next-state set must be contained in one of the maximal compatibles.

	X	
	0	1
(*ABE*)	(*B*)	(*CD*)
(*CD*)	(*E*)	(*A*)

As can be seen, the consistency condition is satisfied by this set of maximal compatibles. Hence, the reduced table will have two states: *P*, corresponding to (*ABE*), and *Q*, corresponding to (*CD*). The reduced table is shown below:

	X	
	0	1
P	*P*/1	*Q*/0
Q	*P*/0	*P*/1

The output is set to 1 or 0, depending on whether any of the next-state transitions corresponding to the maximal compatible as the present state produce a 1 or a 0 output. If all the outputs are don't-cares (i.e., "–"), then the resulting output will also be a don't-care. Thus, the reduced state table generally can also be incompletely specified.

Example 6.15

Consider the incompletely specified circuit described by the state table of Figure 6.44(a). The implication chart is shown in (b). The maximal compatibles are derived in (c), and the maximal incompatibles are derived in (d). Note that the derivation of maximal compatibles (and incompatibles) requires an elaborate comparison process.

Since $NS = 8$ and $NC = 5$, the upper bound on the number of states is 5. The lower bound is 4, since the largest maximal incom-

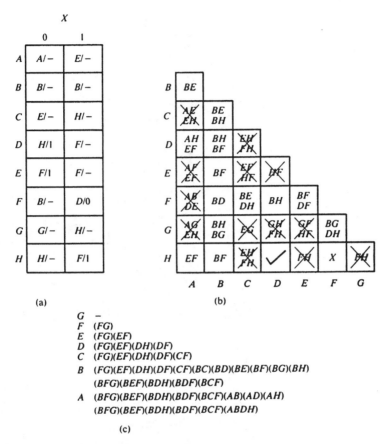

Figure 6.44 State reduction (Example 6.15). (a) State table; (b) implication chart; (c) compatibility classes; (d) incompatibility classes; (e) closure table; (f) reduced table.

patible has four states in it. In order to satisfy the completeness criterion, $(ABDH)$, (BCF), (BEF), and (BFG) must be selected since states A, C, E, and G are each covered, respectively, by these maximal compatibles only.

The closure table is shown in (e). As can be seen, (BEF) makes a transition to (BDF), which is not a selected maximal compatible. To satisfy closure, (BDF) must be included into the chosen set. The reduced table will thus have five states.

The reduced table is shown in (f). Note that the next state of R for $X = 0$ can be any one of the four states Q, R, S, or T, since

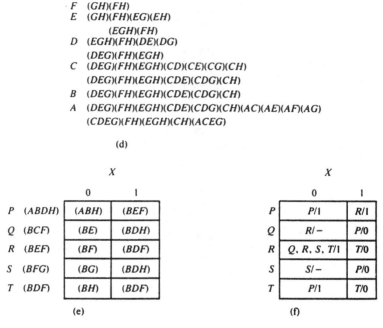

$$
\begin{array}{ll}
G & (GH) \\
F & (GH)(FH) \\
E & (GH)(FH)(EG)(EH) \\
  & (EGH)(FH) \\
D & (EGH)(FH)(DE)(DG) \\
  & (DEG)(FH)(EGH) \\
C & (DEG)(FH)(EGH)(CD)(CE)(CG)(CH) \\
  & (DEG)(FH)(EGH)(CDE)(CDG)(CH) \\
B & (DEG)(FH)(EGH)(CDE)(CDG)(CH) \\
A & (DEG)(FH)(EGH)(CDE)(CDG)(CH)(AC)(AE)(AF)(AG) \\
  & (CDEG)(FH)(EGH)(CH)(ACEG) \\
\end{array}
$$

(d)

	X	
	0	1
P $(ABDH)$	(ABH)	(BEF)
Q (BCF)	(BE)	(BDH)
R (BEF)	(BF)	(BDF)
S (BFG)	(BG)	(BDH)
T (BDF)	(BH)	(BDF)

(e)

	X	
	0	1
P	$P/1$	$R/1$
Q	$R/-$	$P/0$
R	$Q, R, S, T/1$	$T/0$
S	$S/-$	$P/0$
T	$P/1$	$T/0$

(f)

Figure 6.44 *Continued*

(BF)—the next-state set of (BEF) in the closure table—is covered by four of the maximal compatibles. Note also that some of the outputs remain unspecified.

This completes our discussion of state reduction for incompletely specified circuits. The remaining steps in the design (e.g., state assignment, flip-flop input equation derivation) remain the same as in the case of completely specified sequential circuits.

6.11 Summary

The functional and timing characteristics of the four popular types of flip-flops used in synchronous sequential circuits were described in this chapter. The analysis of synchronous sequential circuits by tracing the signals through the circuit diagram is useful for simple circuits. The classical analysis and design procedures introduced here are applicable to circuits of any complexity, although for a circuit with a large number of states, the procedures become unwieldy. These procedures can be programmed to provide automatic tools for design and analysis. It is important to note that in practice, digital circuits tend to have a large number of states, thus making the application of classical design procedures impractical. In such cases, the circuit should be partitioned into smaller subsystems that can be handled by these procedures.

Several other state assignment and state reduction techniques, in addition to the ones described in this chapter, are available in the literature. (See the references listed below.) The advances in IC technology have lessened the importance of these classical minimization techniques. The emphasis now is on fabrication techniques that allow denser integration of devices on the chip.

In addition to the random logic implementation techniques discussed in this chapter, use of programmable logic devices (PLD) has become a preferred method of sequential logic implementation. Chapter 9 provides the details on this implementation mode.

References

Dietmeyer, D. *Logical Design of Digital Systems*. Boston: Allyn and Bacon, 1971.

Ercegovac, M. D., and Lang, T. *Digital Systems and Hardware/Firmware Algorithms*. New York: John Wiley & Sons, 1985.

Hartmanis, J. "On the State Assignment Problem for Sequential Machines I." *IRE Transactions on Electronic Computers*, EC-10 (June 1961): 157–165.

Karp, R. M. "Some Techniques of State Assignment for Synchronous Sequential Machines." *IEEE Transactions on Electronic Computers*, EC-13, No. 5 (October 1964): 507–518.

Kohavi, Z. *Switching and Finite Automata Theory*. New York: McGraw-Hill, 1970.

Mano, M. M. *Digital Design*. Englewood Cliffs, N.J.: Prentice-Hall, 1991.

National Advanced Bipolar Logic Data Book, Santa Clara, Calif.: National Semiconductor, 1995.

Paul, M. C., and Unger, S. H. "Minimizing the Number of States in Incompletely Specified Sequential Switching Functions." *IRE Transactions on Electronic Computers*, EC-8, No. 3 (September 1959): 356–357.

Shiva, S. G., and Nagle, H. T. A series of articles on computer-aided logic design. *Electronic Design*, 22 (October 11, October 25, and November 8, 1974).

Stearns, R. E., and Hartmanis, J. "On the State Assignment Problem for Sequential Machines II." *IRE Transactions on Electronic Computers*, EC-10 (December 1961): 593–603.

Story, J. R., Harrison, H. J., and Reinhard, E. A. "Optimum State assignment for Synchronous Sequential Circuits." *IEEE Transactions on Computers*, C-21, No. 12 (December 1972): 1,365–1,373.

TTL Data Manual. Sunnyvale, Calif.: Signetics, 1986.

Weiner, P., and Smith, E. J. "Optimization of Reduced Dependencies for Synchronous Sequential Machines." *IEEE Transactions on Electronic Computers*, EC-16, No. 6 (December 1967): 835–847.

Problems

6.1   Replace the NOR gates in Figure 6.4 with NAND gates and analyze the circuit for its operation as an SR flip-flop.

6.2   For the special flip-flops shown below, derive the (a) state table; (b) next-state equation; (c) characteristic table; and (d) excitation table:

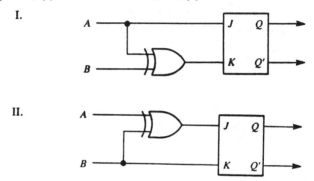

I.

II.

6.3   You are given a *JK* flip-flop. Design the circuitry around it to convert it into a (a) *T* flip-flop; (b) *D* flip-flop; and (c) *SR* flip-flop. (Hint: A flip-flop is a sequential circuit. Start the design with the state table of the flip-flop required. Use a *JK* flip-flop in the design.)

6.4   A *set-dominate* flip-flop is similar to an *SR* flip-flop, except that an input of $S = R = 1$ will result in setting the flip-flop. Draw the state table and derive the next-state equation, characteristic table, and the excitation table for the flip-flop.

6.5   Using the following state table, derive the output sequences for the input sequence shown below, with the starting states as A, B, and C:

$$X = 00100111$$

	X	
	0	1
A	B/1	C/1
B	D/1	D/0
C	B/0	C/1
D	C/0	A/1

6.6   For the circuit shown below:

   (a)   Complete the timing diagram, starting with $Y_1Y_2 = 00$ (at time = 0).
   (b)   Derive excitation tables.
   (c)   Derive the state table. Use

$$Y_1Y_2:\qquad 00 = A,\qquad 01 = B,\qquad 11 = C,\qquad 10 = D$$

Assume that the flip-flops are triggered by the rising edge of the clock.

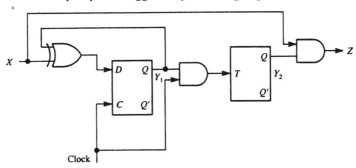

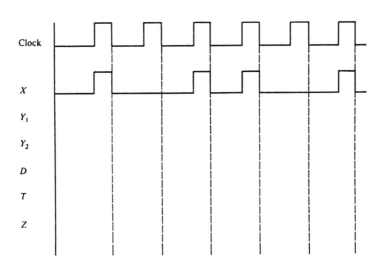

6.7   The circuit shown below gave an output sequence of $Z = 11011111$ for an input sequence $X = 01101010$. What was the starting state?

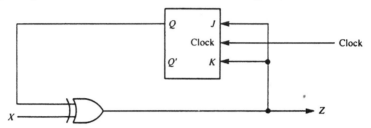

6.8   For the circuit shown below, assume that the flip-flops are triggered by the 1-to-0 edge (negative-edge) transition of the T input.

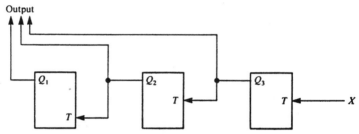

Note that the flip-flops are not clocked. Assume the transitions of X to be as shown below and complete the timing diagram for Q_1, Q_2, and Q_3. (Hint: This is a ripple counter.)

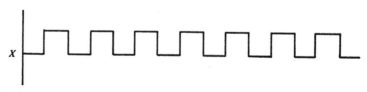

6.9   Draw a circuit diagram for the synchronous sequential circuit described by the following equations:

$$\text{Output: } Z = XY_1Y_2$$
$$\text{Flip-flop input: } J_1 = XY_2 \qquad K_1 = X'$$
$$J_2 = X' \qquad K_2 = X' + Y_1'$$

6.10   (a)   Derive the circuit diagram for a BCD counter that counts from 0 through 9 (and returns to 0) using clocked SR flip-flops. The counter increments only when the input is 1; otherwise, the count remains the same.
       (b)   Compare your design with the TTL BCD counter IC from the IC catalogues.

6.11 Construct a state diagram for a 1101 sequence detector. The sequences may overlap. Typical input and output sequences are shown below:

$$\text{Input } X = 001101101011011$$

$$\text{Output } Z = 000001001000010$$

6.12 Construct state diagrams for sequence detectors that can detect the following sequences:

(a) 11011 (sequences may overlap)
(b) 11011 (overlap not allowed)
(c) 1101 or 1001 (overlap allowed)
(d) 1101 or 1001 (overlap not allowed)

6.13 Derive the state diagram for an odd parity checker. The input arrives on a single input line, X, one bit at a time. The circuit should produce an output of 1 if the number of 1s in the input sequence of four bits is odd. The circuit should reset to the starting state every four bits on the input.

6.14 Design a minimum synchronous sequential circuit to detect the sequence 1001. Sequences may overlap. Use SR flip-flops.

6.15 Design the circuit for a soft drink machine. Each drink costs 30¢. The machine accepts quarters, dimes, and nickels. Assume that a coin sorter that accepts the coins and produces three signals, one corresponding to each of the three coins, is available. Assume that the signals on the output lines of the coin sorter are separated far enough apart to enable other circuits to make a state transition between the pulses on any two of these lines. The sequential circuit should generate signals to release the soft drink bottle and the correct change. The coin sorter releases the appropriate coin when it receives a coin release signal. Use JK flip-flops in the design.

6.16 Reduce the following state table by inspection:

	\multicolumn{2}{c}{X}	
	0	1
A	$B/0$	$E/0$
B	$E/0$	$D/0$
C	$D/1$	$A/0$
D	$C/1$	$E/0$
E	$B/0$	$D/0$

6.17 Reduce the following state tables using (a) equivalence partitioning and (b) an implication chart:

I.

	X	
	0	1
A	D/0	A/1
B	A/1	F/0
C	D/0	C/1
D	D/0	D/1
E	C/1	D/1
F	C/1	G/1
G	G/0	G/1
H	G/1	H/0

II.

	X	
	0	1
A	B/0	C/0
B	E/1	A/0
C	B/0	C/0
D	F/1	C/0
E	C/0	D/0
F	D/1	F/0

6.18 From the following reduced state table, derive a minimum circuit using (a) *SR* flip-flops; (b) *JK* flip-flops; and (c) *T* flip-flops. Use the following assignment: $A = 00$, $B = 01$, and $C = 10$.

	X_1X_2		
	00	01	11
A	A/0	B/0	C/0
B	B/0	C/1	A/0
C	C/0	A/0	B/1

6.19 Examine the operation of the circuit designed in problem 6.18(c), if it starts with a state of 11. Does any of the three allowed input combinations, take it to a valid state? If not, examine and change the usage of don't-cares in the design, to make the circuit transition to a valid state.

6.20 What happens to the circuit designed in problem 6.18, if an input combination of 10 is applied when in states *A*, *B*, or *C*?

6.21 Design a counter that counts in the sequence 0, 1, 4, 3, 6 using *T* flip-flops. Examine its operation starting from the remaining states (2, 5, 7).

6.22 To add two positive numbers, *A* and *B*, *A* can be incremented by 1, *B* times. Design the circuit to perform the addition of two four-bit numbers, *A* and *B*. *A* and *B* registers can each be an up-down counter. Stop incrementing *A* when *B* reaches 0. Assume that *B* is positive.

6.23 Repeat 6.22 for *B*, a negative number, represented in the twos complement system.

6.24 Find an optimal state assingment for each of the following state tables:

I.

	X	
	0	1
A	$B/0$	$E/0$
B	$A/1$	$C/1$
C	$B/0$	$C/1$
D	$C/0$	$E/0$
E	$D/1$	$A/0$

II.

	X	
	0	1
A	$D/0$	$E/0$
B	$C/0$	$E/1$
C	$A/1$	$D/0$
D	$B/1$	$C/1$
E	$A/0$	$D/1$
F	$B/1$	$C/0$

6.25 For each of the following incompletely specified state tables, derive the (a) implication chart; (b) maximal conpatibles; (c) maximal incompatibles; and (d) reduced state table:

I.

	X	
	0	1
A	-/-	$A/0$
B	$B/0$	-/-
C	$E/0$	$C/1$
D	-/-	$C/-$
E	$H/1$	$B/-$
F	$B/0$	-/-
G	-/-	$G/0$
H	$H/1$	-/-

II.

	$X_1 X_2$			
	00	01	11	10
A	$A/0$	$B/-$	$A/0$	$C/-$
B	$B/0$	$D/1$	$C/-$	$A/-$
C	$C/0$	$B/1$	$C/0$	$E/-$
D	$D/0$	$E/0$	$E/-$	-/-
E	$E/0$	-/-	$C/1$	$E/-$

III.

	$X_1 X_2$			
	00	01	11	10
A	-/-	$C/1$	$E/1$	$B/1$
B	$E/0$	-/-	-/-	-/-
C	$F/0$	$F/1$	-/-	-/-
D	-/-	-/-	$B/1$	-/-
E	-/-	$F/0$	$A/0$	$D/1$
F	$C/0$	-/-	$B/0$	$C/1$

7

Popular Sequential Circuits

7.1 Introduction

Sequential circuits are built out of memory elements and combinational logic devices. Flip-flops are the most common memory elements. Any circuit consisting of at least one flip-flop is thus a sequential circuit. In addition to flip-flop ICs, several other MSI and LSI components that are useful in sequential circuit design are available. As in combinational logic design using MSI components, the design of sequential circuits is also first partitioned into as many subsystems as can be directly implemented using off-the-shelf ICs. The classical sequential circuit design procedure discussed in Chapter 6 can be used to implement the remaining subsystems to complete the design. This chapter introduces the popular ICs and illustrates their utility in sequential circuit design. Based on their function, these ICs can be classified into the following categories:

1. Flip-flops
2. Latches
3. Registers (or multiple flip-flops)
4. Shift registers
5. Register files (or multiple registers)
6. Counters
7. Programmable logic sequencers and
8. Memories

Flip-flop ICs usually contain two individual flip-flops that can be used in the implementation of any sequential circuit. (Examples of such ICs were given in Chapter 6.)

Multiple flip-flop ICs contain a set of flip-flops with the common terminals (e.g., Clock, Preset and Clear) of each flip-flop connected to the corresponding terminal of other flip-flops. Such an interconnection of flip-flops is called a *Register*. A register is a storage device that is capable of holding binary information. Because each flip-flop can store one bit, an *n*-bit register made up of *n* flip-flops can store any binary information consisting of *n* bits. In addition to flip-flops, a register typically consists of some combinational logic that controls when and how the binary data enter and leave the register.

While the flip-flops in a register are either of the edge-triggered or the master-slave type, a *latch* is a flip-flop that is level-sensitive. The data can enter a latch throughout the duration of the controlling clock's pulse, whereas the data can enter a flip-flop only at a clock edge. A latch can be replaced by a flip-flop. If a latch is used in the circuit, care must be taken to ensure that the output of the latch is not connected to the inputs of other flip-flops that are triggered by the same clock controlling the latch.

Shift registers are special registers that contain additional control inputs and combinational logic circuitry that enable the data stored in them to be shifted right, left, or in either direction.

Register file ICs contain an array of registers and control signals that enable entering (i.e., writing) data into or retrieving (i.e., reading) data from one of the selected registers. In fact, with some of these devices it is possible to write data into one register while reading data from another.

Examples of counters were given in Chapter 6. A counter is a special register that goes through a predetermined sequence of states upon the application of the input pulse. Various types of counters are available.

The application of programmable logic sequencers (PLSs) in the design of sequential circuits will be illustrated in Chapter 9. Also in Chapter 9, we illustrate the use of one type of memory device, the read-only memory (ROM) in the implementation of combinational logic. In its broad definition, a flip-flop is a 1-bit memory; an *n*-bit register is a memory with one *n*-bit *word*, and a register file with *m* *n*-bit registers is a memory with *m* words, each *n* bits wide. The description of memory devices is deferred to Chapter 10. We will examine various types of registers and counters next.

7.2 Registers

Figure 7.1 shows a four-bit register built out of *D* flip-flops. There are four input lines—IN_1 through IN_4—each of which is connected to the *D*

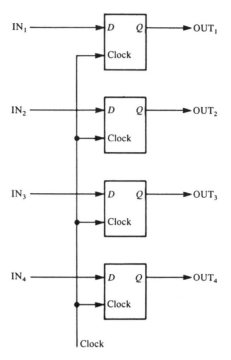

Figure 7.1 A four-bit register.

input of the corresponding flip-flop. At the positive edge of the clock, the data from IN_1–IN_4 enter the register. The clock thus *loads* the register. Because all the bits enter the register simultaneously, the register is said to be loaded in *parallel*. Q outputs of flip-flops are connected to the output lines OUT_1–OUT_4, and hence all four bits of data in the register (i.e., the *contents* of the register) are available (in *parallel*) simultaneously. Thus, this is a parallel-input, parallel-output register. (Shift registers with serial-input and serial-output capabilities are described in the next section.) At each clock pulse, the four-bit data enter the register from the input lines and are available on the output lines until the next clock edge, at which point the new data enter the register.

Figure 7.2 shows an arrangement in which the entry of the data is controlled by the Load control signal. The Load signal must be high for the data to enter the register; if it is low, the clock edge is inhibited from reaching the clock input of the flip-flops. The Clear input must be low for the regular operation of the register. If the Clear input is high when load is high, zeros enter the register, thus clearing it.

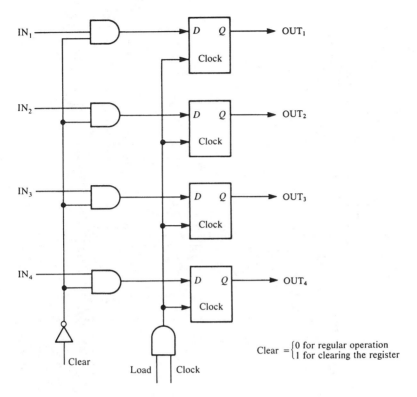

Figure 7.2 A four-bit register with CLEAR and LOAD.

Clearing the register is a very common operation. Usually, the direct Reset input of the flip-flop is used for this purpose, as shown in Figure 7.3. Also note that the Load control is now on the data input lines rather than the clock input, and the outputs of the flip-flops are fed back to the corresponding inputs, in order to preserve the data in the register when the Load signal is low.

Usually data control operations are not performed with the clock, since including combinational logic in the clock circuitry may result in different path delays, thereby resulting in the arrival of a clock edge at different times at the flip-flop inputs (i.e., *clock skewing*), which spoils the synchronous operation of the register. (Clock skewing is discussed further in Section 7.8.2.)

Figure 7.4(a) shows a four-bit register built out of *JK* flip-flops. In this circuit, two AND gates are used for the Load control at each flip-flop input. The common tendency is to eliminate one of these gates by implementing

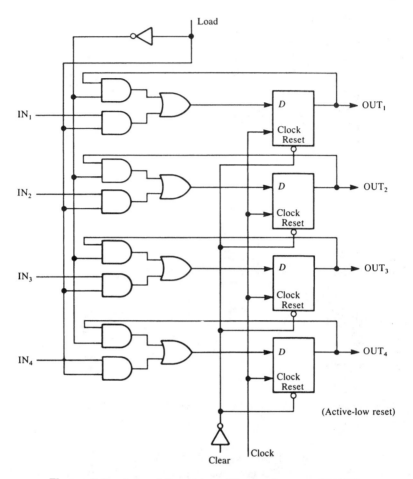

Figure 7.3 A four-bit register with asynchronous CLEAR.

the Load control with one AND gate at the IN line, as shown in (b). However, this arrangement is not desirable, as it results in unequal delays on the J and K input lines because of the extra NOT gate delay on the K line.

Figure 7.5 shows the TTL 74175, a four-bit register. Both the true and complemented outputs are available. The clock (CP) and master-reset (MR) inputs are buffered in order to reduce the load on the master clock and master-clear signals when several such devices are used in the circuit. The state of each D input (one setup-time before the positive clock edge) enters the register. Larger registers can be formed by concatenating the required number of four-bit registers, as shown in Figure 7.6.

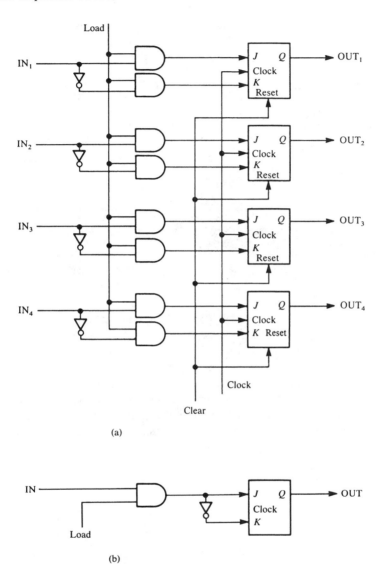

(a)

(b)

Figure 7.4 A four-bit register using *JK* flip-flops. (a) Equal delay in *J* and *K* lines; (b) unequal delay in *J* and *K* lines.

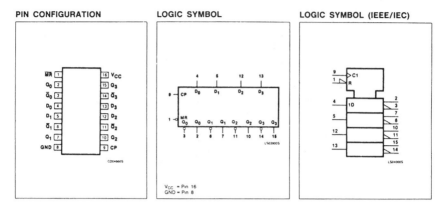

Figure 7.5   A four-bit register (TTL 74175) (Courtesy of Signetics Corporation).

7.3 Latches

Figure 7.7 shows the details of a TTL 74279. This IC contains four independent SR latches. The function table summarizes the operation of each latch. When R is high, the output of the latch will be high if one of the S inputs is low. The output is low when both S inputs are high and R is low. If all the inputs become high simultaneously, the output is indeterminate.

Figure 7.8 compares a latch's operation with that of a flip-flop through a timing diagram consisting of a clock and an input signal. The outputs of an edge-triggered D flip-flop and a latch corresponding to these input signals are also shown. For the latch, the clock is used as the enable input.

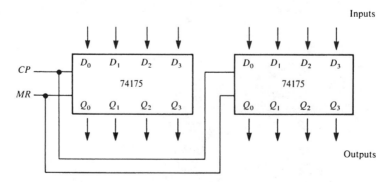

Figure 7.6   An eight-bit register using two 74175s.

TYPE	TYPICAL PROPAGATION DELAY	TYPICAL SUPPLY CURRENT (TOTAL)
74279	13ns	18mA

FUNCTION TABLE

INPUTS			OUTPUT
$\bar{S}_1$	$\bar{S}_2$	$\bar{R}$	Q
L	L	L	h
L	X	H	H
X	L	H	H
H	H	L	L
H	H	H	No change

L = LOW voltage level.
H = HIGH voltage level.
X = Don't care.
h = The output is HIGH as long as $\bar{S}_1$ or $\bar{S}_2$ is LOW. If all inputs go HIGH simultaneously, the output state is indeterminate; otherwise, it follows the truth table.

ORDERING CODE

PACKAGES	COMMERCIAL RANGE V_{CC} = 5V ±5%; T_A = 0°C to +70°C
Plastic DIP	N74279N
Plastic SO-16	N74279D

NOTE:
For information regarding devices processed to Military Specifications, see the Signetics Military Products Data Manual.

INPUT AND OUTPUT LOADING AND FAN-OUT TABLE

PINS	DESCRIPTION	74
All	Inputs	1ul
Q	Output	10ul

NOTE:
A 74 unit load (ul) is 40μA I_{IH} and −1.6mA I_{IL}.

PIN CONFIGURATION	LOGIC SYMBOL	LOGIC SYMBOL (IEEE/IEC)
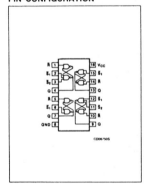	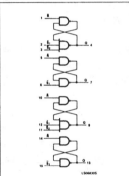	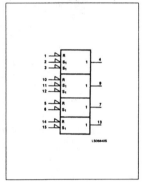

Figure 7.7   A four-bit latch (TTL 74279) (Courtesy of Signetics Corporation).

7.4   Shift Registers

There are two types of shift registers: *static* and *dynamic*. Static shift registers are built with flip-flops, and dynamic shift registers are built using capacitors as storage devices.

7.4.1   Static Shift Registers

Figure 7.9 shows a four-bit shift register built out of D flip-flops. The output of each flip-flop is connected to the D input of the flip-flop to its right. The clock inputs of all the flip-flops are connected together and form the Shift input. At each Shift pulse, the content of d_1 moves to d_2; the content of d_2 moves to d_3, and that of d_3 moves to d_4, simultaneously. That is, the four-

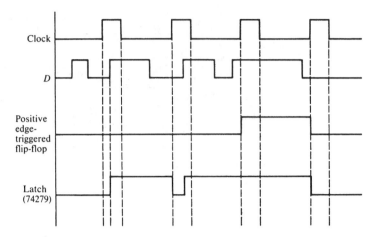

Figure 7.8   Comparison of latch and flip-flop outputs.

bit data are shifted right once for each shift pulse. The data on the INPUT line enters d_1 and the OUTPUT is the data in d_4. Thus, this is a *serial-input, serial-output* device. If the INPUT is set to 1, a 1 is entered into d_1 at each shift pulse. Similarly, a 0 is entered by setting the INPUT to 0. In order to enter the four-bit data into the register, we need four shift pulses. The data are loaded by setting the INPUT to the corresponding bit value at each shift pulse. For example, to load the register with 1101, the INPUT is first set to 1 and a shift pulse is applied, followed by a 0 and shift, 1 and shift, and finally 1 and shift.

The circuit in Figure 7.9 is a unidirectional shift register, since it can shift in only one direction. If the data are loaded from the least significant (rightmost) to the most significant (leftmost) bit, as in the example above, the shift register shifts the data right at each shift pulse. The same circuit

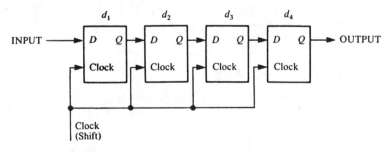

Figure 7.9   A four-bit shift register.

can also be used as a left shift register by loading the data from the most significant bit (MSB) to the least significant bit (LSB)—that is, 1011 in the example. If it is necessary to shift in both directions, a bidirectional shift register (shown in Figure 7.10) can be used. In this circuit, the Direction control input determines the shift direction (left or right).

When an *n*-bit shift register is shifted right *n* times, all the *n* bits of the data are shifted out. The new data in the register depend on the value of the left input at each shift pulse. The shift register can be cleared by setting the left input to 0, or all 1s can be loaded by setting it to 1. Alternatively, new *n*-bit data can be entered while the current data are being shifted out. If the data are to remain intact in the register after the *n*-bit shift (in cases in which the data are needed in the register for further operations), a *circulating shift register*, shown in Figure 7.11, can be used. Here, the contents of the LSB enters the MSB at each shift pulse. Thus, after *n* shifts, the shift register contains the original data.

Loading the *n*-bit shift register in a serial mode consumes *n* shift pulse times. To increase operating speed, a parallel load feature can be added to the shift register. Figure 7.12 shows a three-bit shift register built out of SR flip-flops. The shift register can be loaded in parallel by setting the Mode control to 0. The serial load operation is performed when the Mode is set at 1. Both serial and parallel outputs are available.

Figure 7.13 shows a register with serial input and output, parallel output, circulate (left or right), and shift (left or right) capabilities. The Direc-

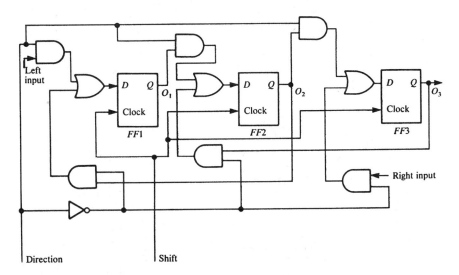

Figure 7.10 Bidirectional shift register.

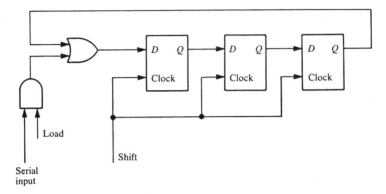

Note: Load = 1 only during data input

Figure 7.11   A circulating shift register.

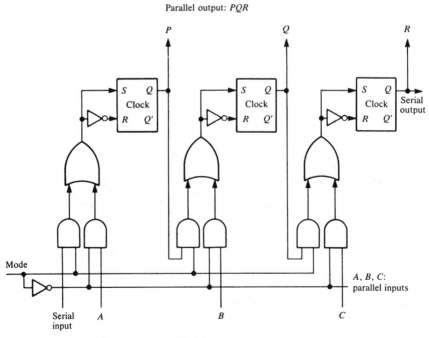

Figure 7.12   A three-bit shift register.

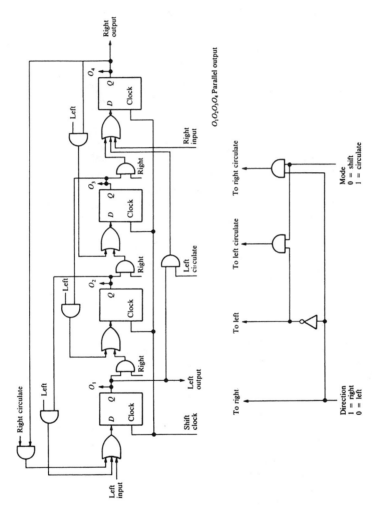

Figure 7.13 A four-bit universal shift register.

tion signal determines the shift direction. Since the Right and Left signals are complements of each other, the shift register can shift in only one direction at any time. The shift register performs the shift or circulate functions depending on the Mode signal. When in shift mode, the data on the left input enter the register if the Direction signal is 1, and the data on the right input enter the register if it is 0. The contents of the register can be output in parallel, through $O_1O_2O_3O_4$, or in serial mode, through O_4.

Figure 7.14 shows the TTL 7495, a four-bit shift register with serial and parallel synchronous operating modes. It has serial data (D_s) and parallel data $(D_0 - D_3)$ inputs and parallel outputs $(Q_0 - Q_3)$. The Mode select input (S) controls the serial or parallel mode of operation. There are two clock inputs (CP_1 and CP_2). The serial (shift right) or parallel data transfers occur synchronously with the high-to-low transition of the selected clock input.

When S is high, CP_2 is enabled. A high-to-low transition on CP_2 loads the data from the parallel inputs into the register. When S is low, CP_1 is enabled, and a high-to-low transition on CP_1 loads the data from the serial input (shift right). To perform a left shift, we would need external connections from Q_3 to D_2, Q_2 to D_1, and Q_1 to D_0, with S set high. In regular operations, S should change states only when both clock inputs are low. However, changing S from high to low when CP_2 is low, or changing S from low to high when CP_1 is low, will not cause any changes on the register outputs.

The following examples illustrate the utility of shift registers in sequential circuit design.

Example 7.1: Serial Adder

The parallel binary adder (discussed in Chapter 6) used $n - 1$ full adders and one half-adder to generate the sum of two n-bit numbers. The addition is done in parallel, although the carry has to propagate from the LSB position to the MSB position. This carry propagation delay determines the speed of the adder. If a slower addition speed can be tolerated by the system, a serial adder can be employed. The serial adder uses one full adder and two shift registers. The bits to be added are brought to full adder inputs, and the sum output of the full adder is shifted into one of the operand registers, while the carry output is stored in a flip-flop and is used in the addition of the next most significant bits. The n bit addition is thus performed in n cycles (i.e., n clock pulse times) through the full adder.

Figure 7.15 shows the serial adder for six-bit operands stored in shift registers A and B. The addition follows the stage-by-stage addition process (as done on paper) from the LSB to the MSB. The carry flip-

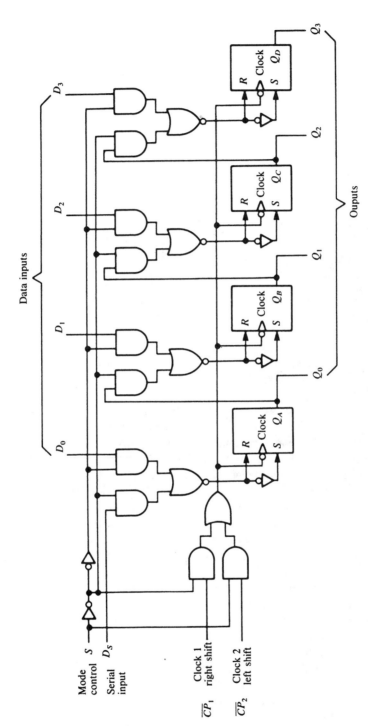

Figure 7.14 A four-bit parallel-access shift register (TTL 7495) (Courtesy of Texas Instruments, Inc.).

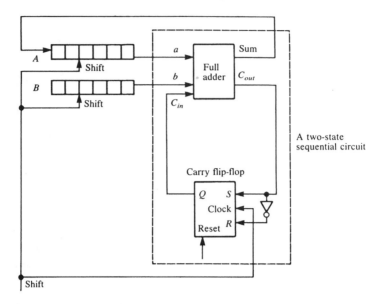

Figure 7.15   Serial adder.

flop is reset at the beginning of addition since the carry into the LSB position is 0. The full adder adds the LSBs of A and B with C_{in} and generates SUM and C_{out}. During the first shift pulse, C_{out} enters the carry flip-flop, SUM enters the MSB of A, and the A and B registers are shifted right, simultaneously. Now the circuit is ready for the addition of the next stage. Six pulses are needed to complete the addition, at the end of which the least significant n bits of the SUM of A and B will be in A, and the $(n + 1)$th bit will be in the carry flip-flop. Operands A and B are lost at the end of the addition process.

If the LSB output of B is connected to its MSB input, B will become a circulating shift register. If that is the case, the contents of B will be unaltered as a result of addition, since the bit pattern in B after the sixth shift pulse will be same as that before addition began. If the value of A also must be preserved, A should be converted into a circulating shift register and the SUM output of the full adder must be fed into a third shift register.

The circuit enclosed by dotted lines in Figure 7.15 is a sequential circuit with one flip-flop and hence two states, two input lines (a and b), and one output line (SUM). C_{in} is the present-state vector and C_{out} is the next-state vector.

Example 7.2: Serial Twos Complementer

A serial twos complementer follows the copy/complement algorithm for twos complementing the contents of a register. The algorithm examines the bits of the register, starting from the LSB. The zero bits from the LSB until and including the first nonzero bit are retained as they are (i.e., copied), and the remaining bits until and including the MSB are complemented, to convert a number into its twos complement. An example is given below.

```
1 0 1 1 0 1 0 | 1 0 0 0     An 11-bit number
↑_____ | ←_____|
Complement    | Copy

0 1 0 0 1 0 1 | 1 0 0 0     Its twos complement
```

There are two distinct operations in this algorithm: copy and complement. Furthermore, the transition from a copying to a complementing mode is brought about by the first nonzero bit. The serial complementer circuit must be a sequential circuit, since the mode of operation at any time depends on whether or not the nonzero bit has occurred. There will be two states and hence one flip-flop in the circuit; one input line on which bits of the number to be complemented are entered, starting with the LSB; and one output line, which is either the copy or the complement of the input. The circuit starts in the copy state and changes to the complement state when the first nonzero bit enters through the input line. At the beginning of each twos complement operation, the circuit must always be set to the copy state.

Figure 7.16 shows the design of this circuit. From the state diagram in (a), the state table in (b) is derived, followed by the output table in (c). The input tables for the *SR* flip-flop shown in (e) are derived from the next-state table (d). The circuit is shown in (f). The circuit is set to the copy state by resetting the flip-flop before each complementation.

7.4.2 Dynamic Shift Registers

The operation of static circuits is based on the switching of transistors between on and off states and the corresponding output voltage levels, determined by the flow of current through load devices. Dynamic circuits use the parasitic capacitance inherent in NMOS circuits. The 1 and 0 states depend on the charge on the capacitor.

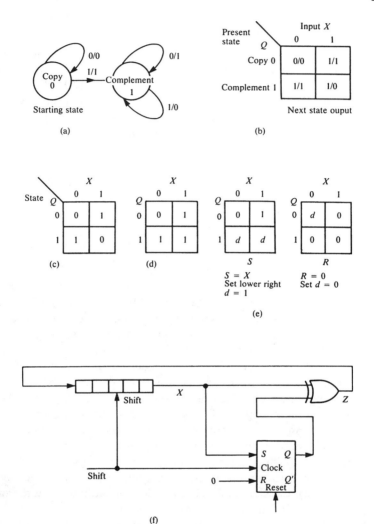

Figure 7.16 Serial twos complementer. (a) State diagram; (b) state table; (c) output table; (d) next-state table; (e) flip-flop input tables; and (f) circuit diagram.

Consider the circuit shown in Figure 7.17. Here, several inverter stages are coupled by *pass transistors* (or *transmission gates*). A pass transistor conducts when its gate voltage is high. Thus, Q_1 connects X to the input of inverter I_1 when ϕ_1 is high. C_1 through C_4 are the input parasitic capacitances of the inverters I_1 through I_4, respectively. A two-phase nonoverlapping

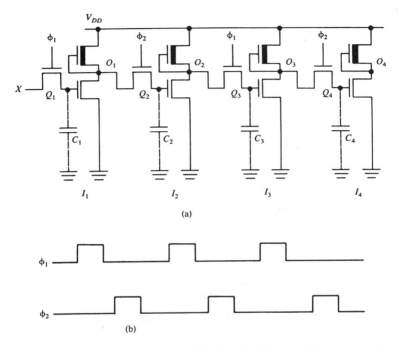

Figure 7.17 Dynamic shift register. (a) Circuit; (b) two-phase nonoverlapping clock.

clock, shown in (b), is required for the operation of this circuit. (A two-phase clock consists of two clock signals; one is the complement of the other. In a two-phase nonoverlapping clock, there is a region in which neither of the clocks will be active.)

During ϕ_1 of the clock, Q_1 and Q_3 conduct, transferring X and O_2 and hence charging C_1 and C_3, respectively. During ϕ_2, Q_2 and Q_4 conduct, charging C_2 and C_4, respectively. The charge on C_2 and C_4 thus corresponds to that on C_1 and C_3, respectively. The circuit thus behaves like a shift register, shifting right once for each clock phase. Note that the data are stored as a charge on the gate capacitance of each inverter in the circuit, which in turn depends on the values of X loaded into the circuit.

The charge on the capacitors decays with time. At each shift the capacitors are charged by the data from the adjacent cell. Thus, so long as the shift frequency is greater than the discharge time of the capacitors, no special refresh (or recharge) circuitry is needed.

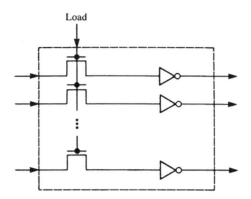

Figure 7.18   A dynamic register.

Figure 7.18 shows a dynamic register built out of pass transistors and inverters similar to those in Figure 7.17. In this circuit, the data must be reloaded periodically in order to retain the charge on the capacitors. A register with selective loading is shown in Figure 7.19. Here, if load is 1, a new data value is loaded during ϕ_1 of the clock. If load is 0, the previous data value is fed back to the input, thereby recharging the capacitor.

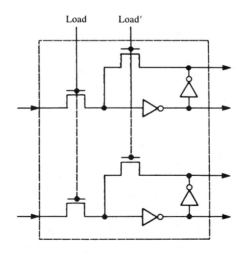

Figure 7.19   Selective loading dynamic register.

The advantage of dynamic circuits is that the currents flow only when the clock is active, and hence there is low power consumption. Furthermore, all the transistors can be of the smallest size since it is unnecessary to retain the ratio between the pull-up and pull-down transistors. Thus, compared with static circuits, dynamic circuits enable the fabrication of denser circuits. The disadvantage of dynamic circuits is that they are slower than static circuits and require a clock even for implementing combinational circuits. Several dynamic MOS shift registers are available off-the-shelf.

7.5 Register Transfer Logic

Manipulating data in most digital systems involves the movement of data between registers. This data movement can be accomplished either in *serial* or in *parallel* mode. Transferring *n*-bit data from one register to the other (each of *n* bits) takes *n* shift pulses if done in serial mode, while it is done in one pulse in parallel mode. A data path that can transfer one bit between the registers is sufficient for serial mode operation. This path is used repeatedly to transfer all *n* bits one at a time. In a parallel transfer scheme, *n* such data paths are needed. Thus, the serial transfer scheme is less expensive (in terms of hardware) and slower than the parallel scheme. Figure 7.20 shows both serial and parallel data transfer schemes.

The transfer circuit shown in Figure 7.20 is usually called a "jam transfer" circuit. The other method of transfer shown in Figure 7.21 is the "clear-and-copy" transfer. This circuit shows a one-bit transfer path from the source register to the destination register. Clear-and-copy transfer requires two clock pulses and is slower than jam transfer, which requires one clock pulse.

All data processing done in the processing unit of a computer is accomplished by one or more register transfer operations. Data in one register often must be transferred into several other registers or a register must receive its inputs from one or more other registers. Figure 7.22 shows two schemes for transferring the contents of either register *A* or register *B* into register *C*. In (a), when the *control* signal "*A* to *C*" is on, the contents of *A* are moved into *C*. When the "*B* to *C*" signal is on, the contents of *B* are moved into *C*. Only one control signal can be active at any time. This can be accomplished by using the true and complement of the same control signal to select one of the two transfer paths. Figure 7.22(b) shows the use of a four-line, two-to-one multiplexer to accomplish the register transfer required in (a). In this figure, the number shown next to a slash mark in-

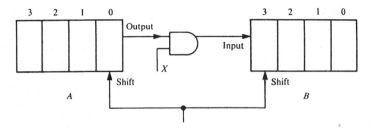

A and B are four-bit shift registers

X: Transfer control

X = 1: Transfer contents of A into B

(a)

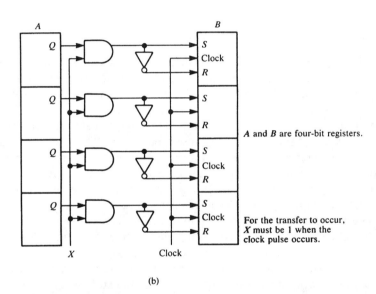

A and B are four-bit registers.

For the transfer to occur, X must be 1 when the clock pulse occurs.

(b)

Figure 7.20 Register transfers (a) serial; (b) parallel.

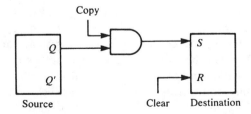

Figure 7.21 Clear-and-copy transfer scheme. The destination register is first cleared and then the source is copied. Two clock pulses are thus required for each transfer operation.

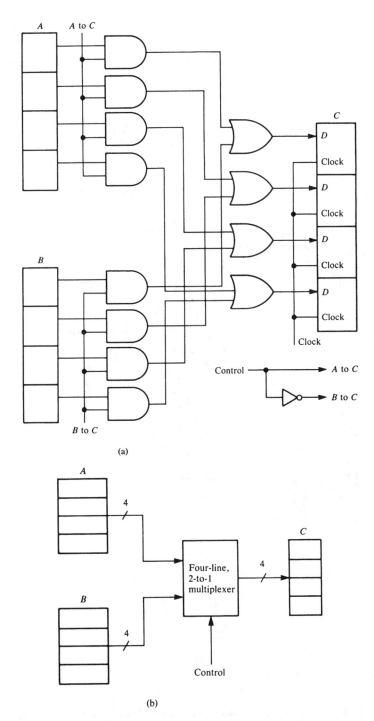

Figure 7.22 Transfer from multiple source registers. (a) Using individual control signals; (b) using a multiplexer.

dicates the number of bits (lines). This is a common convention used to represent multiple bits of any signal in a circuit diagram.

7.5.1 Register Transfer Schemes

When data must be transferred between several registers to complete a processing sequence in a digital computer, two transfer schemes are generally used: (1) point-to-point and (2) bus. In a point-to-point scheme, there will be one transfer path between each of the two registers involved in the data transfer. In a bus scheme, one common path is time-shared for all register transfers.

Point-to-Point Transfer   The hardware required for a point-to-point transfer between three three-bit registers (*A*, *B*, and *C*) is shown in Figure 7.23.

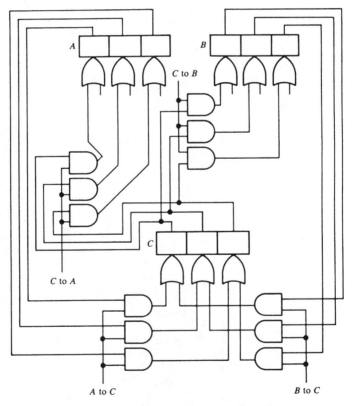

(Paths from *A* to *B* and *B* to *A* are not shown.)

Figure 7.23 Point-to-point transfer. (Paths from *A* to *B* and *B* to *A* are not shown.)

Only a few of the paths are shown. *"A* to *C"* and *"B* to *C"* are control signals used to bring about the data transfer. This scheme allows more than one transfer to be made simultaneously (in parallel), since independent data paths are available. For example, the control signals *"A* to *C"* and *"C* to *B"* can both be enabled at the same time. The disadvantage of the scheme is that the hardware required for the transfer grows rapidly as additional registers are included, since the new register must be connected to other registers through newer data paths. This growth makes the scheme too expensive, and hence a point-to-point scheme is used only when rapid parallel operation is desired.

Bus Transfer Figure 7.24 shows a bus scheme for the transfer of data between three three-bit registers. A bus is a common data path (highway) into which each register feeds the data or from which the data are taken. When a register is feeding the data to the bus, we say the register is on-the-bus; when it is receiving the data from the bus, we say the register is off-the-bus. At any time, only one register can be putting the data on the bus. This requires that the bits in same position in each register be ORed and connected to the corresponding bit (line) of the bus. Figure 7.24(b) shows typical timing for the transfer from *A* to *C*. The control signals *"A* to Bus" and "Bus to *C"* must be 1 simultaneously for the transfer to take place. Several registers can receive the data from the bus simultaneously. But only one register can put the data on the bus at any time. This characteristic makes the bus transfer scheme slower than the point-to-point scheme. But the hardware requirements are considerably less. Furthermore, additional registers can be added onto the bus structure simply by adding two paths: one each from the bus to the register and the register to the bus. The bus scheme is the most commonly used data transfer scheme.

In practice, many registers are connected to a bus. The transfer control circuitry is implemented by using either open collector or tristate gates, as shown in Figure 7.25. In (b), the bus signals are inverted since tristate inverters are used. These signals must be reinverted in paths from the bus to each register (by using NAND gates in place of AND gates).

7.5.2   Register Transfer Languages

Because register transfer is the basic operation in a digital computer, several register transfer notations have evolved over the past decade. These notations are sufficiently complete to describe any digital computer at the register transfer level and hence have come to be known as register transfer languages. Because they are used to describe the hardware structure and be

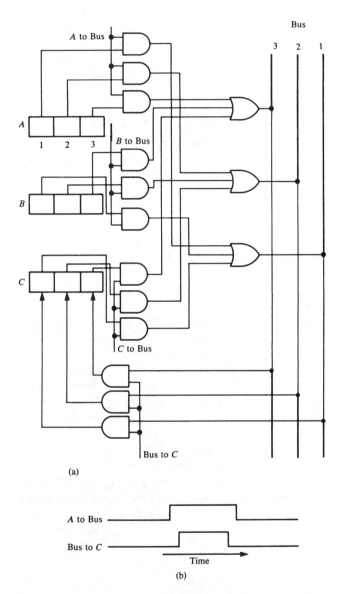

(a)

(b)

Figure 7.24   Bus transfer. (a) Bus structure (Bus to A and Bus to B are not shown); (b) timing.

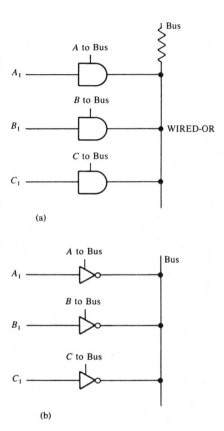

Figure 7.25   Special devices for a Bus interface. (a) Open collector gates; (b) OR using Tristates.

havior of digital systems, they are more generally known as Hardware Description Languages (HDLs).

Here, we will use a primitive HDL to describe register transfers. See the books by Shiva, Hill and Peterson, and Dietmeyer listed in the references at the end of this chapter for further details on HDLs.

Tables 7.1 and 7.2 show the basic operators and constructs of our HDL. The general format of a register transfer is

Destination ← Source.

TABLE 7.1   HDL Operators

Operator	Description	Examples
Left arrow ←	Transfer operator	$Y \leftarrow X$ Contents of register X are transferred to register Y.
Plus +	Addition	$Z \leftarrow X + Y$.
Minus −	Subtraction	$Z \leftarrow X - Y$.
¢	Concatenation	$C \leftarrow A$ ¢ B.
Quote′	Complement	$D \leftarrow A'$.
∧	Logical AND	$C \leftarrow A \wedge B$.
∨	Logical OR	$C \leftarrow A \vee B$.
SHL	Shift left one bit Zero filled on the right	$A \leftarrow$ SHL (A).
SHR	Shift right one bit Copy the MSB on the left	$A \leftarrow$ SHR (A).

TABLE 7.2   HDL Constructs

Construct	Description	Examples	
Capital letter strings	Denote registers	ACC, A, MBR	
Subscripts	Denote a bit or a range of bits of a register	A_0, A_{15}	Single bit
		A_{5-15}	Bits numbered left to right, bits 5 through 15
		A_{5-0}	Bits numbered right to left, bits 0 through 5
Parentheses ( )	Denote a portion of the register (subregister)	IR (ADR)	ADR portion of the register IR—a symbolic notation to address a range of bits
Colon :	Control function delimiter	ADD:	Terminates the control signal definition
Comma ,	Separates register transfers; implies that transfers are simultaneous	$Y \leftarrow X, Q \leftarrow P$.	
Period .	Terminates register transfer statement	$Y \leftarrow X$.	

where Source is a register or an expression consisting of registers and operators and Destination is a register or a concatenation of registers. The number of bits in Source and Destination must be equal.

A transfer controlled by a control signal has the format

Control: transfer.

Multiple transfers controlled by a control signal are indicated by

Control: $transfer_1$, $transfer_2$, . . . , $transfer_n$.

The transfers are simultaneous.

The general format of a conditional register transfer is:

IF condition THEN $transfer_1$ ELSE $transfer_2$.

where

condition is a Boolean expression,

$transfer_1$ occurs if condition is TRUE (or 1),

$transfer_2$ occurs if condition is FALSE (or 0).

The ELSE clause is optional. Thus,

IF condition THEN transfer.

is valid.

A control signal can be associated with a conditional register transfer:

Control: IF condition THEN $transfer_1$
ELSE $transfer_2$.

The following examples illustrate the features of HDLs.

Example 7.3

$B \leftarrow A.$ *A* and *B* must have the same number
                                           of bits.

$C \leftarrow A + B' + 1.$ The twos complement of *B* is added
                                           to *A* and transferred to *C*.

$B \leftarrow A, A \leftarrow B.$ *A* and *B* are exchanged. (*A* and *B*
                                           must be formed using master-slave
                                           flip-flops to accomplish this ex-
                                           change.)

$T_1: B \leftarrow A.$

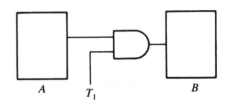

$A \leftarrow C \not{\mathrel{\mkern1mu}} D.$ The total number of bits in *C* and *D*
                                           must be equal to that in *A*.

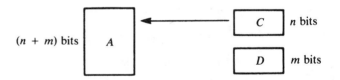

$C \not{\mathrel{\mkern1mu}} D \leftarrow A.$ This is the reverse operation of the
                                           previous line.

$T_1:$ IF *C* THEN $B \leftarrow A$ This is equivalent to
         ELSE $B \leftarrow D.$ $T_1 \wedge C: B \leftarrow A.$ and
                                           $T_1 \wedge C': B \leftarrow D.$

The transfers in Figure 7.22 can be described by the statement

$$\text{IF CONTROL THEN } C \leftarrow A$$
$$\text{ELSE } C \leftarrow B.$$

Example 7.4

The following is the HDL description of a six-bit serial twos complementer circuit:

1   REGISTER, $R(1-6)$, S

2   SWITCH, SW(ON, OFF)

3   REGISTER, COUNT$(2-0)$, T

4   CLOCK, P

5   SW(ON): $T \leftarrow 1$, COUNT $\leftarrow 0$, $S \leftarrow 0$.

6    $T*P$: IF $(S = 0)$ THEN $(S \leftarrow R(6), R \leftarrow R(6) \notin R(1-5))$ ELSE

$$R \leftarrow R(6)' \notin R(1-5),$$
$$\text{IF (COUNT} = 5) \text{ THEN T} \leftarrow 0 \text{ ELSE}$$
$$\text{COUNT} \leftarrow \text{COUNT} + 1.$$

R is the six-bit register containing the data to be twos complemented. Its bits are numbered 1 through 6, left to right. S and T are single-bit registers (flip-flops). COUNT is a three-bit register whose bits are numbered 0 through 2, right to left. SWITCH has an ON and an OFF position. As described by line 5, when SWITCH is ON, T is set to 1 and COUNT and S are initialized to 0. As described by line 6, when T is 1 and a CLOCK pulse occurs, R is circulated right once if $S = 0$. If $S = 1$, R is circulated with bit 6 complemented. Note that $R(6)$ enters S when $S = 0$. Simultaneously, COUNT is incremented by 1 if it has not already reached a value of 5. When COUNT $= 5$, T is set to 0 thereby halting the complementation process.

7.6 Register Files

Register file ICs contain one or more registers and control circuitry to enable such common operations as data selection from more than one source and simultaneous input and output of data to different registers in a file.

For example, TTL 74298 (Figure 7.26) contains a four-bit register with two *ports*. Each port corresponds to a four-bit input, one of which is selected by the S input.

Figure 7.27 shows a 16-bit register file (TTL 74170) organized as four words of four bits each (i.e., a 4 × 4 register file). Simultaneous input (writing) into one word (selected by write address inputs W_A and W_B) and output (reading) from another word (selected by read address inputs R_A and R_B) are possible. In order to write the data into a word, the W_A and W_B inputs are set to one of the four *addresses* (00, 01, 10, or 11), and the input lines $(D_0 - D_3)$ are set with the data. The write enable (WE) is set low. When WE is set high, data and write address inputs are inhibited. To read the contents of one of the registers, the read address is set on read address lines R_A and R_B, and the read enable (RE) is set low. The contents of the addressed word

LOGIC DIAGRAM

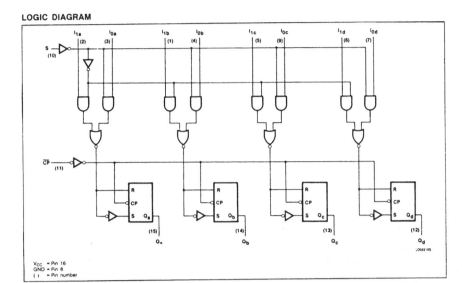

MODE SELECT — FUNCTION TABLE

OPERATING MODE	INPUTS				OUTPUTS
	$\overline{CP}$	S	I_0	I_1	Q_n
Load	↓	l	l	X	L
Source "0"	↓	l	h	X	H
Load	↓	h	X	l	L
Source "1"	↓	h	X	h	H

H = HIGH voltage level
h = HIGH voltage level one set-up time prior to the HIGH-to-LOW clock transition
L = LOW voltage level
l = LOW voltage level one set-up time prior to the HIGH-to-LOW clock transition.
X = Don't care
↓ = HIGH-to-LOW clock transition.

PIN CONFIGURATION   **LOGIC SYMBOL**   **LOGIC SYMBOL (IEEE/IEC)**

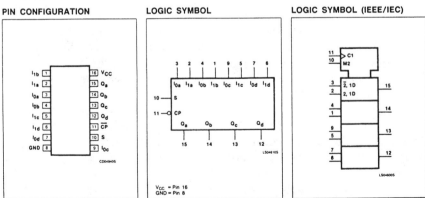

Figure 7.26 A two-port four-bit register (TTL 74298) (Courtesy of Signetics Corporation).

LOGIC DIAGRAM

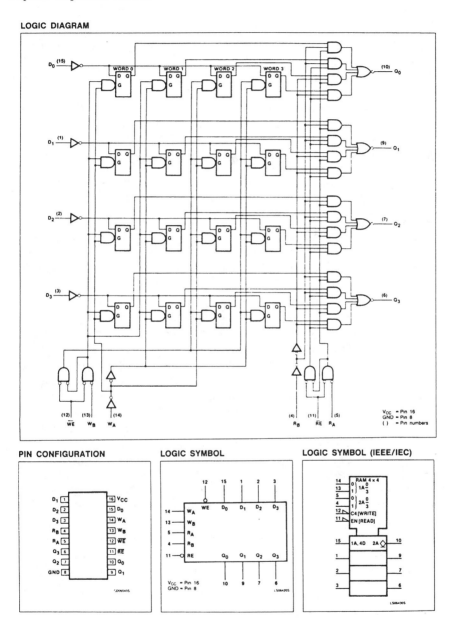

Figure 7.27 A 4 × 4 register file (74170) (Courtesy of Signetics Corporation).

appear at the outputs. The outputs are inhibited and remain high when the RE is high. Simultaneous reading and writing thus are possible, to enable a high-speed operation.

The open collector outputs can be tied together to allow stacking of up to 256 devices, to increase the number of words to 1,024. Word length is increased by driving the enable and address inputs of each device in parallel. Figure 7.28 shows an 8 × 8 register file built out of four 74170s.

A flip-flop is a one-bit register and is usually referred to as a *memory cell*. A register is a *memory word*. A register file is a *memory* with several memory words.

7.7 Counters

Counters are the least complex type of sequential circuits and are used very commonly in digital circuit design, both in arithmetic and control applications. There are two types of counters: *synchronous* and *ripple* (or *asynchronous*). The up-down counter described in Chapter 6 is an example of a synchronous counter. In a synchronous counter, all the flip-flops change state simultaneously in response to the common clock input pulse. The counter follows the predetermined sequence of states in response to the clock input, advancing one state for each pulse. In addition to the clock, there may be other control inputs, such as the up-down control, to determine the direction of the count. In a ripple counter, the state change of one flip-flop triggers the next flip-flop in line. The count pulse forms the triggering input to the first flip-flop, and the state change ripples through the set of flip-flops. Hence, ripple counters are asynchronous circuits.

Ripple counters are more difficult to design than synchronous counters. But if designed properly, they tend to use less hardware than synchronous counters do. In this section, we will examine the design of both types of counters. Descriptions of representative counter ICs available off-the-shelf are also included.

A *modulo*-n counter is a sequential circuit with one input X (usually called clock) and n states. If the states are labeled with the integers 0, 1, 2, ..., $(n - 1)$, as shown in Figure 7.29, the state transitions can be described by

$$S(t + 1) = (S(t) + X) \text{ modulo } n$$

In most counters, the output corresponds to the state of the counter, and hence no other output circuitry is needed. The number of states in counter

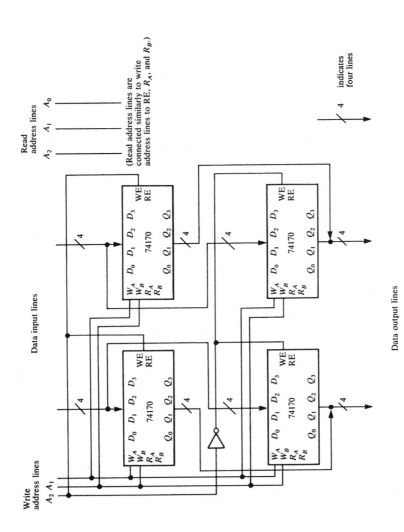

Figure 7.28 An 8 × 8 register file using 74170s.

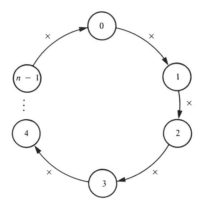

Figure 7.29   Modulo-*n* counter.

modules differs. Counters are classified into the following groups (based on the code in which the output is represented):

1. A *binary* counter counts in the binary numerical sequence. An *n*-bit (or *n*-stage) binary counter counts from 0 through $(2^n - 1)$ and returns to 0. This is also called a modulo-*n* counter.
2. A *decade* counter (or a *decimal* counter) counts from 0 through 9 and returns to 0. Any of the codes described in Chapter 1 can be used to represent the count sequence (e.g., BCD, Excess-3).
3. A *gray-code* counter has 2^n states and *n* output variables. The count sequence follows the gray code, in which only one variable changes at each count.
4. A *ring* counter has *n* states and *n* output variables. The output corresponds to integers from 0 through $(n - 1)$ using a ''1 out of *n*'' code (i.e., only one 1 out of *n* variables). Ring counters are implemented by *n*-bit circulating shift registers.
5. A *twisted-tail* ring counter has $2n$ states and *n* state variables. It is implemented by *n*-bit shift registers in which the complement output of the $(n - 1)$th bit is connected to the input of the 0th bit.

Figure 7.30 shows the two types of ring counters. Table 7.3 shows typical count sequences. Although these are the popular types of counters, other counters can be designed to count in an arbitrary (i.e., nonbinary) sequence.

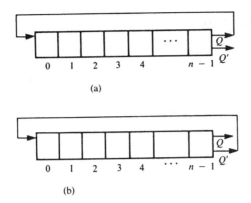

Figure 7.30 Ring counters. (a) Ring counter; (b) twisted-tail ring counter.

Single-mode counters count in either up or down direction. *Multimode* counters, however, have control signals to allow more than one counting sequence. The up-down counter is a simple example of a multimode counter.

7.7.1 Design of Synchronous Counters

The classical design procedure of Chapter 6 can be adopted to design synchronous counters. We will illustrate the design procedure through the following examples and show that binary counters can be designed more directly without resorting to the classical design procedure.

TABLE 7.3 Typical Counting Sequences

Count	Binary	Ring	Tail-ring	Gray
0	000	00000001	0000	000
1	001	00000010	0001	001
2	010	00000100	0011	011
3	011	00001000	0111	010
4	100	00010000	1111	110
5	101	00100000	1110	111
6	110	01000000	1100	101
7	111	10000000	1000	100

Example 7.5: Three-Stage Binary Counter

The state table for a three-bit binary counter is shown below:

Present state			Next state		
A	B	C	A	B	C
0	0	0	0	0	1
0	0	1	0	1	0
0	1	0	0	1	1
0	1	1	1	0	0
1	0	0	1	0	1
1	0	1	1	1	0
1	1	0	1	1	1
1	1	1	0	0	0

Here, the input is not specifically shown. The state change occurs in response to the clock pulse. The binary counter counts from 000 through 111 and resets to the 000 value at the next pulse. The circuit's output is the state (count) itself, and hence no other output circuitry is needed.

Let us use clocked T flip-flops in the design. We will use one flip-flop for each stage. That is, three flip-flops will be needed for the design of this counter. The clock inputs of all three flip-flops are tied together and form the count pulse input to the circuit. The Q outputs of the flip-flops (designated as A, B, and C, respectively) form the present state and the circuit output. To complete the circuit, we need to derive the flip-flop input circuits, each of which is a function of the present state ABC. Using the T flip-flop characteristic table, the flip-flop excitation tables are derived below:

Present state			Next state			Flip-flop inputs		
A	B	C	A	B	C	I_A	I_B	I_C
0	0	0	0	0	1	0	0	1
0	0	1	0	1	0	0	1	1
0	1	0	0	1	1	0	0	1
0	1	1	1	0	0	1	1	1
1	0	0	1	0	1	0	0	1
1	0	1	1	1	0	0	1	1
1	1	0	1	1	1	0	0	1
1	1	1	0	0	0	1	1	1

For example, in the first row, flip-flop A makes a transition from 0 to 0, and hence $T_A = 0$, while flip-flop C makes a transition from 0 to 1, and hence $T_C = 1$. The other input conditions are derived similarly.

The flip-flop input equations can be derived by an inspection of the table above. As can be seen, T_C always equals 1; $T_B = C$, and $T_A = A'BC + ABC = BC$. The circuit diagram is shown below:

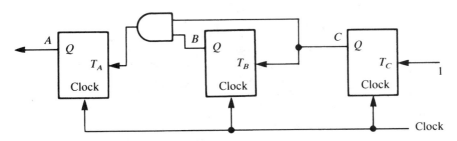

In fact, these input equations can be derived directly by a close inspection of the state table. Note that C changes state every count pulse. Hence, its input T_C must always be set to 1. B complements its state only when C is 1 and a count pulse arrives; hence, $T_B = C$. A complements its state only when both B and C are at 1 and a count pulse arrives. Hence, $T_A = BC$.

This observation can be generalized to a binary counter of any number of stages: the ith flip-flop in an n-stage binary counter complements its state only when all the flip-flops in stages $(i - 1)$ through 0 are at 1, where the counter stages are numbered from $(n - 1)$ to 0 from the MSB to the LSB. Hence, a binary counter can be implemented by using trigger-type (JK or

T) flip-flops in each stage. The trigger input at any stage is derived by the AND of the *Q* outputs of the earlier (less significant) stages.

Figure 7.31 shows a five-stage binary counter built out of *T* flip-flops using the above scheme.

Figure 7.32 illustrates the design of a three-stage binary counter using *SR* flip-flops. The flip-flop input equations can be derived in the usual manner, as shown in the figure, or can be determined from the flip-flop input equations for *T* flip-flop design. Note that in order for an *SR* flip-flop to complement its state, the *S* input must be set to Q' and *R* must be set to *Q*. Since $T_C = 1$, $S_C = 1 \cdot C' = C'$ and $R_C = 1 \cdot C = C$. Similarly, $S_B = B' \cdot C$ and $R_B = B \cdot C$; $S_A = A' \cdot BC$ and $R_A = A \cdot BC$.

The counter circuit in this example is an *up counter.* A *down counter* can be realized by the same circuit by using the Q' outputs of the flip-flops rather than the *Q* outputs.

Example 7.6

Design a three-stage counter for the count sequence shown in the following state diagram:

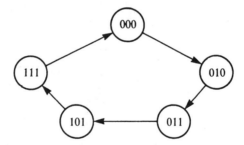

Figure 7.33 shows the design and the circuit diagram using *T* flip-flops. Since only five out of the possible eight states are used in the

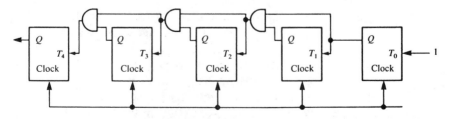

Figure 7.31   Five-stage counter.

Present State	Next State	Flip-flop inputs		
A B C	A B C	$S_A R_A$	$S_B R_B$	$S_C R_C$
0 0 0	0 0 1	0d	0d	10
0 0 1	0 1 0	0d	10	01
0 1 0	0 1 1	0d	d0	10
0 1 1	1 0 0	10	01	01
1 0 0	1 0 1	d0	0d	10
1 0 1	1 1 0	d0	10	01
1 1 0	1 1 1	d0	d0	10
1 1 1	0 0 0	01	01	01

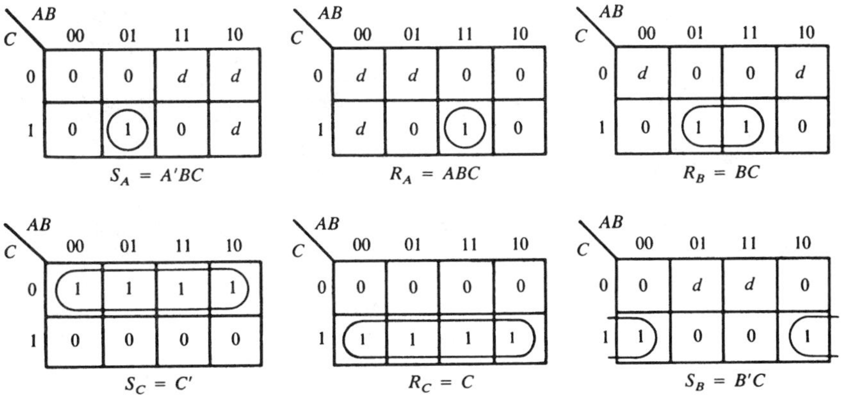

S_A map:

C \ AB	00	01	11	10
0	0	0	d	d
1	0	(1)	0	d

$S_A = A'BC$

R_A map:

C \ AB	00	01	11	10
0	d	d	0	0
1	d	0	(1)	0

$R_A = ABC$

R_B map:

C \ AB	00	01	11	10
0	d	0	0	d
1	0	(1	1)	0

$R_B = BC$

S_C map:

C \ AB	00	01	11	10
0	(1	1	1	1)
1	0	0	0	0

$S_C = C'$

R_C map:

C \ AB	00	01	11	10
0	0	0	0	0
1	(1	1	1	1)

$R_C = C$

S_B map:

C \ AB	00	01	11	10
0	0	d	d	0
1	(1)	0	0	(1)

$S_B = B'C$

Figure 7.32 Three-stage binary counter using *SR* flip-flops.

count sequence, the other three are designated as don't-cares, to optimize the circuit.

7.7.2 Design of Ripple Counters

In ripple counters, the count pulse triggers the first flip-flop, and the flip-flops in subsequent stages are triggered by the transition occurring in other flip-flops. Binary ripple counters are built using complementing flip-flops (*T* or *JK*) and connecting the output of one flip-flop to the clock input of the next. These are the easiest ripple counters to build. If the count sequence required is other than binary, the design of ripple counters becomes cumbersome. Because the state change of flip-flops depends on the order in which the signals change rather than the input values at a clock edge, the classical procedures for synchronous circuit design are not applicable. The design often depends on the designer's ingenuity.

| t | $(t+1)$ | | | |
$A\,B\,C$	$A\,B\,C$	T_A	T_B	T_C
0 0 0	0 1 0	0	1	0
0 0 1	$d\,d\,d$	d	d	d
0 1 0	0 1 1	0	0	1
0 1 1	1 0 1	1	1	0
1 0 0	$d\,d\,d$	d	d	d
1 0 1	1 1 1	0	1	0
1 1 0	$d\,d\,d$	d	d	d
1 1 1	0 0 0	1	1	1

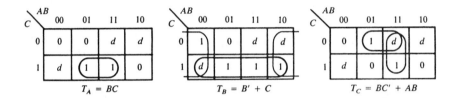

$$T_A = BC \qquad T_B = B' + C \qquad T_C = BC' + AB$$

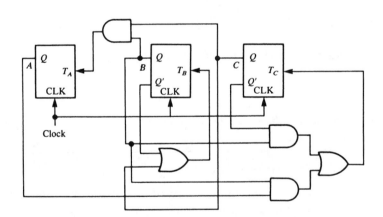

Figure 7.33   An arbitrary-sequence counter.

The following examples illustrate the design procedure.

Example 7.7: Binary Ripple Counter

Figure 7.34 shows a four-stage ripple counter using negative edge-triggered T flip-flops. The flip-flop at the least significant stage (D) is triggered by the count pulse and hence complements at each clock pulse. Flip-flop C is triggered by the 1-to-0 transition of the output of D, which occurs only every other count pulse. Thus, C complements every two clock pulses. Similarly, B complements every four count pulses, and A complements every eight clock pulses. We can trace the circuit for typical transitions to verify its operation.

When the count is $ABCD = 0000$, the count pulse triggers flip-flop D, and hence the count changes to 0001. The other flip-flops will not change, since they cannot see a 1-to-0 transition.

When the count is 0011 and a count pulse arrives, flip-flop D changes to 0; the 1-to-0 transition of D triggers flip-flop C to 0, which in turn provides a 1-to-0 edge to flip-flop B, changing it to 1. Flip-flop A is not affected. Thus, the new count is 0100.

When the count is 1111 and the count pulse arrives, all the flip-flops receive a 1-to-0 edge, thereby changing the count to 0000. Thus, it can be verified that the counting sequence is binary, starting at 0000, going through 1111, and returning to 0000.

The four-stage counter requires in a worst-case scenario four delays through the flip-flops before achieving the new state. If TTL 7473 JK flip-flops were used in the design, the result would be a total delay of $4 \times 40 = 160$ ns (the clock-to-output delay of the 7473 is a maximum of 40 ns). However, a synchronous counter built out of these flip-flops can be triggered every 55 ns, assuming a gate delay of 15 ns for other gates in the circuit. As the number of stages increases, the ripple counter slows down, while the synchronous counter becomes

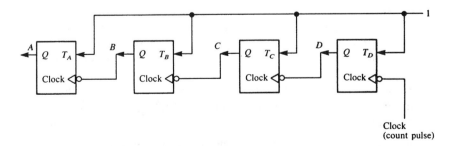

Figure 7.34 A four-stage ripple counter.

more complex. A compromise between these two parameters is needed to determine the appropriate counter for a particular design.

Just as in the case of a synchronous binary counter, a down counter can be realized by using the Q' outputs of the flip-flops. Also note that if each flip-flop in the circuit of Figure 7.34 were replaced by a positive edge-triggered flip-flop, the circuit would behave as a down counter. Verification of the down-counter operation is left as an exercise.

Example 7.8

We will now examine the design of another ripple counter. The following state table illustrates the count sequence:

Present state			Next state		
A	*B*	*C*	*A*	*B*	*C*
0	0	0	0	0	1
0	0	1	0	1	0
0	1	0	0	1	1
0	1	1	1	0	0
1	0	0	1	0	1
1	0	1	0	0	0

The count terminates at 101 and returns to 000. One possible way to realize this counter is to use a three-stage binary ripple counter. Provisions must then be made to reset the counter after the count of 101 and before the next count pulse occurs. One approach to this is shown in Figure 7.35. This circuit is derived by the following observation.

From the state table, it is seen that C complements every count pulse. B complements its state only if C goes from 1 to 0 and if $A = 0$. Flip-flop A sets if C goes from 1 to 0 while $A = 0$ and $B = 1$. This gives the condition for J input of the flip-flop A. Flip-flop A resets if C goes from 1 to 0 while $A = 1$ and $B = 0$. This determines the K input of flip-flop A. The circuit of Figure 7.35(a) satisfies these conditions. The timing diagram shown in (b) verifies the circuit's operation.

As can be seen by this design example, designing ripple counters that count in nonbinary sequences is not straightforward. It is a trial-and-error

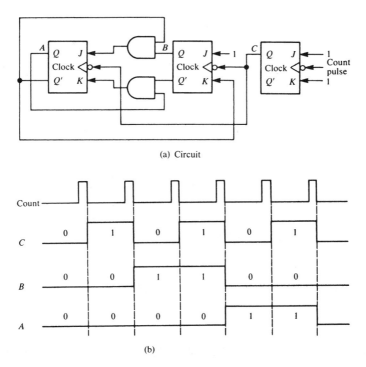

(a) Circuit

(b)

Figure 7.35 Ripple counter (Example 7.8). (a) Circuit; (b) timing diagram.

process that depends entirely on the designer's ability to recognize the appropriate operating conditions.

7.7.3 Divide-by-*N* Circuits

As its name indicates, a circuit that divides the clock frequency by N is called a *divide-by-N* circuit. That is, the frequency of the output of the circuit is $1/N$ of its input clock (or the count pulse).

Consider the timing diagram for the three-stage binary counter shown in Figure 7.36. Here, the B flip-flop changes its state every two count pulses, and hence the frequency of its output will be half that of the clock input to the circuit. That is, the B output *divides* the clock input by 2. Similarly, the A output divides the clock by 4. For each additional stage, the frequency is halved again. Thus, a *divide-by-2^k* circuit can be realized using a binary counter with (k + 1) stages.

Divide-by-*N* circuits where N is not a power of 2 are designed by the classical procedure shown above. In fact, the counter of Example 7.8 is a

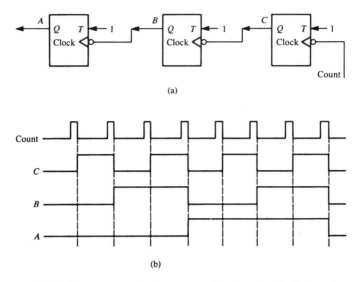

(a)

(b)

Figure 7.36   Three-stage ripple counter. (a) Circuit; (b) timing diagram.

divide-by-six counter since it has six states. A divide-by-three circuit is shown in Example 7.9.

Example 7.9: Divide-by-Three Counter

The following state table illustrates the operation of a divide-by-three counter:

Present state	Next state
00	01
01	10
10	00

Using two *JK* flip-flops, the following input table can be derived:

Present state		Next state		Flip-flop inputs	
A	*B*	*A*	*B*	J_1K_1	J_2K_2
0	0	0	1	$0d$	$1d$
0	1	1	0	$1d$	$d1$
1	0	0	0	$d1$	$0d$

The flip-flop input equations can be derived from the above input table, using the state vector 11 as a don't-care condition, thus contributing the last row of all don't-cares in the input columns above. The flip-flop input equations are

$$J_1 = B \qquad K_1 = 1$$

and

$$J_2 = A' \qquad K_2 = 1$$

The circuit diagram is shown below:

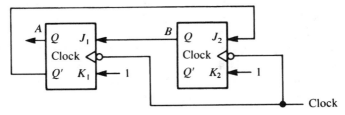

As we can see by the following timing diagram, the circuit performs a divide-by-three operation on the clock.

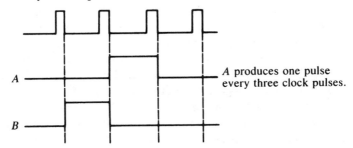

A produces one pulse every three clock pulses.

7.7.4 IC Counters

Both synchronous and ripple counter ICs are available in various configurations. In this section, we will describe representative IC counters.

Figure 7.37 shows the TTL 7493, a four-stage binary ripple counter. The counter has two sections: a divide-by-two section triggered by the count pulse $\overline{CP_0}$ and a divide-by-eight section triggered by $\overline{CP_1}$. Q_0 is connected to CP_1 externally to form a four-stage counter. Thus, the IC can be used as a divide-by-two, a divide-by-eight (three-stage counter), or a divide-by-16 (four-stage) counter. When the reset inputs MR_1 and MR_2 are both high, the counter is reset (to 0000). These asynchronous reset inputs override the clock. State changes are initiated by the high-to-low transition of the clock.

It is important to remember the ripple characteristic of asynchronous counters when designing with them. A common operation with the counters is to decode the count at a particular value to initiate some activity. For example, the circuit of Figure 7.38 shows a decoding count 5. Because the 7493 is a ripple counter, the output (Z) of the AND gate will not be high until the ripple is complete and the counter state is stabilized. During the ripple, spikes (glitches) occur at Z, and hence Z should not be used for clocks or strobes. One way to avoid glitches at Z is to use the clock as an input to the decoding gate, as shown by the dotted line in Figure 7.38. In this scheme, Z will be high only during the clock pulse after the count of 5 has been reached, thus avoiding the spikes but introducing one clock pulse delay into Z. This scheme is useful only when such a delay can be tolerated by the rest of the circuit.

Figure 7.39 shows a four-bit ripple decade counter (TTL 7490). The operation of this IC is similar to that of 7493, except that this is a decade (or BCD) counter. A decade counter counts from LLLL through HLLH and returns to LLLL. This IC consists of four master-slave flip-flops that are internally connected to form a divide-by-two and a divide-by-five section. The function of asynchronous reset (MR_1 and MR_2) and set (MS_1 and MS_2) inputs are defined in the mode selection function table. The counter can be reset to LLLL or set to HLLH by using these inputs.

If the Q_3 output is externally connected to the $\overline{CP_0}$ input and the count pulse is applied to $\overline{CP_1}$, a divide-by-10-square wave is obtained at the output Q. Verification of this mode of operation is left as an exercise.

Figure 7.40 shows the details of a synchronous four-bit binary counter IC (TTL 74163). The state change is triggered by the positive going edge of the clock. The clock input is buffered. There are four operating modes for this IC, as described in the function table. The counter is cleared when a low voltage is applied to the synchronous master reset ($\overline{MR}$) input, one setup time before the positive going edge of the clock. When it is brought

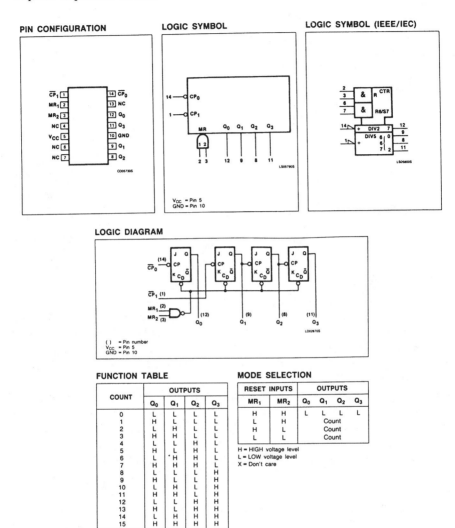

Figure 7.37 A four-stage binary counter (TTL 7493) (Courtesy of Signetics Corporation).

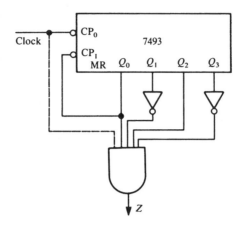

Figure 7.38   Decoding count 0101 on a 7493.

low one setup time before the positive edge of the clock, the parallel enable ($\overline{PE}$) input loads the four-bit data from the D inputs into the counter. Thus, the count can be started at any value. For the counter to count, the count enable inputs (CEP and CET) must both be high one setup time before the positive clock edge. The counter will be in a hold (do-nothing) mode when the CEP or the CET input is low.

Because of the synchronous reset feature, the counter can be reset after any maximum count using only one NAND gate. Figure 7.41 shows an arrangement that clears the counter after the count of 6. That is, the counting sequence will be from 0 through 6 and back to 0.

The counter can be started at any count by loading the appropriate data into the counter. For example, the circuit in Figure 7.42 loads a 1100 into the counter after a count of 1111. When the counter reaches 1111, the carry output TC is 1, which sets the PE to low, thereby loading the 1100 value from the D inputs. The count then continues from 1100 through 1111.

Because of the TC output, 74163s can be cascaded to form counters of larger numbers of bits. Figure 7.43 shows a multiple-bit counter built out of 74163s. When the least significant counter reaches 1111, the TC will be 1, enabling the other IC. At the next pulse, the least significant IC goes to 0000 and the other IC counts up from 0000.

More elaborate counting schemes can be designed using the clear, load, carry, and data inputs of the 74163.

TTL 74160 through 74163 are all four-bit synchronous counter ICs with the same pin-out as 74163. The 74160 and 74162 are decade counters, and the 74161 and 74163 are binary counters. The 74160 and 74161 have

PIN CONFIGURATION

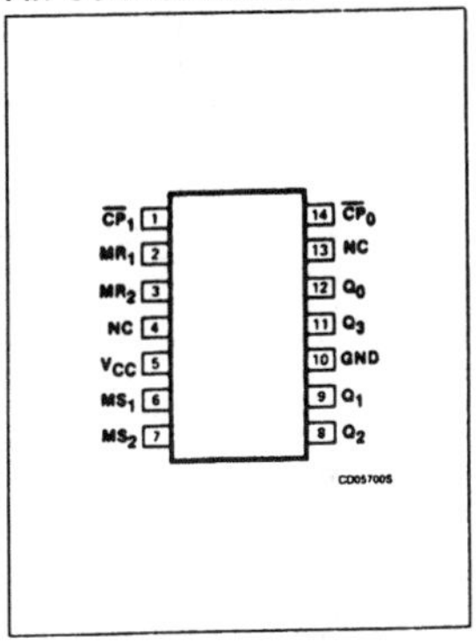

LOGIC SYMBOL

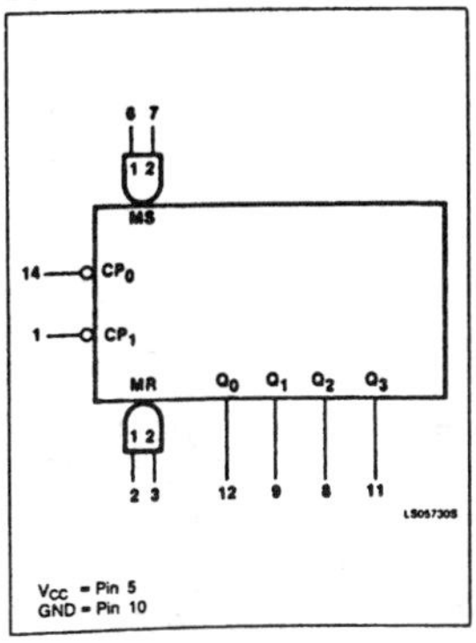

LOGIC SYMBOL (IEEE/IEC)

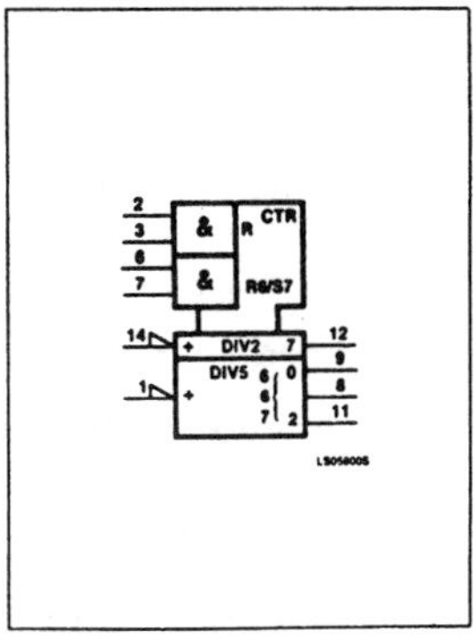

LOGIC DIAGRAM

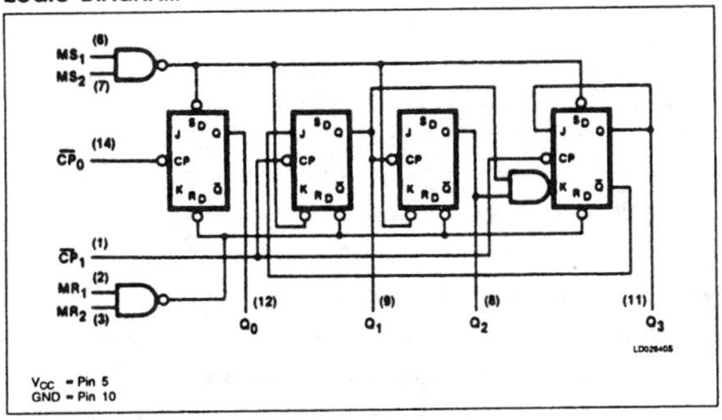

ABSOLUTE MAXIMUM RATINGS (Over operating free-air temperature range unless otherwise noted.)

PARAMETER		74	74LS	UNIT
V_{CC}	Supply voltage	7.0	7.0	V
V_{IN}	Input voltage	−0.5 to +5.5	−0.5 to +7.0	V
I_{IN}	Input current	−30 to +5	−30 to +1	mA
V_{OUT}	Voltage applied to output in HIGH output state	−0.5 to $+V_{CC}$	−0.5 to $+V_{CC}$	V
T_A	Operating free-air temperature range	0 to 70	0 to 70	°C

NOTE:
V_{IN} is limited to +5.5V on $\overline{CP}_0$ and $\overline{CP}_1$ inputs on the 74LS90 only.

MODE SELECTION — FUNCTION TABLE

RESET/SET INPUTS				OUTPUTS			
MR_1	MR_2	MS_1	MS_2	Q_0	Q_1	Q_2	Q_3
H	H	L	X	L	L	L	L
H	H	X	L	L	L	L	L
X	X	H	H	H	L	L	H
L	X	L	X	Count			
X	L	X	L	Count			
L	X	X	L	Count			
H	L	L	X	Count			

H = HIGH voltage level
L = LOW voltage level
X = Don't care

BCD COUNT SEQUENCE — FUNCTION TABLE

COUNT	OUTPUTS			
	Q_0	Q_1	Q_2	Q_3
0	L	L	L	L
1	H	L	L	L
2	L	H	L	L
3	H	H	L	L
4	L	L	H	L
5	H	L	H	L
6	L	H	H	L
7	H	H	H	L
8	L	L	L	H
9	H	L	L	H

NOTE:
Output Q_0 connected to input $\overline{CP}_1$.

Figure 7.39 A four-bit ripple decode counter (TTL 7490) (Courtesy of Signetics Corporation).

PIN CONFIGURATION

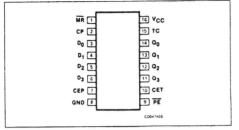

C0047405

LOGIC SYMBOL

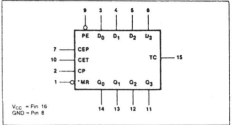

V_{CC} = Pin 16
GND = Pin 8

MODE SELECT — FUNCTION TABLE, '162, '163

OPERATING MODE	INPUTS						OUTPUTS	
	$\overline{MR}$	CP	CEP	CET	$\overline{PE}$	D_n	Q_n	TC
Reset (clear)	l	↑	X	X	X	X	L	L
Parallel load	$h^{(f)}$	↑	X	X	l	l	L	L
	$h^{(f)}$	↑	X	X	l	h	H	(d)
Count	$h^{(f)}$	↑	h	h	$h^{(f)}$	X	count	(d)
Hold (do nothing)	$h^{(f)}$	X	$l^{(e)}$	X	$h^{(f)}$	X	q_n	(d)
	$h^{(f)}$	X	X	$l^{(e)}$	$h^{(f)}$	X	q_n	L

H = HIGH voltage level steady state.
L = LOW voltage level steady state.
h = HIGH voltage level one set-up time prior to the LOW-to-HIGH clock transition.
l = LOW voltage level one set-up time prior to the LOW-to-HIGH clock transition.
X = Don't care.
q = Lower case letters indicate the state of the referenced output prior to the LOW-to-HIGH clock transition.
↑ = LOW-to-HIGH clock transition.

NOTES:
(a) The TC output is HIGH when CET is HIGH and the counter is at Terminal Count (HHHH for '161 and HLLH for '160).
(b) The HIGH-to-LOW transition of CEP or CET on the 74161 and 74160 should only occur while CP is HIGH for conventional operation.
(c) The LOW-to-HIGH transition of $\overline{PE}$ on the 74161 and 74160 should only occur while CP is HIGH for conventional operation.
(d) The TC output is HIGH when CET is HIGH and the counter is at Terminal Count (HLLH for '162 and HHHH for '163).
(e) The HIGH-to-LOW transition of CEP or CET on the 74163 should only occur while CP is HIGH for conventional operation.
(f) The LOW-to-HIGH transition of $\overline{PE}$ or $\overline{MR}$ on the 74163 should only occur while CP is HIGH for conventional operation.

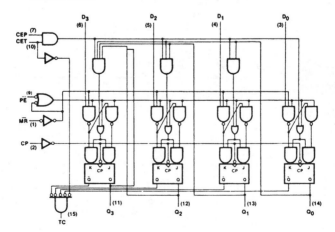

Figure 7.40 A four-bit synchronous binary counter (TTL 74163) (Courtesy of Signetics Corporation).

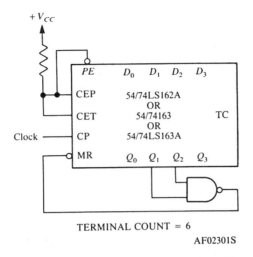

TERMINAL COUNT = 6

AF02301S

Figure 7.41 TTL 74163 with a terminal count of 6 (Courtesy of Signetics Corporation).

an asynchronous reset feature, and the 74162 and 74163 have a synchronous reset feature. Other types of counters are also available, such as up-down or bidirectional counters (74168, 74169, MC14510) and up-down counters with three-state output (74568, 74569). For further details on these ICs see the manufacturers' manuals.

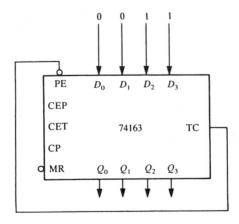

Figure 7.42 TTL 74163 arranged to count from 1100 to 1111.

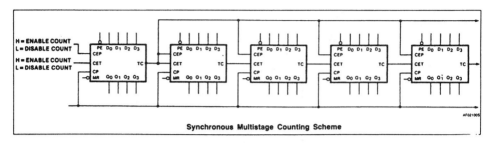

Figure 7.43   Multistage counter using 74163s (Courtesy of Signetics Corporation).

7.8   Designing with ICs

As illustrated by the examples shown above, it is not always necessary to follow the classical design procedure to design sequential circuits. In fact, the number of states in a practical circuit can become so large that the classical design procedure becomes impractical. In such cases, the circuit functions are partitioned into those that can be implemented using off-the-shelf ICs. Therefore, a thorough familiarity with the available ICs is necessary. We will now examine two examples of sequential circuit design using ICs.

Example 7.10: Divide-by-128 Circuit

A divide-by-128 circuit can be built using a cascade of seven flip-flops (by the method shown in Section 7.7), since $128 = 2^7$. Using dual flip-flop ICs, such an implementation requires four ICs. An alternative design scheme would be to use two 7493s. Since $128 = 16 \times 8$, one 7493 is configured as a divide-by-16 circuit and the other as a divide-by-eight circuit. The two ICs are cascaded to form the divide-by-128 circuit, as shown in Figure 7.44.

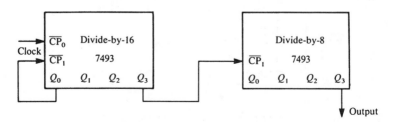

Figure 7.44   Divide-by-128 circuit.

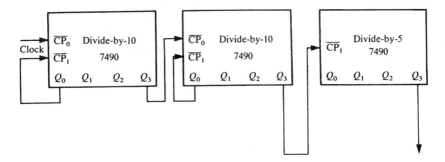

Figure 7.45 Divide-by-500 circuit.

Example 7.11: Divide-by-500 Circuit

Figure 7.45 shows a divide-by-500 circuit built out of three 7490s. The first two ICs are configured as divide-by-10 circuits, and the last one is configured as a divide-by-five circuit. The cascade of the three ICs thus forms a divide-by-500 ($10 \times 10 \times 5$) circuit.

7.8.1 Generation of Timing Signals

When synchronous control units are used to control the operation of a digital system, each event is controlled (or brought about) by a clock pulse. A sequence of events is invoked by a sequence of clock pulses, each pulse activating an event. For example, Figure 7.46 shows a sequence of eight clock pulses, each on a different line but spaced equally apart in time. The clock generator circuit is activated by the master clock and produces the sequence of required clock pulses.

Two approaches, shown in Figure 7.47, can be used to design such timing circuits. In (a), the timing circuit is built out of a three-bit binary counter and a three-to-eight decoder. The master clock is used to increment

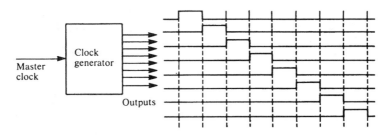

Figure 7.46 A timing-signal generator.

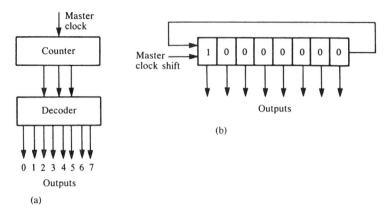

Figure 7.47 Methods of timing signal generation. (a) Counter-decoder method; (b) circulating shift register.

the counter. At each count, only one of the decoder outputs is active, thus producing the required sequence of eight pulses. In (b), an eight-bit circulating shift register is used. The shift register is first initialized to 10000000, and at each shift pulse the 1 shifts from bit to bit, thus producing the sequence of pulses required.

Figure 7.48 shows the counter-decoder circuit using the 74163 counter and a 74138 decoder. The total delay through the circuit is 67 ns (the sum of the counter delay, which is a maximum of 29 ns, and the decoder delay, which is a maximum of 38 ns). Hence, the output pulse width will be a minimum of 67 ns, and the maximum master clock frequency is $(1/67) \times 10^6$ Hz.

Figure 7.49 shows a timing circuit built out of a 74164 shift register. The worst-case delay of the shift register (clock to output) is 37 ns and that of the 7430 is 11 ns; hence, the output pulse width will be a minimum of $37 + 11 = 48$ ns, and the master clock frequency is limited to $1/48 \times 10^6$ Hz. (Draw the timing diagram for this circuit and compare it with that of Figure 7.46.)

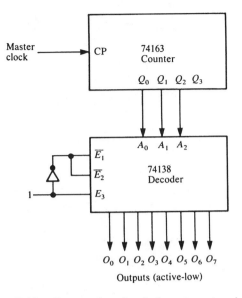

Figure 7.48   Counter-decoder timing generator circuit.

Example 7.12

A circuit is required to produce a sequence of four pulses on four lines $(P, R, S,$ and $T)$, with the following timing:

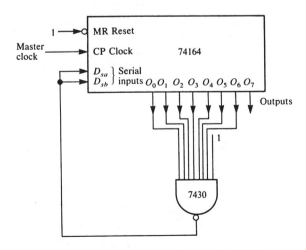

Figure 7.49   Timing circuit using a TTL 74164.

P: from 15 to 30 μs

R: from 20 to 40 μs

S: from 30 to 55 μs

T: from 45 to 65 μs

The sequence must repeat every 65 μs.

Figure 7.50 shows a circuit that generates the above pulses using two 74164 shift registers. In this circuit, a 1 is shifted into the register at the beginning of the cycle. As the register is shifted, the 1 is propagated through the shift register outputs. When the sequence of pulses is completed, the shift register is cleared to start a new sequence.

Since these pulse widths are multiples of 5 μs, a 200 kHz clock is used to shift the register. Because the serial inputs are set to high, a 1 is always shifted into the register. The 1 appears at the A_1 output at 5 μs, at the B_1 output at 10 μs, and so on, until it appears on the E_2 output at 65 μs. Since E_2 is inverted and connected to the CLEAR input, the shift registers are cleared at 65 μs.

In order to realize signal P, note the C_1 remains 0 from 0 μs and goes to 1 at 15 μs, and F_1 goes to 1 at 30 μs, which are the limits of the P signal. By EXCLUSIVE-ORing C_1 and F_1, the P line stays at 1 only during the 15-to-30-μs interval. The other three signals are derived similarly.

7.8.2 Clock Skewing

In order to maintain synchronous operation, all the flip-flops involved in the register transfer or state change must be clocked simultaneously. But if the paths from the clock source to flip-flop inputs contain some other circuitry, clock skewing might occur as a result of differing delays introduced by those circuits. Clock skewing may not always be hazardous, but it can result in the improper operation of the circuit, especialy when there are shift registers in the circuit.

Consider the circuit of Figure 7.51(a). As shown by the timing diagram in (b), flip-flop 2 receives the clock at t_2, while flip-flop 1 receives it at t_1. Although both the flip-flops will be in their correct states after t_2, during the transition time between t_1 and t_3 a glitch may appear in the sequential circuit's output. This glitch may not be a problem if the circuit's output is also gated by the clock.

Consider the two-bit shift register shown in (c). Now, flip-flop 1 changes its state at t_1, to that on I, and at t_2 this new state is transferred to flip-flop 2. Thus, the state of flip-flop 1 before the occurrence of the clock pulse is lost. In this case, the delays in clock paths must be equalized. This is done by including sufficient additional gates into the path of clock 1.

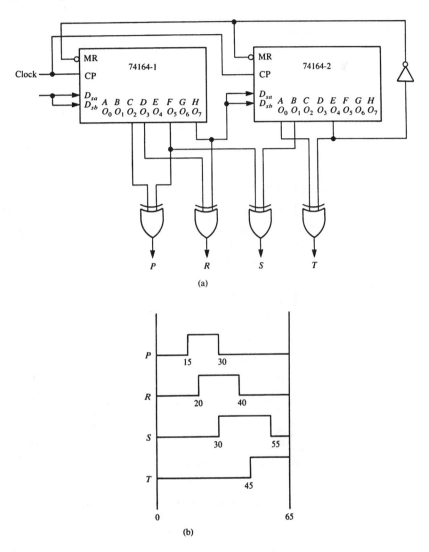

Figure 7.50 A timing signal generator circuit. (a) Circuit; (b) timing diagram.

Example 7.13: Parallel-to-Serial Data Converter

It is required to design a parallel-to-serial data converter that accepts four-bit data in parallel and produces as its output a serial bit stream of the data input into it. The input consists of the sign bit (a_0) and three magnitude bits ($a_1 a_2 a_3$). The serial device expects to receive the

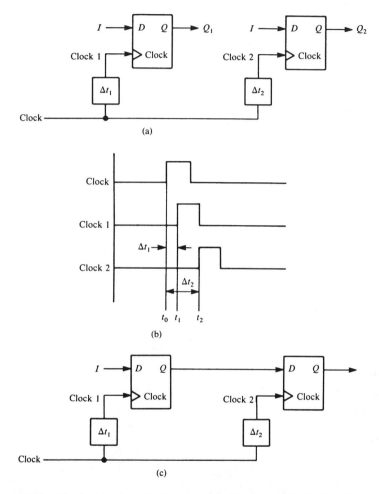

Figure 7.51 Clock skewing. (a) Independent flip-flops; (b) timing diagram; (c) shift register.

sign bit a_0 first, followed by the three magnitude bits in the order a_3, a_2, a_1, as shown in Figure 7.52(a).

Note that the output bit pattern can be obtained by circulating the input data right three times and then shifting right one bit at a time. Performing this requires a four-bit shift register that can be loaded in parallel and shifted right.

The TTL 7495 shown in Figure 7.14 is a four-bit shift register. When the mode select input (S) is 1, the shift register is loaded in parallel. When S is 0, the serial input and right shift are active. The Q_3 output must be connected to the serial input (D_s) in order to circulate the data.

Figure 7.52(b) shows the operation of the circuit being designed. The circuit operation requires eight steps, designated 0 through 7. Two idle steps, 8 and 9, are shown since a decade counter (7490) is available to count from 0 through 9.

The circuit diagram is shown in (c). The output of the decade counter (7490) is decoded by a BCD-to-decimal decoder (7442). Because the outputs of the 7442 are low-active, its output 0 will have a low value when it is active at step 0 and a high value during other times. Therefore, it can be used as the mode select signal for 7495. The CP_1 and CP_2 inputs of the 7495 are tied together and form the shift clock input. Output 3 of the 7442 is used to alert the serial device for data acceptance, and output 8 indicates the idle state.

Note that the 7490 and 7442 introduce a delay into the clock line, which is also used as the shift clock for the 7495. In order to compensate for this delay, two NAND gates are inserted between the clock and the 7495. The computation of the exact delay and the corresponding compensation is left as an exercise.

7.9 Design of Control Circuits

The function of a digital circuit is to transform the data input into the desired output form. In this transformation, the circuit uses various logic elements such as gates, flip-flops, registers, counters, etc., which constitute the "processing" part of the circuit. This processing part is guided by the "controlling" part of the circuit composed of signals such as load, clear, count, etc. These control signals need to be generated by a control circuit, in appropriate order, to make the processing part of the circuit perform its function properly.

Figure 7.53 shows the general model of a digital system consisting of a controller and a processor. The processor receives the data from the input devices or the memory and converts them into the required form and outputs them to either the output devices or the memory. The controller receives the external control signals such as start, clear, and add, and produces a sequence of control signals for the processor. This sequence of control signals directs the processor through the sequence of operations needed to complete the processing function. In addition to the external inputs, the controller also receives the status signals from the processor. These signals provide such status information as overflow and parity error, among others.

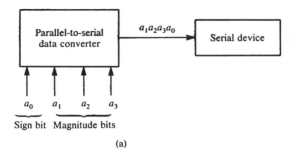

(a)

Count	Event	Action	Mode
0	Load the register (Parallel)	Parallel in	1
1	↑	Shift-circular	0
2	Circulate	Shift-circular	0
3	↓	Shift-circular	0
4	↑	Shift right	0
5	Serial output	Shift right	0
6	↓	Shift right	0
7		Shift right	0
8	Idle	Idle	d
9	Idle	Idle	d

(b)

Figure 7.52 A parallel-to-serial data converter. (a) Requirements; (b) operation of circuit design; (c) circuit.

The controller is a sequential circuit whose internal states dictate the control signals (commands) for the processor. Its inputs are the external control inputs and the status signals from the processor. These inputs take the controller from the present state to the next state. At each state, the controller produces a set of control signals as its outputs. The state transitions and the corresponding control signals are defined by means of a *hardware algorithm*, which is a procedure for implementing the processing function using the hardware available in the processing unit.

For simple digital systems, the controller can be treated as a classical sequential circuit design problem starting from a state diagram derived from the hardware algorithm. As the system becomes more complex, the number of states required in the controller increases, thereby making the classical technique cumbersome. In such cases an Algorithmic State Machine (ASM)

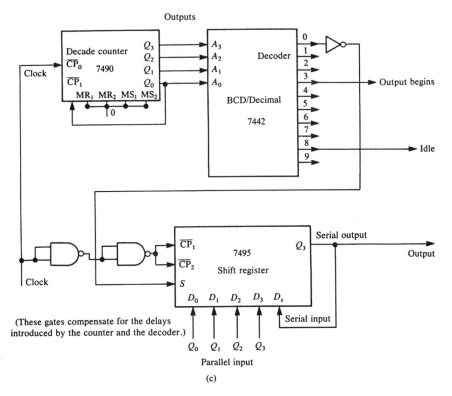

Figure 7.52 Continued

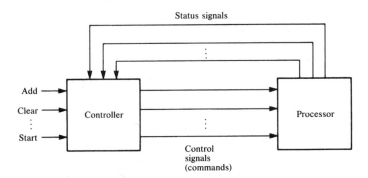

Figure 7.53 Controller/processor interaction.

chart (or simply the SM chart) can be used to describe the operation of the controller. *State Machine* is another term for a sequential circuit.

The concept of hardware algorithm and the corresponding operation of the controller are illustrated further by the following example.

Example 7.14

Consider the digital computer described in Chapter 1 (Figure 1.5). In order to perform the addition of two integers located at A and B in the memory, leaving the result at B, the controller initiates the following sequence of actions:

1. Fetch the ADD A to B instruction from the memory.
2. Decode the instruction to derive the addresses of A and B and to derive the operation (i.e., ADD) to be performed on these operands.
3. Execute the sequence of operations to complete the addition.

Thus, the controller will have three internal states, and the state sequence is shown by the following diagram:

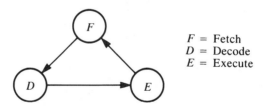

F = Fetch
D = Decode
E = Execute

Let us now expand the description of the controller operation. Two other registers that are commonly found in a digital computer are the program counter (PC) and the instruction register (IR). The digital computer retrieves instructions in the program one at a time, in the sequence in which they are stored in the memory. The PC always contains the address of the instruction to be retrieved next. This means that once an instruction is fetched by the control unit, the address in the PC is incremented to point to the next instruction in sequence. The control unit fetches the instruction into the IR, where it is decoded by the circuitry around the IR during the decode state of the controller. Let us also include a memory address register (MAR) and a memory buffer register (MBR) into the system. The digital system is shown in Figure 7.54.

The fetch state can now be described by the following sequence of register transfer operations:

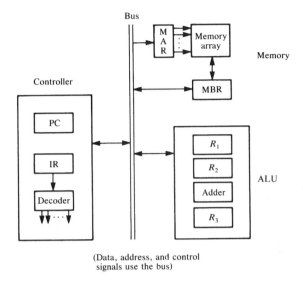

Bus

Memory

Controller

ALU

(Data, address, and control
signals use the bus)

Figure 7.54   Digital computer system.

MAR ← PC	The instruction address is transferred to MAR.
READ memory	The instruction is brought into MBR.
IR ← MBR	The instruction is transferred to the IR.
PC ← PC + 1	The PC is incremented to point to the next instruction.

This sequence is the hardware algorithm for the fetch phase. The controller generates one control signal corresponding to each step. The control signal is used to gate the register transfers. For instance, the control signal MAR ← PC is used to bring about the transfer of data from the PC to the MAR, as shown below:

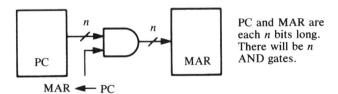

MAR ← PC

PC and MAR are
each n bits long.
There will be n
AND gates.

The READ memory control signals enables the memory read input. IR ← MBR enables another data path from the MBR to the IR, and PC $\leftarrow$ PC $+$ 1 is used to increment the PC, either through an adder or by realizing the PC as a counter.

Let us assume that the IR is subdivided into three fields: OPERATION, OPERAND$_1$, and OPERAND$_2$, as shown below:

OPERATION	OPERAND$_1$	OPERAND$_2$

During the decode phase, the decoder connected to the OPERATION field of the IR enables the signal corresponding to the operation called for. The controller uses the enabled operation signal during the execute phase to bring about the appropriate sequence of operations. Thus, for each operation allowed in the instruction set of the digital computer, there will be one hardware algorithm for the controller to follow. For instance, if the operation field calls for an addition, the execute phase consists of the following sequence of register transfers:

MAR $\leftarrow$ OPERAND$_1$
READ memory
$R_1 \leftarrow$ MBR   OPERAND$_1$ is now in R_1.
MAR $\leftarrow$ OPERAND$_2$
READ memory
$R_2 \leftarrow$ MBR   OPERAND$_2$ is in R_2.
ADD   Command to the adder; the result is in
                                               R_3.

MBR $\leftarrow R_3$
WRITE memory   Note that the MAR still contains the address of OPERAND$_2$, which will be replaced by the result.

The control unit produces this sequence of signals when in the execute phase of an add operation.

The fetch and decode phases will be common to all the instructions in the instruction set, but the execute phase differs for each instruction. Thus, in order to complete the controller hardware algorithm, the execute phase of each instruction must be analyzed to derive the register transfer sequence similar to the above. Note that although we defined only three states for the controller, each state again corresponds to a series of states, since the controller is required to make transitions in the sequence specified above in order to generate the control signals at appropriate time.

We will leave this example at this level since the design of a complete digital computer is beyond the purposes of this book. For further details, refer to the author's *Computer Design and Architecture*, listed in the references at the end of this chapter.

Example 7.15

Consider the serial twos complementer of Example 7.2. Figure 7.55 shows the control and processing parts of this circuit, as per the HDL description in Example 7.4. The shift register R and the flip-flop S form the processing section. The controller consists of the counter (count) and the control flip-flop T. The inputs to the controller are the clock P and the switch SW. The controller outputs a load signal that is used to load the six-bit data into R (in parallel) and to clear S, and a shift signal that initiates a right shift of R.

Note from the HDL description that the control circuit has two states, as indicated by the flip-flop T. The state diagram is also shown in the figure. The initial state corresponds to $T = 0$. When the switch is turned on, the circuit makes a transition to the $T = 1$ state, producing the load output for the processing section and CLEAR COUNT and SET T signals for internal components. As long as $T = 1$, the controller produces SHIFT and INCREMENT COUNT signals at each clock pulse. When the count reaches 5, the controller transfers to the $T = 0$ state.

The controller can be implemented as a synchronous sequential circuit (with one input and two outputs) by the methods discussed earlier in this Chapter.

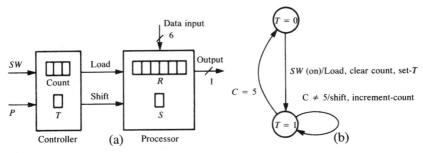

Figure 7.55 Serial twos complementer. (a) Circuit; (b) state diagram.

7.9.1 ASM Charts

As the digital system becomes increasingly complex, the number of states in the controller increases and the state diagram becomes more cumbersome. ASM charts are useful in such cases. An ASM chart is similar to the conventional flowcharts used to describe the program flow, except that the timing relationships between the various steps in the chart are defined more explicitly. The conventional flowchart describes only the sequence of events.

An ASM chart is composed of three types of elements, as shown in Figure 7.56. The rectangular box indicates a state. The register transfers and outputs occurring while in that state are listed within the rectangle. The name of the state and the binary vector assigned to it are usually listed next to the rectangle.

The diamond-shaped decision box shows an input condition and two or more exit paths based on the input condition. When the condition is binary, one exit path (marked with a 1) is taken if the condition is true and the other (marked with a 0) is taken if it is false.

The state and decision boxes are similar to those in conventional flowcharts. The oval-shaped conditional box is unique to ASM charts. The input path to a conditional box must always come from one of the exit paths of a decision box. The register transfers and the outputs listed within the conditional box occur during a given state if the input conditions are satisfied.

In the ASM chart of Figure 7.57, there are two states S_1 and S_2. While in S_1, $T \leftarrow 0$ is performed and if Q is 0, $S \leftarrow 0$ is performed and the next state is S_2. While in S_1 if Q is 1, $T \leftarrow 0$ is performed and the next state of S_2 is reached.

An *ASM block* is a structure consisting of one state box and all the decision and conditional boxes associated with it. It has one input path and any number of exit paths. The structure shown by the dotted lines in Figure 7.57 is an ASM block. A simple ASM block consists of only a state box.

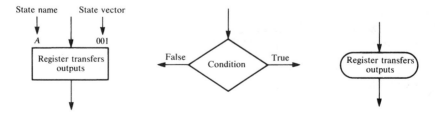

Figure 7.56 Components of an ASM chart. (a) State box; (b) decision box; (c) conditional box.

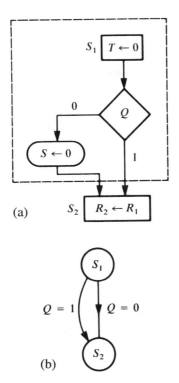

(a)

(b)

Figure 7.57   An ASM chart example. (a) ASM chart; (b) state diagram.

Each ASM block represents the state of the system during one clock-pulse interval. All the operations indicated by the block are performed during that clock-pulse interval, and the system moves to one of the next states indicated by the exit paths of the block. Thus, the ASM chart of (a) is equivalent to the state diagram of (b). Note that the unconditional and conditional operations are not indicated on the state diagram.

Figure 7.58 shows the ASM chart for the serial twos complementer example. This chart has two blocks, corresponding to the two states in the twos complementer control circuit. From the chart, the controller can be readily implemented. We will illustrate the implementation of this circuit using a PLD in Chapter 9. This is one implementation technique. There are other methods to implement the circuit described by the ASM chart. Refer to books by Mano and Stone listed in the references for such methods.

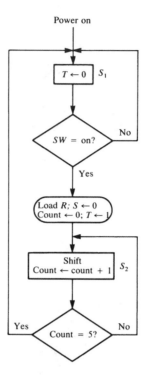

Figure 7.58 ASM chart for the serial twos complementer.

7.10 Summary

Registers and counters are the most common components used in sequential circuit design. The design of sequential circuits using these components was illustrated in this chapter. Details of representative ICs were given. This chapter also served as an introduction to asynchronous sequential circuit design through the design of asynchronous counters.

Several other ICs, such as oscillators and timers that serve as sources of clock signals, were not described here. (See the IC manufacturers' manuals for details on such ICs. The books by Greenfield and Blakeslee, listed as references, also provide descriptions of these components.)

A brief introduction to the design of control circuits and the uses of ASM charts in such designs was provided. Books on computer design and architecture listed in the references provide further details on this topic.

References

Blakeslee, T. R. *Digital Design with Standard MSI and LSI.* 1975. New York, John Wiley, & Sons, 1979.

CMOS Data Manual. Austin, Tex.: Motorola, 1978.

Computer (special issues on hardware description languages) (December 1974); 10, (June 1977).

Dietmeyer, D. *Logic Design of Digital Systems.* Boston: Allyn & Bacon, 1971.

Ercegovac, M. D. and Lang, T. *Digital Systems and Hardware/Firmware Algorithms.* New York: John Wiley & Sons, 1985.

Fletcher, W. I. *An Engineering Approach to Digital Design.* Englewood Cliffs, N.J.: Prentice-Hall, 1980.

Greenfield, J. D. *Practical Digital Design Using ICs.* New York: John Wiley & Sons, 1983.

Hill, F. J. and Peterson, G. R. *Introduction to Switching Theory and Logic Design.* New York: John Wiley & Sons, 1981.

Mano, M. *Digital Design.* Englewood Cliffs, N.J.: Prentice-Hall, 1994.

National Advanced Bipolar Logic Databook, Santa Clara, Calif. *National Semiconductor*, 1995.

Pasahow, E. J. *Learning Digital Electronics Through Experiments.* New York: McGraw-Hill, 1982.

Roth, C. H. *Fundamentals of Logic Design.* St. Paul, Minn.: West Publishing, 1985.

Shiva, S. G. *Computer Design and Architecture.* Glenview, Harpercollins, 1991.

TTL Data Manual. Sunnyvale, Calif.: Signetics, 1986

Problems

7.1 Design a four-bit register using TTL 7474 D flip-flops. Include a LOAD control input. The data should enter the register when LOAD is high and at the (a) positive edge of the clock and (b) negative edge of the clock.

7.2 Compare the complexity of the circuit in problem 7.1 with that of the TTL 74173.

7.3 Design an eight-bit shift register using 74109 JK flip-flops. Parallel and serial input, serial output, and right-shift features are required.

7.4 Compare the complexity of the shift register in problem 7.3 with that of the TTL 74199.

7.5 The content of a four-bit shift register is 1001. Show the content of the register after each shift for five (a) right shifts and (b) left shifts. Assume that the serial input is 10101 in either case.

7.6 Use two TTL 7495s to implement an eight-bit shift register capable of left and right shifts, serial and parallel inputs, and serial and parallel outputs.

7.7   Design a complete circuit using off-the-shelf ICs to load an eight-bit data into a register, circulate it right five bits, and output in parallel.

7.8   Design a four-bit synchronous binary counter using (a) D flip-flops and (b) JK flip-flops.

7.9   Design a decade counter using (a) D flip-flops and (b) JK flip-flops.

7.10   Design a four-bit ripple counter using SR flip-flops.

7.11   Design a synchronous counter to count in the sequence 0000-0101-1100-1001-1110-1111-0000, using T flip-flops. Draw the timing diagram to show this sequence.

7.12   Show the state transition sequence of the counter in problem 7.11 if the starting state is 0001.

7.13   A flip-flop has a p ns delay from the clock transition until its output changes. Assume a gate delay of g ns for each gate used in the circuit. Determine the maximum frequency at which an n-bit counter can operate if the counter is (a) synchronous and (b) ripple.

7.14   Design divide-by-N circuits where N is 5, 40, and 64. Use available ICs if possible.

7.15   There are two four-bit registers, A and B, built out of SR flip-flops. There is a control signal C. The following operations are needed:

> If $C = 0$, send contents of A to B.
> If $C = 1$, send ones complement of contents of A to B.

Draw the circuit to perform these functions:

(a) In parallel mode.
(b) In serial mode.

7.16   Design the circuit in problem 7.15(a) for a twos complement transfer.

7.17   There are 3 two-bit registers A, B, and C. Design the logic to perform

$$\text{AND: } C \leftarrow A \land B.$$
$$\text{OR: } C \leftarrow A \lor B.$$

AND and OR are control signals. Each bit in the register is a D flip-flop.

7.18   A two-bit counter C controls the register transfers shown below:

$$C = 0: B \leftarrow A.$$
$$C = 1: B \leftarrow A'.$$
$$C = 2: B \leftarrow A + B.$$
$$C = 3: B \leftarrow 0.$$

A and B are two-bit registers. Draw the circuit. Use 4-to-1 multiplexers in your design. Show the details of register B.

7.19 Draw a bus structure to perform the operations in problem 7.18.

7.20 Connect four five-bit registers A, B, C, D using a bus structure capable of performing the following:

$$C_0: B \leftarrow A. \qquad C_4: A \leftarrow C + D.$$
$$C_1: C \leftarrow A \vee B. \qquad C_5: B \leftarrow C' \wedge D'.$$
$$C_2: D \leftarrow A \wedge B. \qquad C_6: D \leftarrow A + C.$$
$$C_3: A \leftarrow A + B'. \qquad C_7: B \leftarrow A + C'.$$

C_0 through C_7 are control signals. Use a three-bit counter and a decoder to generate these signals. Assume tristate outputs for each register.

7.21 Assume regular outputs for each register in problem 7.20. How many OR gates are needed to implement the bus structure?

7.22 It is required to transmit the eight-bit data in a register over a serial line. The receiver expects the eight data bits followed by a parity bit. Odd parity is used. The data bits are loaded into the source register in parallel. Design the data formatter/converter circuit and interface it to the source and destination registers. Use ICs from the catalogues.

7.23 Design a serial subtractor by modifying the serial adder circuit of this chapter.

7.24 Design a serial adder circuit to add two BCD digits. Each digit is in a four-bit shift register. The sum should occupy a five-bit shift register. Remember the Add-6 correction needed.

7.25 Design the control circuit for the serial full-adder. Draw an ASM chart.

7.26 Derive the ASM chart for the multiplier circuit that uses the shift-and-add multiplication algorithm of Chapter 1.

7.27 Assume that a digital system can perform the following instructions:

$$\text{SUM } A, B \qquad A \leftarrow A - B$$
$$\text{TCA } A \qquad A \leftarrow A' + 1$$

Include the features needed to implement these instructions and draw an ASM chart depicting the operation of the system.

8

Asynchronous Circuits

8.1 Introduction

In a synchronous sequential circuit, the change of an internal state occurs in response to clock pulses. Asynchronous sequential circuits do not use clock pulses, and the change in their internal state occurs in response to input changes. Asynchronous sequential circuits are used when operating speed is important. Recall that clock frequency in a synchronous circuit is determined by the slowest memory element. If the application requires that the circuit respond immediately after a change in the input without waiting for a clock pulse to arrive, asynchronous sequential circuits are more appropriate. Furthermore, in small digital systems with very few components, the overhead of clock circuitry can be eliminated by using asynchronous sequential circuits.

For example, consider the case of serial data transmission over telephone lines. The source and the destination subsystems each may be a synchronous circuit with its own clock, but the two clocks are not directly connected. As and when the source produces the data stream, the destination subsystem starts its internal clock and receives the data stream. Once the data transfer is complete, the two subsystems are disconnected. This is called an asynchronous data interface and is the most common mode of data communication between two remote devices.

Because the asynchronous circuit does not require a clock pulse to activate its processing activity, it can respond to changes in input conditions as and when they occur. Thus, an asynchronous circuit, if designed properly, will be faster than a synchronous circuit.

Earlier in this book, we examined two simple asynchronous circuits: the SR latch formed out of two cross-coupled NOR (or NAND) gates, described in Chapter 6, and the ripple counters, described in Chapter 7. (Although the ripple counters were triggered by a clock edge in that earlier discussion, a clock is not necessary for their operation. The rising or falling edge of an input signal would serve as a trigger.) We will use the SR latch as an example again later in this chapter.

Consider the model of an asynchronous sequential circuit shown in Figure 8.1. It is identical to the model of the synchronous circuit except that there is no clock. The combinational logic portion accepts the external inputs $(x_1, x_2, x_3, \ldots, x_n)$ and the present state of the circuit as represented by the *secondary variables* $(y_1, y_2, \ldots, y_p)$ and produces the outputs $(z_1, z_2 \ldots, z_m)$ and the next-state information represented by the *excitation variables* $(Y_1, Y_2, \ldots, Y_p)$. There are p memory elements. They respond to a change in the excitation variables Y. Because there is no clock, the secondary variables y assume the state brought about by Y, after a certain delay (Δt) required by the memory elements to make the transition. Thus, the inputs should not change faster than Δt if circuit operation is to be predictable. The input signals can be in the form of *pulses* or *levels*. Asynchronous circuits with pulse inputs are said to operate in *pulse mode*, while those with level inputs are said to operate in *fundamental mode*.

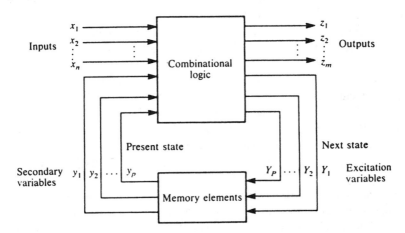

Figure 8.1 Block diagram of an asynchronous circuit.

Because of the absence of a clock in asynchronous circuits, the probability of two pulses simultaneously occurring on two signal lines is small. Hence, simultaneous pulses on two or more input lines to a pulse-mode circuit are not allowed. Furthermore, the time between two pulses on the input lines must be at least equal to the transition time of the slowest memory element. The input signals occur only in the uncomplemented form in these circuits, since an input either carries a pulse or is at 0. Pulse-mode circuits use unclocked flip-flops (latches) as memory elements. The analysis and synthesis of pulse-mode circuits is very similar to that of synchronous circuits, except for the above restrictions on the inputs.

In fundamental-mode circuits, the inputs are restricted so that only one of them changes its level at any time. In practice, it is unlikely that two signals will change their levels exactly at the same time, since there is no synchronizing clock present. Again, for predictable circuit operation, the time between two input changes should be at least equal to the transition time of the slowest memory element. These circuits use delay lines as memory elements. Delay lines introduce a delay of Δt into the excitation variables. Thus, $y(t + \Delta t) = Y(t)$. In practice, the use of physical delay lines may not be necessary, since the delay introduced by the gates in the feedback path may be sufficient for the proper operation of the circuit. Thus, these circuits can be treated as combinational circuits with feedback. We will assume that each memory element has the same transition time (Δt). In practice, this may not be true, especially when there are no physical delay lines present, since each feedback path may have a different number and type of gates, thereby introducing different delays. These unequal delays result in *hazards* and *race* conditions. Techniques for avoiding these unequal delay effects are described later in this chapter.

It is important to note that in asynchronous circuits, the state transitions occur in response to changes in input rather than because of a clock edge, as in synchronous circuits. As soon as the input changes, the excitation variables (Y) assume the new value based on the new input condition and the current value of the secondary-state variables (y), neglecting the propagation delays introduced by the combinational logic. After the delay Δt, the secondary variables y assume the value of Y. For the circuit to be *stable*, $y(t + \Delta t)$ must equal $Y(t)$. Thus, if $y(t + \Delta t) \neq Y(t)$, the circuit becomes *unstable*. That is, a change in y after Δt from the change in Y causes a further change in Y and the circuit will not become stable. In general, an asynchronous circuit will have several stable and unstable states. The change from one stable state to the other occurs as a result of a change in an input. During this transition, the circuit may go through one or more unstable states.

8.2 Pulse-Mode Circuits

The analysis and synthesis of pulse-mode circuits are very similar to that of synchronous sequential circuits, as illustrated by the following examples.

Example 8.1

Consider the sequential circuit shown in Figure 8.2. The circuit uses two latches: an SR and a D type. There are two input lines, x_1 and x_2, and one output, z. The excitation variable Y_1 corresponds to (S_1, R_1), and Y_2 corresponds to D_2. The secondary variables y_1 and y_2 correspond to the outputs of latches.

From an analysis of the circuit diagram, the input or excitation functions are

$$S_1 = x_1 y_1' y_2$$
$$R_1 = x_1 y_1' y_2' + x_1 y_1 y_2 + x_2 y_1' y_2 + x_2 y_1 y_2$$
$$D_2 = x_1 y_1',$$

and the output function is

$$z = x_1 y_1 y_2$$

S_1, R_1, and D_2 are shown in map form in (b). Here, the columns are identified by x_1 and x_2 rather than all four combinations of $x_1 x_2$. x_1 and x_2 indicate the occurrence of a pulse on the corresponding input line. These are the only two conditions possible. From the S_1 and R_1 tables and using the excitation table for the SR latch (which is same as that of the SR flip-flop—see Chapter 6), we can derive the next-state table $y_1(t + 1)$, shown in (c). Similarly, the next-state table $y_2(t + 1)$ for the second latch is derived from the table for D_1.

The transition table in (d) is derived by merging the $y_1(t + 1)$ and $y_2(t + 1)$ tables, entry by entry. The output table is shown in (e). Using an arbitrary state assignment of $A = 00$, $B = 01$, $C = 11$, and $D = 10$, the state table is shown in (f). The state diagram is shown in (g). As can be seen from the state diagram, this is an $x_1 - x_1 - x_1$ sequence detector. That is, the output pulse occurs only when three consecutive pulses occur on input x_1. Note that state D is not entered by the circuit, starting from the initial state A. The circuit's timing diagram for a typical input sequence is shown in (h).

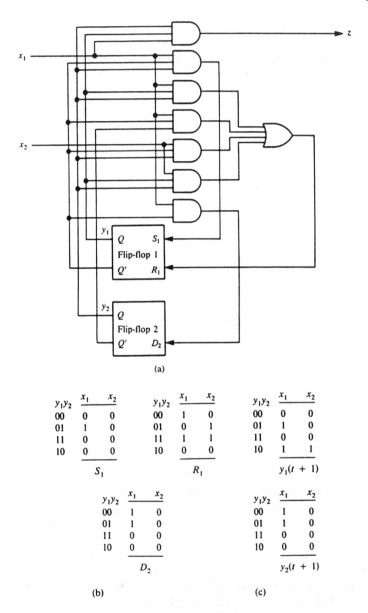

Figure 8.2 A pulse-mode circuit. (a) Circuit; (b) excitation tables; (c) transitions; (d) transition table; (e) output table; (f) state table; (g) state diagram; (h) timing diagram.

	y_1y_2	x_1	x_2
$A =$	00	01	00
$B =$	01	11	00
$C =$	11	00	00
$D =$	10	10	10

(d)

y_1y_2	x_1	x_2
00	0	0
01	0	0
11	1	0
10	0	0

(e)

	x_1	x_2
A	$B/0$	$A/0$
B	$C/0$	$A/0$
C	$A/1$	$A/0$
D	$D/0$	$D/0$

(f)

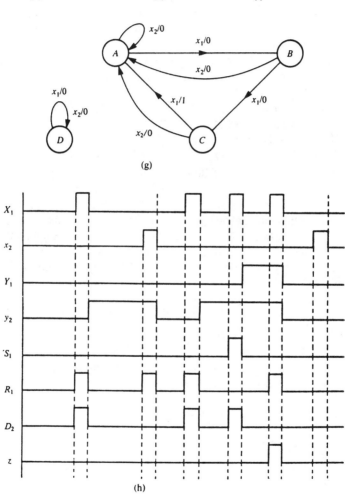

(g)

(h)

Figure 8.2 Continued

Example 8.2

Design an asynchronous circuit with three input lines (x_1, x_2, and x_3) and an output line (z) that can detect the sequence of pulses $x_1 - x_3 - x_2 - x_2$.

Because the output coincides with the last pulse in the sequence to be detected, a Mealy-type circuit is used. The state diagram is shown in Figure 8.3(a). A is the starting state. B indicates that x_1 has occurred; C indicates that $x_1 - x_3$ has occurred; D indicates that $x_1 - x_3 - x_2$ has occurred. When another x_2 pulse arrives, the sequence is complete and the circuit returns to state A with an output of 1. The state diagram is completed by including transitions as a result of other inputs possible at each state in the circuit. The state table is shown in (b). Using an arbitrary assignment of $A = 00$, $B = 01$, $C = 11$, and $D = 10$, the transition table is shown in (c).

We will use an SR latch and a D latch. Using the transitions of the first latch in transition table (c) and the characteristic table for an SR flip-flop (from Chapter 6), the excitation conditions S_1 and R_1 shown in (d) are derived. Note that the tables for S_1 and R_1 are not K-maps, since the entries in rows (in two physically adjacent columns) are not logically adjacent. Nevertheless, the physically adjacent entries in each column are logically adjacent. Thus, we can combine terms in each column only. The input equations derived from these tables are

$$S_1 = x_3 y_1' y_2$$
$$R_1 = x_1 + x_2 y_2' + x_3 y_1$$

The transitions of the second latch and the corresponding input conditions are shown in (e). The input equation is

$$D_2 = x_1 + x_3 y_1' y_2$$

From the output map of (f), the output equation is

$$z = x_2 y_1 y_2'$$

The circuit is shown in (g).

Example 8.3

An asynchronous pulse-mode circuit must be designed with two inputs (x_1 and x_2) and an output, z. When z is 0, if an input sequence of

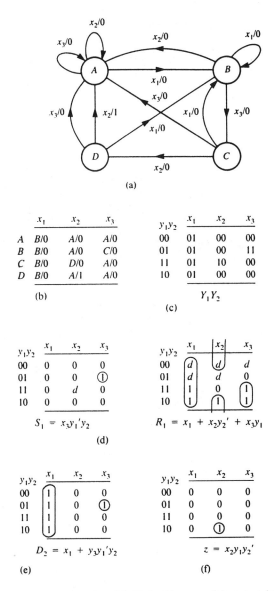

(a)

(b)

	x_1	x_2	x_3
A	B/0	A/0	A/0
B	B/0	A/0	C/0
C	B/0	D/0	A/0
D	B/0	A/1	A/0

(c)

y_1y_2	x_1	x_2	x_3
00	01	00	00
01	01	00	11
11	01	10	00
10	01	00	00

Y_1Y_2

(d)

y_1y_2	x_1	x_2	x_3
00	0	0	0
01	0	0	①
11	0	d	0
10	0	0	0

$S_1 = x_3y_1'y_2$

y_1y_2	x_1	x_2	x_3
00	d	d	d
01	d	d	0
11	1	0	1
10	1	1	1

$R_1 = x_1 + x_2y_2' + x_3y_1$

(e)

y_1y_2	x_1	x_2	x_3
00	1	0	0
01	1	0	①
11	1	0	0
10	1	0	0

$D_2 = x_1 + y_3y_1'y_2$

(f)

y_1y_2	x_1	x_2	x_3
00	0	0	0
01	0	0	0
11	0	0	0
10	0	①	0

$z = x_2y_1y_2'$

Figure 8.3 $x_2-x_3-x_2-x_2$ detector. (a) State diagram; (b) state table; (c) transition table; (d) excitation for latch 1; (e) excitation for latch 2; (f) output; (g) circuit.

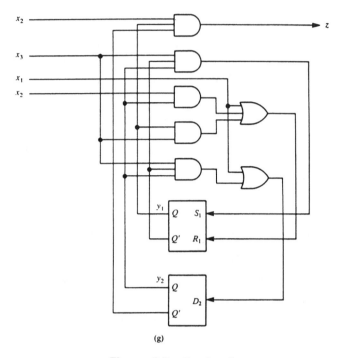

(g)

Figure 8.3 Continued

$x_1 - x_1 - x_2$ occurs, then z changes to 1. z remains at 1 until a sequence $x_1 - x_2$ occurs.

Because a level output is required, as implied by the above conditions, a Moore-type circuit is suitable for this application. The state diagram is shown in Figure 8.4(a). The starting state is A. The circuit makes a transition from A through B and C to D while maintaining an output of 0, in response to the input sequence $x_1 - x_1 - x_2$. Once in D, the circuit attains an output of 1, and Z remains at 1 until an input sequence of $x_1 - x_2$ occurs to take the circuit back to A by way of state E. The state diagram is completed by accounting for transitions resulting from x_1 and x_2 at each state. The state table is shown in (b).

The development of the circuit diagram is very similar to that in Example 8.2. We will leave that as an exercise. Note that in deriving latch input equations, each column in this example corresponds to four variables (an input and y_1, y_2, and y_3). Thus, entries in each column should be rearranged into a four-variable K-map to accommodate simplification.

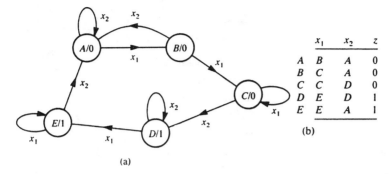

	x_1	x_2	z
A	B	A	0
B	C	A	0
C	C	D	0
D	E	D	1
E	E	A	1

(b)

(a)

Figure 8.4 A pulse-mode circuit. (a) State diagram; (b) state table.

As we can see by these examples, the analysis and design of pulse-mode circuits are very similar to those of synchronous sequential circuits.

8.3 Fundamental-Mode Circuits

8.3.1 Analysis

The analysis of fundamental-mode sequential circuits is illustrated through the following examples.

Example 8.4

Consider the *SR* latch of Chapter 6, formed by cross-coupling two NOR gates. The circuit and the truth table of its operation are shown in Figure 8.5. In terms of the terminology used to describe fundamental-mode circuits, the excitation variable Y corresponds to the Q output of the latch and the secondary variable y is one of the inputs to Gate 1. The feedback of Y to the input of Gate 1 makes this circuit an asynchronous sequential circuit. The truth table in (b) was derived in Chapter 6 by tracing through the signals from the inputs to the outputs of the circuit. By analyzing the circuit, we can derive the *excitation equation* as

$$Y = (R + y')'$$
$$= (R + (S + y)')'$$
$$= R' \cdot (S + y)$$
$$= R'S + R'y$$

The excitation equation is represented on a K-map in (c). This K-map is called the *excitation table* for the circuit. Each row of this map

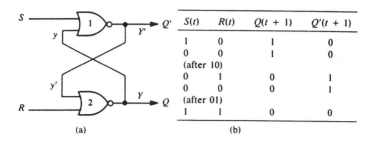

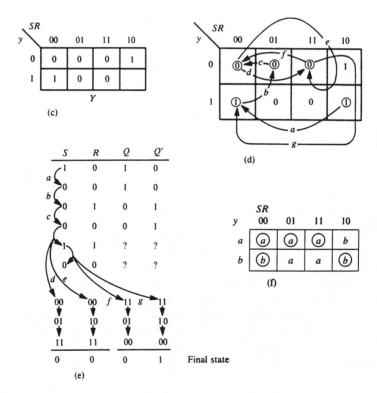

Figure 8.5 Cross-coupled NOR gate circuit. (a) Circuit; (b) truth table (partial); (c) excitation table; (d) transition table; (e) typical transitions; (f) flow table.

corresponds to a secondary state, and each column corresponds to a combination of input values. The entries in the table correspond to the values of Y, the excitation variable. For the circuit to be stable, Y must be equal to y. The circuit is said to be unstable if $Y \neq y$. The circled entries in table (d) correspond to $y = Y$ and hence are the *stable states*. This table is called a *transition table*.

Let us consider various transitions to understand the behavior of this circuit. Initially, the circuit is in state 1 (i.e., $Q = 1$) and $S = 1$, $R = 0$. This corresponds to the first row of the table (e) and the stable state represented by the bottom right-most cell of (d). The input condition (x) combined with the secondary state of the circuit (y) is called the *total state* and is represented as (x, y). In the case of a cross-coupled NOR circuit, the total state is thus indicated by (SR, y). While in total state (10, 1), if the S input goes to 0, the circuit moves to the stable state (00, 1), as shown by the transition marked ''a.'' If the R input now goes to 1, the circuit moves to the unstable state (01, 1) and then makes a transition to the stable state (01, 0), as shown by transition b. If R now goes to 0, the circuit moves to the stable state (00, 0), as shown by c. Now if the inputs were to change to $SR = 11$, two possibilities would result, depending on the order in which the inputs change. That is, the input transitions would either correspond to 00-01-11, as depicted by d, or 00-10-11, as depicted by e. Note that in either case, the final state achieved is (11, 0). Now let us change the inputs to $SR = 00$. Again, there are two possibilities: 11-01-00, as depicted by f, and 11-10-00, as depicted by g. In the former case, we reach the state (00, 0) and in the latter (00, 1). Thus, when both inputs are at 1, circuit operation is not predictable. In general, when SR latches made out of cross-coupled NOR gates are used, care must be taken to ensure that one of the inputs is 0. That is, $S \cdot R$ should always equal 0.

The transition table in (d) is converted into the *flow table* shown in (f) by encoding each state with a symbol. Both transition and flow tables essentially represent the same information on the circuit. In the transition table, the states are represented by binary variables, while in the flow table, the states are represented by symbols.

Example 8.5

A fundamental-mode circuit using a delay line memory element is shown in Figure 8.6(a). There are two inputs, x_1 and x_2, an output z, an excitation variable Y (input to the delay line), and the secondary variable y (the output of the delay line). In addition, $y(t + \Delta t) = Y(t)$.

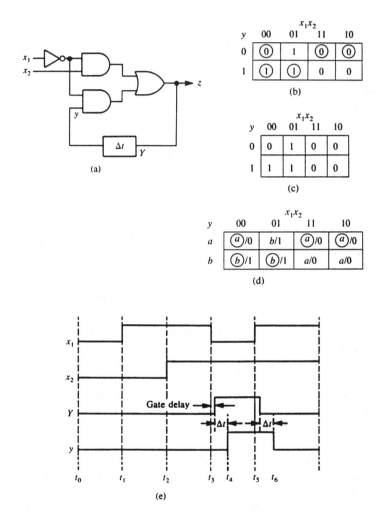

Figure 8.6 Fundamental-mode circuit. (a) Circuit; (b) Y; (c) z; (d) flow table; (e) timing diagram.

From the circuit diagram, we can see that

$$Y = x_1'x_2 + x_1'y$$

and

$$z = x_1'x_2 + x_1'y$$

The excitation and output tables for the circuit are shown in (b) and (c), respectively. Since $y(t + \Delta t) = Y(t)$, the excitation table is also the transition table. The stable states are circled in the transition table. The flow table shown in (d) is derived by encoding the states in the transition table (by $a = 0$ and $b = 1$) and merging the output table.

The timing diagram in (e) shows typical input transitions. At t_0, the circuit is in the stable state $(x_1x_2, y) = (00, 0)$. At t_1, x_1 changes to 1, resulting in no changes in other variables. At t_2, x_2 changes to 1, resulting in no changes to other variables, and the circuit moves to the stable state $(11, 0)$. At t_3, x_1 goes to 0. As a result, Y goes to 1 at t_3, ignoring the propagation delays of the circuit gates. y goes to 1 at t_4, Δt later from t_3. The circuit thus goes through the unstable state $(01, 0)$ to the stable state $(01, 1)$. At t_5, x_1 goes to 1, sending Y to 0 and y to 0 at t_6.

Example 8.6

Consider the fundamental-mode circuit shown in Figure 8.7(a). By analyzing the circuit, we can derive the excitation equations as

$$Y_1 = x'y_1 + x'y_2'$$

and

$$Y_2 = y_1y_2 + xy_1'$$

The output equation is

$$z = x'y_1y_2 + xy_1'$$

The tables for Y_1 and Y_2 are shown in (b) and (c), respectively. They are merged with the output table shown in (d) to obtain the transition table shown in (e) for the circuit. The stable states are circled. Note that the circuit has only three stable states.

Note from the transition table that if the initial state is $(1, 01)$ and x changes to 0, the circuit goes through the unstable states 00-10 before reaching the stable state $(0, 10)$. The output changes from 1 to 0. Similarly, while in state $(0, 11)$, if the input is changed to 1, the circuit goes through the unstable state 01 to the stable state $(1, 01)$. But then the output goes through a momentary 0.

Example 8.7

A fundamental-mode circuit using an SR latch as the memory element is shown in Figure 8.8(a). From the circuit, we can see that the latch input functions are

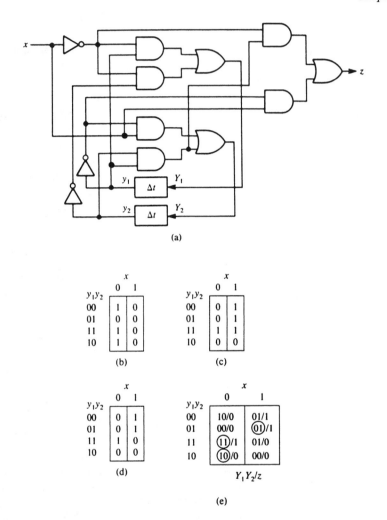

Figure 8.7 Fundamental-mode circuit (Example 8.6). (a) Circuit; (b) Y_1; (c) Y_2; (d) z; (e) transition table.

$$S = x_1y' + x_2'y'$$
$$R = x_1'x_2 + x_1'y$$

and the output function is

$$z = x_1x_2y$$

Note that $S \cdot R = 0$. Hence, the restriction on the use of the *SR* latch is

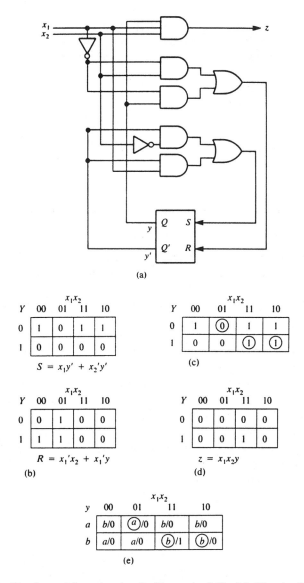

Figure 8.8 Fundamental-mode circuit (Example 8.7). (a) Circuit; (b) excitation tables; (c) transition table; (d) output table; (e) flow table.

obeyed. The latch input tables derived from the above functions are shown in (b). The transition table for the latch derived from the S and R tables is shown in (c), in which the stable states are circled.

The output table of (d) is merged with the transition table, and by assigning values of $a = 0$ and $b = 1$, we can obtain the flow table of (e).

The above examples suggest the following analysis procedure for fundamental-mode circuits:

1. From the circuit diagram, determine the circuit excitation and output equations.
2. Construct the excitation and output tables.
3 (a) If the circuit uses delay lines as memory elements (or has direct feedback with no other memory elements), then the excitation table is the transition table.
   (b) If the memory elements in the circuit are unclocked flip-flops, derive the transition table using the excitation table for each flip-flop derived in step 2 above and the characteristic tables of flip-flops from Chapter 6 (this step is similar to that for synchronous circuits). Merge the transition tables for individual flip-flops entry by entry to form the circuit transition table.
4. Identify and circle all stable states in the transition table by using the condition $Y = y$ for a stable state.
5. Assign a unique symbol to each row of the excitation table.
6. Construct the flow table by replacing the state variables with corresponding symbols and merging the excitation and the output tables, entry by entry.

8.3.2 Design

To design a fundamental-mode circuit, follow the steps in the analysis procedure in reverse order. The procedure is outlined below:

1. From the given design specifications for the circuit, derive a flow table. This step is similar to deriving state diagrams for synchronous sequential circuits and is the most difficult part of the design. The designer's experience and intuition are critical for this step. In this step, flow tables with one stable state per row are generally obtained. Such tables are called *primitive flow tables*.

2. Reduce the number of rows in the primitive flow table if possible. The implication chart method of Chapter 6 is used for this purpose, since numerous entries in the primitive flow table are usually unspecified.

3. Assign unique binary state variables to each row of the reduced flow table. The state assignment should eliminate any possible *critical races* in the circuit. The concept of race in fundamental-mode circuits and the state assignment procedures are described in Section 5 of this chapter.

4. Specify any unspecified output entries in the reduced flow table, so that when the circuit makes a transition from one stable state to another, output toggling is minimized because of the intermediate unstable states. That is, if the present and the next output values are each 1, the intermediate output value should preferably be 1. Similarly, if the present and the next output values are each 0, the intermediate output value should be a 0. However, if the output changes during the transition, it is immaterial as to when the change occurs and hence a don't-care output is specified for the intermediate state.

5. Obtain the transition table and the output table.

6. Separate the transitions corresponding to each excitation variable and obtain the input equations for each memory element. Remember that $Y = y$ if the memory elements used are delay elements. If the memory element is an SR latch, equations for each S and R input are derived by using the excitation table for the corresponding latch and the characteristic table for an SR flip-flop. (This step is similar to the one in the design of synchronous sequential circuits.) In addition, make sure that $S \cdot R = 0$.

7. Derive the output equatons from the output table.

8. Draw the circuit diagram.

The following examples illustrate the design procedure.

Example 8.8

A clocked T flip-flop needs to be designed. Although the T flip-flop was used in designing synchronous circuits earlier in this book, internally it is an asynchronous circuit. It has two inputs: T and the clock (C). We will designate its output as Z (rather than Q, to avoid confusion). If T is 1 and C changes from 1 to 0, the flip-flop complements its output. In all other conditions, the output of the flip-flop remains unchanged.

We will first derive a *primitive flow table*. A primitive flow table has one stable state per row. For each possible change in the input from the stable state in the row, an unstable state is included in the row and the circuit makes a transition to the new row through this unstable state.

Because there are two inputs in this example, the primitive flow table will have four columns, one corresponding to each combination of inputs. Row 1 corresponds to the initial stable state 1 with $TC = 00$ and $Z = 0$.

$$TC$$

	00	01	11	10
1	①/0			

Now, if C changes to 1, the circuit reaches a stable state 2 via the unstable state 2 in row 1 and the second column. The output at the unstable state 2 is not specified at this time. Stable state 2 is represented by the second row. Because only one input changes at a time, the transition from 00 to 11 will never occur. Hence, the state and output entries in the third column of the first row are unspecified.

$$TC$$

	00	01	11	10
1	①/0	2/—	—/—	
2		②/0		

We will return to the input transition from 00 to 10 later. While in stable state 2, if T changes to 1, there should be no change in the output, though the circuit moves to the stable state 3 via another unstable state. While in state 3, if C changes to 0 (corresponding to the 1-to-0 transition of the clock), the output must change. This transition is represented by state 4.

$$TC$$

	00	01	11	10
1	①/0	2/—	—/—	
2		②/0	3/—	
3			③/0	4/—
4				④/1

From state 4, if the inputs make the transitions from 10 to 11 to 01 to 00, the output remains at 1 and the circuit makes transitions to new states 5, 6, and 7, respectively, as shown below:

TC

	00	01	11	10
1	①/0	2/—	—/—	
2		②/0	3/—	
3			③/0	4/—
4				④/1
5			⑤/1	
6		⑥/1		
7	⑦/1			

While in state 7, if the input changes to 01, there should be no change in the output. Because this total state of (01, 1) is already represented by state 6, no new row in the primitive flow table is needed. Similarly, if the input changes to 10 while in state 7, the circuit moves to state 4.

TC

	00	01	11	10
1	①/0	2/—	—/—	
2		②/0	3/—	
3			③/0	4/—
4				④/1
5			⑤/1	
6		⑥/1		
7	⑦/1	6/—		4/—

We can now complete the remaining entries. The transitions corresponding to a change in more than two inputs from the stable state will never occur. Hence, the entries in each row, two columns away from the stable-state entry, are unspecified. From state 1, if the input changes to 10, the output should be 0. Because there is no stable state corresponding to a 0 output in the last column, a new row (8) is added. From state 8, the two transitions possible take the circuit to either state 1 or state 3. The complete primitive flow table is shown below:

TC

	00	01	11	10
1	①/0	2/—	—/—	8/—
2	1/—	②/0	3/—	—/—
3	—/—	2/—	③/0	4/—
4	7/—	—/—	5/—	④/1
5	—/—	6/—	⑤/1	8/—
6	7/—	⑥/1	5/—	—/—
7	⑦/1	6/—	—/—	4/—
8	1/—	—/—	3/—	⑧/0

We will use the implication chart to reduce the primitive flow table. Because each row in the primitive flow table corresponds to a single stable state, each row can be treated as equivalent to the present state, for the purposes of the implication chart. Thus, we compare two rows at a time, and the concept of compatibility of states is replaced by that of compatible rows. In each column, an unspecified output ("—") is compatible with both 0 and 1; stable state i and unstable state i are compatible; stable state i and unstable state j are compatible if i is compatible with j; unstable state i is compatible with unstable state j if i is compatible with j. The implication chart is shown below:

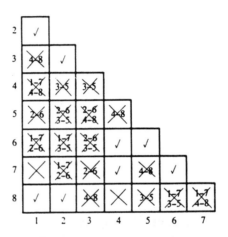

From the implication chart, the compatible pairs are collected and merged if possible to form the maximal compatibles:

2 (1 2)
3 (1 2) (2 3)
4 (1 2) (2 3)
5 (1 2) (2 3)
6 (1 2) (2 3) (4 6) (5 6)
7 (1 2) (2 3) (4 6) (5 6) (4 7) (6 7)
8 (1 2) (2 3) (4 6) (5 6) (4 7) (6 7) (1 8) (2 8)

A *merger diagram* can be used to derive the maximal compatibles from the compatible pairs. A merger diagram plots the states in the flow table along the circumference of a circle; two states are connected by an arc if they are compatible. A group of n states forms

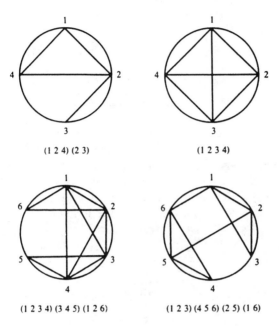

Figure 8.9 Typical merger diagrams.

a compatible set if there is an arc from each state to every other state in the set. Figure 8.9 shows some examples.

The merger diagram for the example at hand is shown below:

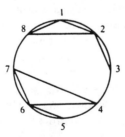

From the diagram, we can see that the maximal compatibles are (1 2 8) (4 6 7) (5 6) (2 3). As in Chapter 6, we select a minimum number of maximal compatibles that satisfy completeness, consistency, and closure criteria. In this example, to satisfy the completeness criterion, all four maximal compatibles are essential. The reduced flow table is shown below:

	00	01	*TC* 11	10
(1 2 8)	①/0	②/0	3/—	⑧/0
(4 6 7)	⑦/1	⑥/1	5/—	④/1
(5 6)	7/—	⑥/1	⑤/1	8/—
(2 3)	1/—	②/0	③/0	4/—

Note that in combining rows to form the reduced table, if at least one of the states in each column is a stable state, the resulting state is also stable; otherwise, the resulting state is unstable. Furthermore, if one of the outputs is specified, the resulting output entry takes that specified value. Renaming the rows as $A = (1\ 2\ 8)$, $B = (4\ 6\ 7)$, $C = (5\ 6)$, and $D = (2\ 3)$, we can derive the following flow table:

	00	01	*TC* 11	10
A	Ⓐ/0	Ⓐ/0	D/—	Ⓐ/0
B	Ⓑ/1	Ⓑ/1	C/—	Ⓑ/1
C	B/—	Ⓒ/1	Ⓒ/1	A/—
D	A/—	Ⓓ/0	Ⓓ/0	B/—

Row 2 appears in two maximal compatibles. This should not present any problems in terms of coding the reduced table. Since the appearance of a 2 in rows 1 and 4 of the reduced table means that each corresponds to a stable state, the 2 in row 1 must be A and that in row 4 must be D.

In this example, we will use an arbitrary state assignment. (Procedures for state assignments are described later in this chapter.) The unspecified outputs are changed so that when the circuit makes a transition from a state with a 0 output to one with a 0 output, the intermediate state also produces an output of 0. Similarly, for a transition with outputs of 1, the intermediate state produces a 1 output. If the outputs at the two states are different, the intermediate state produces a don't-care output, denoted by d. The reduced flow table with the outputs specified is shown below:

	00	01	*TC* 11	10
A	Ⓐ/0	Ⓐ/0	D/0	Ⓐ/0
B	Ⓑ/1	Ⓑ/1	C/1	Ⓑ/1
C	B/1	Ⓒ/1	Ⓒ/1	A/d
D	A/0	Ⓓ/0	Ⓓ/0	B/d

Figure 8.10 shows the transition table using an assignment of $A = 00$, $B = 01$, $C = 11$, and $D = 10$. The output and excitation tables are derived from the transition table, as before. The excitation and output equations are also shown. The circuit diagram is shown in Figure 8.10(b).

    Figure 8.11 shows the excitation tables, equations, and circuit diagram used to realize the above circuit, using two *SR* latches.

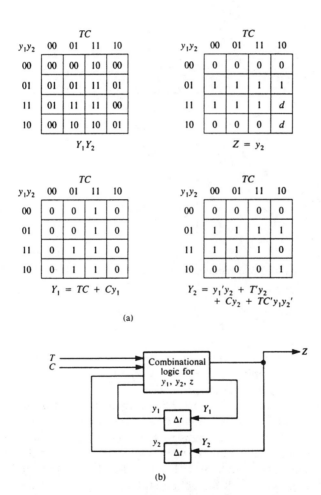

(a)

(b)

Figure 8.10 Fundamental-mode circuit (Example 8.8). (a) Excitations and outputs; (b) circuit.

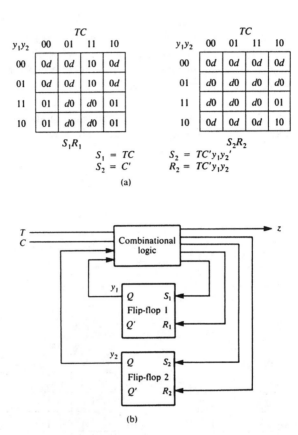

Figure 8.11 Fundamental-mode circuit using SR latch. (a) Excitations; (b) circuit.

Example 8.9

An asynchronous circuit must be designed with one input, x, and one output, z. z complements its state at every 0-1-0 transition of x.

A timing diagram representing the circuit's operation is shown in Figure 8.12(a), and the primitive flow table is shown in (b). Row 1 of the primitive flow table corresponds to the initial state, with $z = 0$ and $x = 0$. Now, if x changes to 1, the circuit moves to state 2, with an output of 0. If the input changes back to 0, the circuit goes to state 3, producing an output of 1. Next, a 0-to-1 transition of x takes the circuit to state 4, with the output still at 1. When x makes a transition to 0, the circuit returns to state 1, complementing z. The output values at unstable states are left unspecified.

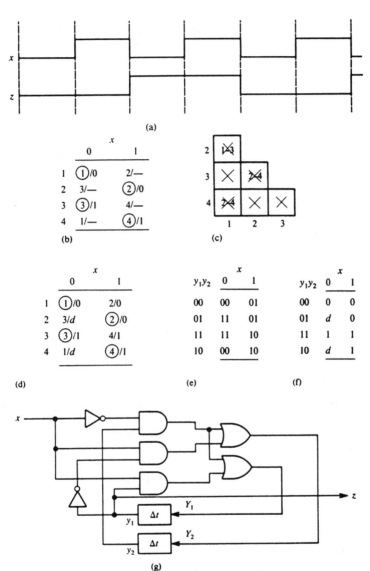

Figure 8.12 Synthesis example. (a) Timing diagram; (b) primitive flow table; (c) implication chart; (d) flow table with outputs defined; (e) transition table; (f) output table; (g) circuit diagram.

The implication chart is shown in (c). Because there are no compatible pairs, the flow table cannot be reduced.

Because the output is 0 in the transition from state 1 to 2, the output corresponding to unstable state 2 is specified to be 0. Similarly, the output at the unstable state 4 must be 1. The outputs at the other two unstable states (1 and 3) can be don't-cares, since the output value is complemented during the transitions through those states. The flow table, with the outputs specified, is shown in (d).

By choosing an arbitrary state assignment of 1 = 00, 2 = 01, 3 = 11, and 4 = 10, we can derive the transition table in (e). From this table, we can derive the excitation equations:

$$Y_1 = x'y_2 + xy_1$$

and

$$Y_2 = x'y_2 + xy_1'$$

The output table is shown in (f), and the output equation is

$$z = y_1$$

The circuit diagram is shown in (g).

8.4   Effects of Component Delays

In our discussion of asynchronous circuits, we have thus far assumed that all the gates are ideal and do not introduce any delay. In practice, gates in the circuit introduce propagation delays, and wires used in the circuit cause transmission delays. As mentioned before, these delays might be sufficient for the circuit to operate properly, thus necessitating no other delay lines in the feedback paths. We will now discuss the effect of these delays on the circuit's operation.

8.4.1   Hazards

A sequential circuit consists of a combinational logic portion and a feedback path, consisting of either delay elements or flip-flops. The combinational logic can introduce both dynamic and static hazards at the excitation and output signals. Such hazards are prevented by including redundant gates in the circuit, as described in Chapter 4.

Another type of hazard that causes the asynchronous circuit to malfunction is an *essential* hazard. An essential hazard is caused by unequal delays along two or more paths that originate from the same input. An excessive delay in a path from the input compared with the delay introduced by the feedback path may cause such a hazard. Essential hazards are thus compensated for by making the delay of the affected feedback path large compared with the delay of other signals originating from the input terminals. Compensation for essential hazards thus depends on the individual circuit characteristics regarding the delays introduced by various paths in the circuit.

Note that the hazards introduced by the combinational logic do not affect the operation of a synchronous circuit, since the transitions are controlled by the clock. However, an asynchronous circuit may make a transition to an incorrect stable state as a result of the momentary incorrect signals produced by hazards. Two methods minimize the effect of hazards: using *inertial* delay lines and using unclocked flip-flops (latches) in the feedback path.

Figure 8.13 shows the characteristics of an inertial delay line. This type of delay line responds to a change in the input only if the change persists for at least Δt. Thus, momentary input changes are ignored. Inertial delay lines are more difficult to realize than pure delay lines.

The most common asynchronous circuit realization technique to minimize the effects of hazards is to use latches in the feedback paths, since the latch output does not change as a result of a momentary change on its inputs.

8.4.2 Cycles and Races

In practice, it is difficult to ensure that the delays introduced by all the feedback paths in an asynchronous circuit are equal. Thus, when two or more secondary variables must be changed simultaneously to bring about a state transition, the secondary variables go through intermediate incorrect values before establishing their final value. Because the circuit is sensitive

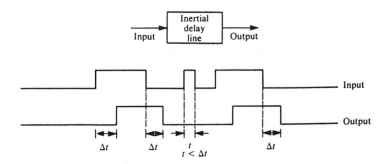

Figure 8.13 Inertial delay line.

to the order in which the signals change, an incorrect stable state might result.

A *race condition* is said to exist in a fundamental-mode circuit if two or more secondary variables must change when the circuit makes a transition from one stable state to another. If the circuit reaches the correct stable state in the presence of the race, the race is said to be *noncritical.* The race is *critical* if the circuit does not reach the correct stable state. Although critical races must always be avoided, the designer may often use noncritical race conditions to his or her advantage in the design of fundamental-mode circuits. Critical races are avoided through proper state assignment. The following examples further illustrate the concept of race. Race-free state assignment procedures are described following these examples.

Example 8.10

Consider the transition table of Figure 8.14(a). Each column has only one stable state. The possible transitions are

$$\textcircled{01} \rightarrow 11 \rightarrow 10 \rightarrow \textcircled{00}$$
$$\textcircled{00} \rightarrow 10 \rightarrow 11 \rightarrow \textcircled{01}$$

There are two unstable states in each transition. A circuit is said to have a *cycle* when it goes through a unique sequence of unstable states. Although the transition table indicates cycles in each transition, the number of unstable states in each transition depends on the order in which secondary variables change. For instance, when the circuit is in the total state $(0, 01)$ and x is changed to 1, if y_2 changes faster than y_1, the circuit reaches the state $(1, 00)$ without a cycle. The circuit goes through the cycle only if y_1 changes faster than y_2. Similar conditions exist for the other transition as well. The order in which the secondary variables change depends on the relative delays introduced by the feedback paths. In this case, since each column has only one stable state, the circuit always reaches the correct stable state.

The transition table in (b) shows another example of cycles. Again, the circuit in this case always reaches the correct stable state regardless of the order in which the secondary variables change. Because both the secondary variables change in their value at each transition between two stable states, race conditions exist. However, these race conditions are noncritical, since the circuit always reaches the correct stable state.

The transition table in (c) is an example of an unstable circuit. Here, as soon as x is changed to 1, the circuit goes into a cycle and oscillates between the unstable states $(11 \rightarrow 10 \rightarrow 00 \rightarrow 01 \rightarrow 11$ ...). This is because there is no stable state corresponding to $x = 1$.

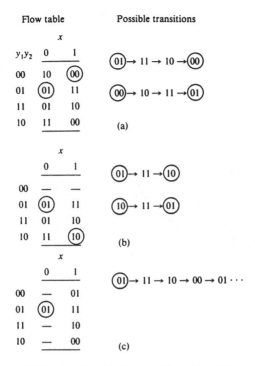

Figure 8.14 Cycles. (a) Stable; (b) stable; (c) unstable.

Example 8.11

As seen by Example 8.10 there must be at least one stable state under each anticipated input condition to ensure that the circuit does not oscillate in a cycle of unstable states. As long as there is a single stable state in each column of the transition table, all the race conditions will be noncritical. This fact is further illustrated by the transition table of Figure 8.15(a). All possible transitions are shown.

Figure 8.15(b) shows a transition table that exhibits critical race conditions. Consider the transition from 00 to 11. If both the variables change exactly at the same time (which is very unlikely), the circuit reaches the stable state 11. However, if y_2 changes more rapidly than y_1, the circuit goes through the unstable state 01 and reaches the stable state 11. If y_1 changes more rapidly than y_2, the circuit reaches a stable state of 10. Thus, there is a critical race condition. As shown in the figure, the transition from 10 to 01 also results in a critical race.

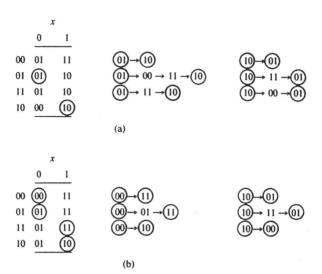

Figure 8.15 Races. (a) Noncritical races; (b) critical races.

Critical races can be avoided by ensuring that only one secondary variable changes during all the transitions between two stable states, through proper state assignment. For a flow table with a small number of rows, each transition can be examined to determine the proper state assignment so that only one variable changes during the transition. Nonetheless, this process becomes tedious as the number of rows increases. We will now examine three state assignment procedures that apply to flow tables with any number of rows.

8.5 State Assignment

Once a reduced flow table is derived during the design of a fundamental-mode circuit, state variables must be assigned to each row in the flow table. The assignment of these variables should avoid all critical races. Three assignment procedures are popular: *shared-row method, multiple-row method*, and *one-hot method*. (All these methods are illustrated in the examples that follow.) Flow tables with two rows require a single secondary-state variable. As such, they do not impose any critical race condition. For simplicity, we will examine state assignment procedures through flow tables with three or four rows, though these methods apply to flow tables with any number of rows and columns. The circuit's output portion is not considered in these examples, since it does not influence the state assignment problem.

8.5.1 Shared-Row Method

By including an extra row into the flow table, this method introduces a cycle into each transition that contributes a critical race. This new row is "shared" by the two stable states involved.

Example 8.12

Consider the three-row flow table shown in Figure 8.16(a). We first draw a *transition diagram*. The transition diagram is a graphical illustration of all the transitions between stable states in the flow table. The transition diagram has one node for each row in the flow table. Two nodes are connected by a line if there is a transition between corresponding rows. The input condition corresponding to the transition (i.e., the column of the flow table in which the transition takes place) is also shown on the line. The transition diagram for the flow table in (a) is shown in (b).

Recall that the columns in the flow table with a single stable state do not contribute critical race conditions. As such, all such column designators can be eliminated from the transition diagram, as far as the critical race-free state assignment problem is concerned. If all column designators on a line in the transition diagram are eliminated, the corresponding line is removed from the transition diagram. In the example here, none of the columns contributes critical rates. Hence all

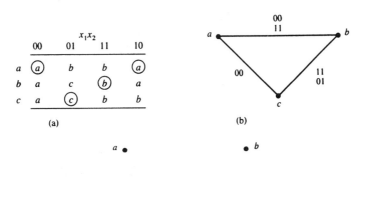

Figure 8.16 Critical race identification. (a) Flow table; (b) transition diagram; (c) reduced transition diagram.

the lines in the transition diagram are removed, resulting in the diagram shown in (c). Thus, the flow table is critical race-free. An arbitrary assignment is sufficient for this circuit.

Example 8.13

A three-row flow table is shown in Figure 8.17(a) and the corresponding transition diagram in (b). Because column 00 is critical race-free, the 00 designation is removed, resulting in the reduced transition diagram in (c). This diagram indicates all the critical transitions. Starting with an arbitrary assignment of 00 to row a, rows b and c must be made adjacent to a. Thus, b can be 01 and c can be 10 (or vice versa). But this results in b and c not being adjacent. Thus, a race-free state assignment cannot be made to these three rows without changing the flow table. In the shared-row method of state assignment, an unstable state is introduced into the flow table. In this example, unstable state d is introduced between b and c because of the nonadjacency of the state variables of these two states. By assigning the remaining combination of 11 to d, all the transitions are made critical race-free. The expanded flow table is shown in (d). All the transitions between b and c now go through d. Note that there cannot be a stable state in row d.

State d could have been introduced in each of the other two transitions in the transition diagram of (c) with the corresponding race-

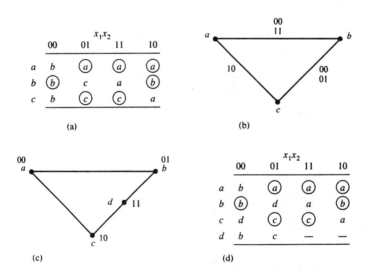

(a)

(b)

(c)

(d)

Figure 8.17 Shared-row state assignment. (a) Flow table; (b) transition diagram; (c) reduced transition diagram; (d) expanded flow table.

free state assignment. Thus, there are three possible race-free assignments for a flow table with three rows.

Example 8.14

Figure 8.18(a) shows a four-row flow table, and the corresponding transition diagram is shown in (b). All the columns in this flow table are critical. All the adjacencies implied by the transition diagram cannot be met by any assignment of two state variables. Hence, additional rows must be introduced, thereby increasing the number of state variables to three. Because there are three independent triangles in the

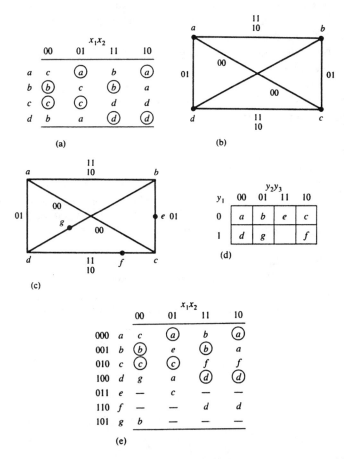

Figure 8.18 State assignment. (a) Flow table; (b) transition diagram; (c) reduced transition diagram; (d) state assignments; (e) expanded flow table.

transition diagram, we need at least three extra states to arrive at a race-free assignment. There are several possible ways to introduce these unstable states. One such scheme is shown in (c). In order to meet all the adjacencies implied by the diagram (c), a three-variable K-map shown in (d) can be utilized. Here, a is arbitrarily entered in cell $y_1y_2y_3 = 000$. States b, c, and d are entered into cells adjacent to that of a. Shared-state e is inserted between b and c, the states sharing it. Similarly, state g is inserted to be shared by states b and d, and state f is shared by c and d.

The modified flow table is shown in (e). In some cases, it is possible to reduce the number of additional rows needed by directing some transitions through the don't-cares in the rows already added. This is accomplished by trial and error until a satisfactory critical race–free assignment is found.

Note that the above assignment is valid for any four-row flow table. As can be seen, the shared-row method depends on the designer's intuitive ability to arrive at the appropriate placement of unstable states and thus can become cumbersome for flow tables with large numbers of rows.

8.5.2   Multiple-Row Method

In this method, each row in the original flow table is replaced by two or more equivalent rows. Thus, there will be two or more assignments of state variables to each state. The behavior of the circuit is the same as that in the original state, regardless of which of these multiple equivalent states the circuit is in.

A possible multiple-row assignment for a four-row flow table is shown in Figure 8.19. Here, there are two state variables for each state, and each is a logical complement of the other. For instance, state a is represented by a_1 and a_2, where a_1 has an assignment of 000, while a_2 is assigned 111. Also note that a_1 is adjacent to b_1, d_1, and c_2, while a_2 is adjacent to b_2, d_2, and c_1.

	y_2y_3			
y_1	00	01	11	10
0	a_1	b_1	c_1	d_1
1	c_2	d_2	a_2	b_2

Figure 8.19   Multiple-row assignment.

| | x_1x_2 | | | |
	00	01	11	10
a_1	c_2	a_1	b_1	a_1
a_2	c_1	a_2	b_2	a_2
b_1	b_1	c_1	b_1	a_1
b_2	b_2	c_2	b_2	a_2
c_1	c_1	c_1	d_1	d_1
c_2	c_2	c_2	d_2	d_2
d_1	b_2	a_1	d_1	d_1
d_2	b_1	a_2	d_2	d_2

Figure 8.20 Expanded flow table.

The flow table in Example 8.14 is expanded in Figure 8.20. Here, each row is replaced by two rows. For instance, row a is replaced by rows a_1 and a_2. In the original flow table, state a makes transitions to states b and c. Correspondingly, in the expanded table, a_1 makes transitions to b_1 and c_2; and a_2 makes transitions to b_2 and c_1, because of the adjacencies implied by the assignments of Figure 8.19. Note also that the stable states in rows a_1 and a_2 are marked appropriately as a_1 and a_2, respectively. The other six rows in the expanded flow table of Figure 8.20 are similarly completed.

If the circuit starts in the total state $(01, a_1)$ and the input sequence is 00-10-00-01-11, the state sequence will be c_2-d_2-b_1-c_1-d_1. This is equivalent to the transition through the state sequence c-d-b-c-d in the original flow table starting from $(01, a)$, as far as the input/output behavior of the circuit is concerned.

Figure 8.21 shows the assignments that are applicable to an eight-row flow table. Figure 8.22 shows an alternate multiple-row assignment for flow tables of various sizes. Here, the assignments are made so that one row in each set of equivalent rows is adjacent to one row in each of the remaining sets of equivalent rows. Hence, race-free transitions can be made between

y_3y_4 \ y_1y_2	00	01	11	10
00	a_1	b_1	c_1	d_1
01	e_1	f_1	g_1	h_1
11	c_2	d_2	a_2	b_2
10	g_2	h_2	e_2	f_2

Figure 8.21 Multiple-row assignments for an eight-row table.

y_1y_2

y_3y_4	00	01	11	10
00	a	b	e	e
01	a	b	f	f
11	c	c	e	f
10	d	d	e	f

y_1y_2

$y_3y_4y_5$	00	01	11	10
000	a	b	c	c
001	a	b	d	d
011	a	a	c	d
010	b	b	c	d
110	e	f	e	e
111	e	f	f	f
101	g	g	g	h
100	h	h	g	h

Figure 8.22   Alternate multiple-row assignments.

any two stable states by properly establishing row-to-row transitions. Because such multiple-row assignments for various-size flow tables are available, the multiple-row assignment method is easier to adopt than the shared-row method. However, the number of rows in the expanded table obtained by this method tends to be larger than that with the shared-row method, thus increasing the circuit's complexity.

8.5.3   One-Hot Method

In this method, if the flow table has n rows, n secondary-state variables are used. Each state is assigned an n-bit vector with all but one of the bits being 0. Hence, the "one hot" assignment. Obviously, this method introduces the largest number of additional rows into the flow table of all the state assignment methods. In order to avoid critical races, cycles are introduced using additional rows, as in the shared-row method.

Example 8.15

Figure 8.23 shows the one-hot assignment for the flow table in Figure 8.18. The original four rows are assigned one-hot four-bit vectors. As shown in the transition diagram (Figure 8.23(b)), additional states are introduced to make the transitions race-free. For instance, state e is introduced between a and b and its state vector is 0011, which is the logical OR of the state vectors of a and b. Other unstable states and corresponding state vectors are obtained similarly.

Although this method is easier to adopt, it results in the most complex circuit of all three state assignment methods. We will now examine a complete design example.

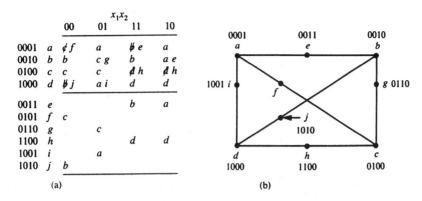

Figure 8.23 One-hot assignment. (a) Extended flow table; (b) transition diagram.

Example 8.16

A fundamental-mode circuit needs to be designed with two inputs, x_1 and x_2 and an output, Z. Z changes from 0 to 1 only when the inputs make the transition from 01 to 11 to 10 and remains at 1 until the next input transition of 00.

The primitive flow table is shown below:

	00	01	11	10
1	①/0	2/—	—/—	7/—
2	1/—	②/0	3/—	—/—
3	—/—	2/—	③/0	4/—
4	1/—	—/—	5/—	④/1
5	—/—	6/—	⑤/1	4/—
6	1/—	⑥/1	5/—	—/—
7	1/—	—/—	3/—	⑦/0

The initial state (RESET) is (00, 1). The circuit makes a transition of states 2, 3, and 4 as the required input sequence is received, and Z changes to 1 when state 4 is reached. From (10, 4) if the input changes to 00, the circuit returns to state 1, with Z changing to 0. The other possibility is for the input to change from 10 to 11 to 01 to 00. In that case, the circuit moves from state 4 to state 1 by way of states 5 and 6. Row 7 is included to handle the transition of inputs from 00 to 10 in row 1.

The implication chart is shown below:

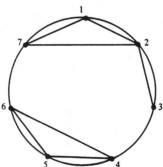

From the implication chart, the following compatible pairs of rows are obtained:

$$(1\ 2)(2\ 3)(4\ 5)(4\ 6)(5\ 6)(1\ 7)(2\ 7)$$

The merger diagram is shown below:

From the merger diagram, the following maximal compatibles are derived:

$$(1\ 2\ 7)(4\ 5\ 6)(2\ 3)$$

As can be seen, all three maximal compatibles are essential to satisfying the completeness criterion. The reduced flow table is shown below:

	00	01	11	10
			x_1x_2	
(1 2 7)	①/0	②/0	3/—	⑦/0
(4 5 6)	1/—	⑥/1	⑤/1	④/1
(2 3)	1/—	②/0	③/0	4/—

The unspecified outputs are now decided upon and by renaming the rows as $a = (1\ 2\ 7)$, $b = (4\ 5\ 6)$, and $c = (2\ 3)$, we can obtain the following flow table:

	x_1x_2			
	00	01	11	10
$(1\ 2\ 7) = a$	$\textcircled{a}/0$	$\textcircled{a}/0$	$c/0$	$\textcircled{a}/-$
$(4\ 5\ 6) = b$	a/d	$\textcircled{b}/1$	$\textcircled{b}/1$	$\textcircled{b}/1$
$(2\ 3)\quad = c$	$a/0$	$\textcircled{c}/0$	$\textcircled{c}/0$	b/d

The transition diagram is shown below:

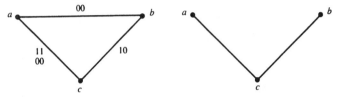

Removing the transition in the noncritical column, we can see that a and b need not be adjacent to obtain a race-free assignment. A state assignment that avoids races is

$$a = 00, \quad b = 11, \quad c = 01$$

Figure 8.24(a) shows the transition, excitation, and output tables and the circuit diagram using delay lines in the feedback path.

Figure 8.24(b) shows another realization using unclocked *JK* flip-flops as the memory elements. Note that because of the grouping on the maps for J_1 and K_1, both inputs will never be 1 at the same time. Thus, a $J = K = 1$ condition is avoided. If not, one of these inputs may change more rapidly than the other, bringing about a state change; this may be followed by a change in the second input, resulting in a further state change, which is not desirable. J_2 and K_2 inputs also satisfy the above condition for correct operation.

8.6 Generation of Timing Signals

In Chapter 7 we illustrated the generation of timing signals using counter and shift-register ICs. In this section, we will generate typical timing signals using one type of off-the-shelf component. These are called *mono-stable multivibrators* (or *one-shots*).

A one-shot is essentially a flip-flop with two outputs (Q and Q') and one stable state. It is normally in the CLEAR ($Q = 0$) state. When a *trigger* is applied, it flips its state to 1 and switches back to 0 after a certain time period. This time period (pulse width or pulse duration) is determined by the values of a capacitor and a resistor connected to the one-shot. The 1

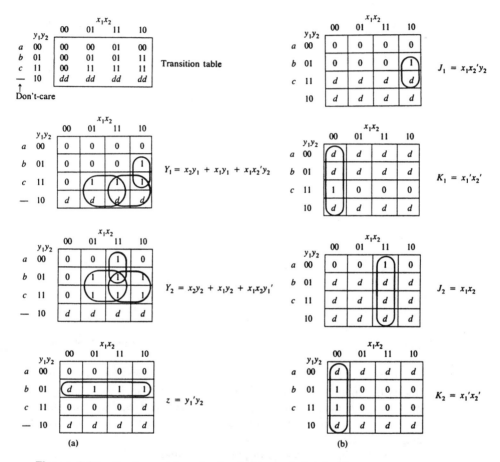

Figure 8.24 Fundamental-mode circuit (Example 8.16). (a) Using delay lines; (b) using *JK* latches.

state of the one-shot is called a *quasi-stable* state since it always returns to its stable state of 0. Two types of one-shots are available: *nontriggerable* and *retriggerable*. A trigger is an input change that "fires" the one-shot into its quasi-stable state. A trigger occurring on a nontriggerable one-shot that is in the quasi-stable state will have no effect on it; while a trigger occurring on a retriggerable one-shot triggers it further, thus continuing the 1 state for another pulse duration from the triggering point.

Nontriggerable one-shots are also limited by their *duty cycle*. After each ON cycle (i.e., 1 state), they need a certain amount of time (OFF) to recover before they can be retriggered. The duty cycle is indicated by

$$\frac{T_{ON}}{T}$$

where $T = T_{ON} + T_{OFF}$ is the period, T_{ON} is the on time, and T_{OFF} is the off time.

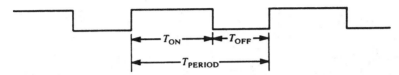

The duty cycle is specified in the data sheets supplied by the one-shot manufacturer. If the trigger is applied so often that it exceeds the duty cycle of the one-shot, the output waveform *jitters*. That is, the width of each pulse does not stay constant.

Example 8.17

Suppose the duty cycle of a one-shot is 70 percent and it is designed to produce a pulse of 100 μs. That is, $T_{ON} = 100$ μs and hence

$$\frac{100}{100 + T_{OFF}} = 0.7$$

Thus, $T_{OFF} = 42.85$ μs. This means that the one-shot cannot be triggered for at least 42.85 μs after the 100-μs on state.

Figure 8.25 shows a nontriggerable TTL one-shot (74121). It has three trigger inputs (A_1, A_2, and B). A_1 and A_2 are used with normal (fast-changing) TTL inputs, which change more rapidly than 1 V/μs. The B input is used with inputs that change as slowly as 1 V/s. This is indicated by the hysteresis symbol (⊐) on the AND gate. (Gates that accept slow-changing signals and respond only when the input signal reaches a threshold value (V_t) are called *Schmitt triggers*. The characteristic of a Schmitt trigger is shown in Figure 8.26. Schmitt triggers are used to smooth out a noisy or an irregular signal.)

The first four rows of the function table for the one-shot indicate the *quiescent state* (i.e., $Q = 0$ and the one-shot is not triggered). The last four rows indicate the inputs required to trigger the one-shot. As can be seen, it is triggered by two input conditions:

1. Making the B input go high when either or both A inputs are low.
2. Either making one of the A inputs go low while the other is high or making both A inputs go low simultaneously (when the B input is high).

PIN CONFIGURATION   **LOGIC SYMBOL**   **LOGIC SYMBOL (IEEE/IEC)**

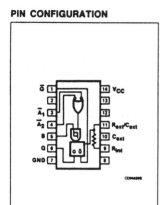 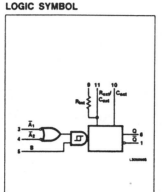 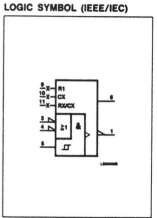

FUNCTION TABLE

INPUTS			OUTPUTS	
$\overline{A}_1$	$\overline{A}_2$	B	Q	$\overline{Q}$
L	X	H	L	H
X	L	H	L	H
X	X	L	L	H
H	H	X	L	H
H	↓	H	⎍	⎍
↓	H	H	⎍	⎍
↓	↓	H	⎍	⎍
L	X	↑	⎍	⎍
X	L	↑	⎍	⎍

H = HIGH voltage level
L = LOW voltage level
X = Don't care
↑ = LOW-to-HIGH transition
↓ = HIGH-to-LOW transition

Figure 8.25 Nontriggerable one shot (TTL 74121) (Courtesy of Signetics Corporation).

The pulse width produced by the one-shot is determined by

$$T_{\rm ON} = 0.7 C_T R_T$$

where R_T and C_T are the timing resistor and capacitor values, respectively. The pulse width from 74121 can be varied from 20 ns to 28 s. When the circuit has no external resistor and capacitor (i.e., the internal resistor is connected to V_{CC} by connecting pins 9 and 14, and there is a stray capacitance between pins 10 and 11), the pulse width will be 30 to 35 ns. Other pulse widths can be obtained by connecting an external capacitor (whose value can range from 10 pF to 10 μF) between pins 10 and 11 and an external resistor (whose value can range from 2 KΩ to 40 KΩ) between

AC ELECTRICAL CHARACTERISTICS $T_A = 25°C$, $V_{CC} = 5.0V$

PARAMETER		TEST CONDITIONS	74 $C_L = 15pF$, $R_L = 400\Omega$		UNIT
			Min	Max	
t_{PLH} t_{PHL}	Propagation delay $\overline{A}$ input to Q & $\overline{Q}$ output	Waveform 1 $C_{ext} = 80pF$, R_{int} to V_{CC}		70 80	ns
t_{PLH} t_{PHL}	Propagation delay B input to Q & $\overline{Q}$ output	Waveform 2 $C_{ext} = 80pF$, R_{int} to V_{CC}		55 65	ns
t_W	Minimum output pulse width	$C_{ext} = 0pF$, R_{int} to V_{CC}	20	50	ns
t_W	Output pulse width	$C_{ext} = 80pF$, R_{int} to V_{CC}	70	150	ns
		$C_{ext} = 100pF$, $R_{ext} = 10k\Omega$	600	800	ns
		$C_{ext} = 1\mu F$, $R_{ext} = 10k\Omega$	6.0	8.0	ms

AC WAVEFORMS

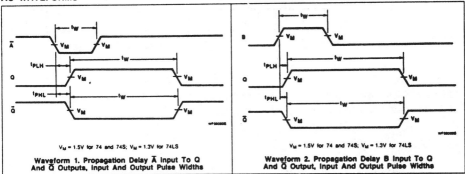

$V_M = 1.5V$ for 74 and 74S; $V_M = 1.3V$ for 74LS

Waveform 1. Propagation Delay $\overline{A}$ Input To Q And $\overline{Q}$ Outputs, Input And Output Pulse Widths

$V_M = 1.5V$ for 74 and 74S; $V_M = 1.3V$ for 74LS

Waveform 2. Propagation Delay B Input To Q And $\overline{Q}$ Output, Input And Output Pulse Widths

AC SET-UP REQUIREMENTS $T_A = 25°C$, $V_{CC} = 5.0V$

PARAMETER		TEST CONDITIONS	74		UNIT
			Min	Max	
t_W	Minimum input pulse width to trigger	Waveforms 1 & 2	50		ns
R_{ext}	External timing resistor range		1.4	40	$k\Omega$
C_{ext}	External timing capacitance range		0	1000	μF
	Output duty cycle	$R_{ext} = 2k\Omega$		67	%
		$R_{ext} = R_{ext}(Max)$		90	%

Figure 8.25 Continued

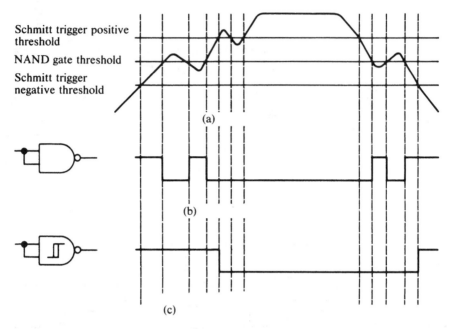

Schmitt trigger positive threshold

NAND gate threshold

Schmitt trigger negative threshold

(a)

(b)

(c)

Figure 8.26 Schmitt trigger characteristics (Courtesy of Signetics Corporation). (a) Input; (b) NAND output; (c) Schmitt trigger output.

pins 11 and 14. The pulse-width values for typical resistor and capacitor values are shown in Figure 8.27.

The duty cycle of the 74121 depends on the timing resistor and ranges from 67% for a 2 KΩ timing resistor to 90% for a 40 KΩ timing resistor.

Example 8.18

Figure 8.28 shows two ways of generating a pulse of 0.5 μs using a 74121. In (a), the 2 KΩ internal resistor is used. The capacitance required is determined by

$$0.5 \times 10^{-6} = 0.7 \times 2 \times 10^{3} \times C_{T}$$

That is, $C_{T} = 357$ pF. In (b) an external resistor (10 KΩ) is used. Hence, the capacitance required is given by

$$0.5 \times 10^{-6} = 0.7 \times 10 \times 10^{3} \times C_{T}$$

That is, $C_{T} = 71.4$ pF.

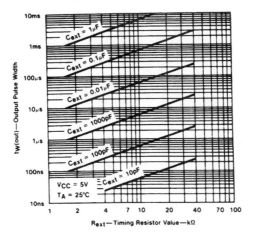

Figure 8.27 Output pulse width versus timing resistor value for the TTL 74121 (Courtesy of Signetics Corporation).

Example 8.19

Figure 8.29(a) shows an *oscillator* circuit built out of two 74121s. Each of these is designed to produce a 0.5-µs pulse. The first 74121 is triggered by throwing the switch to the start position, thus sending the B input high while both the A inputs are low. The falling edge of Q_1 (at the end of the pulse) triggers the second one-shot, which produces a pulse on Q_2. Both one-shots go into a stable state after one pulse each.

If the A_2 input of the first one-shot is tied to V_{CC}, as shown in (b), the pulse sequence repeats. In this example, both one-shots pro-

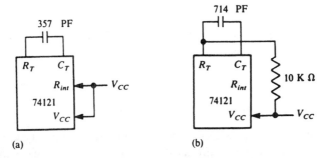

Figure 8.28 Generating a 0.5-µs pulse using the TTL 74121 (Courtesy of Signetics Corporation). (a) Using an internal resistor; (b) using an external resistor.

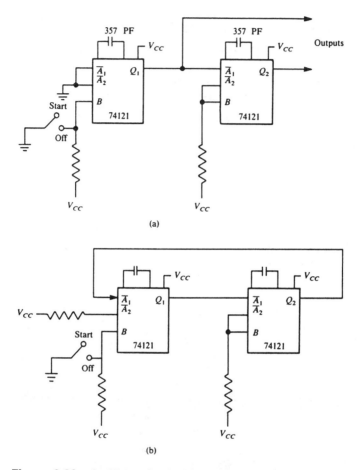

Figure 8.29   Oscillator circuit. (a) Single cycle; (b) continuous.

duce pulses of the same duration. These durations can be changed by the appropriate resistor and capacitor values to obtain required clocks with various timing characteristics.

8.7   Switch Bouncing

In this section, we will examine a very common use for an asynchronous circuit. Mechanical switches are used to start, stop, reset, and also to input data to digital systems. One position of the switch provides a voltage corresponding to logic-1 and the other position corresponds to logic-0. When a switch is thrown from one position to the other, the switch blade goes through several vibrations (bounces) before coming to rest at the latter position. Because of this bouncing action, several spikes occur in the signal produced by the switch. Typically, the switch bounce lasts several milliseconds. Figure 8.30 shows the waveforms produced by a single-pole double-throw switch when it is thrown from one position to the other. Most digital systems require just one transition in the signal from the switch rather than multiple transitions produced by a bouncy switch, since multiple transitions may result in unnecessary triggers to the system. Three popular switch debouncing circuits are shown in Figure 8.31. The output of each of these circuits is a clean pulse corresponding to each toggle of the switch.

In (a), when the switch is in the up position, the flip-flop is set. If the switch is now thrown to the down position, the bounces that occur when the switch leaves the up position do not affect the state of the flip-flop. When the switch blade reaches the down position, the flip-flop is reset by the first pulse. The pulses, as a result of the bounce after the first pulse, merely serve as multiple clear pulses. Thus, the output of the flip-flop produces a clean transition. A similar process occurs when the switch is moved from the down to the up position. The first pulse sets the flip-flop, and the following bounce pulses merely serve as additional set pulses. The debouncing action of circuits shown in (b) and (c) is similar to that of the circuit in (a).

Single-pole single-throw switches have only two contacts. They are either open or closed. The circuits in Figure 8.31 cannot be used to debounce

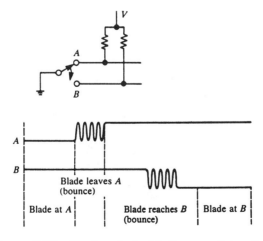

Figure 8.30 Single-pole double-throw swtich action.

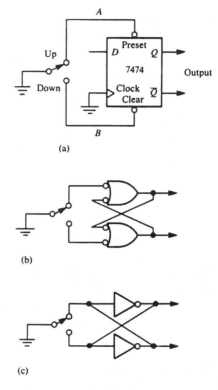

Figure 8.31 Switch debouncing circuits. (a) Using a flip-flop; (b) using cross-coupled NOR gates; (c) using cross-coupled NOT gates.

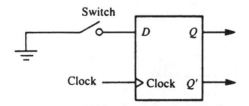

Figure 8.32 Debouncing single-pole single-throw switches using a flip-flop.

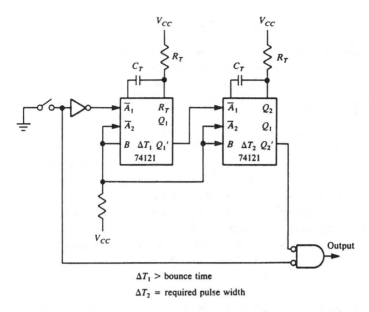

$\Delta T_1 >$ bounce time

$\Delta T_2 =$ required pulse width

Figure 8.33 Debouncing single-pole single-throw switches using one-shots.

such switches. A simple debouncing circuit using a clocked D flip-flop is shown in Figure 8.32. Here, the switch is connected to the D input of the flip-flop and the clock period is chosen to be longer than switch bounce time. Thus, the clock pulse that arrives after the bouncing is complete changes the flip-flop state.

Figure 8.33 shows a debouncing circuit using two one-shots. This circuit is useful when a clock is not available, as required by the previous circuit. The pulse duration of the first one-shot is set to be longer than the switch bounce time. Thus, the switch bounces have no further effect on the first one-shot once it is fired by the first switch pulse. When the first one-shot times out, the second one-shot is triggered by the trailing edge of Q_1'. The pulse duration of the second one-shot is set to the required value. When the second one-shot is active, the switch will be at the low position. Because an output pulse is desired only when the switch is at the low position, the output of the second one-shot (Q_2') is gated by the AND gate as shown. Thus, the output of the circuit provides a pulse only at the high-to-low transition of the switch. A single pulse is produced by this circuit for each switch throw. The capacitor and resistor values for the first one-shot are determined by the bounce time of the switch, while those of the second one-shot are determined by the output pulse duration required (see problem 8.19).

8.8   A Power-On Reset Circuit

For proper operation, synchronous sequential circuits must be initialized to their starting state. Usually an all-zero state vector is used to denote the starting state in order to facilitate resetting the circuit when the power is turned on, thereby bringing the circuit up with its starting state. One way to accomplish this is to include a master-clear switch that is connected to the asynchronous reset terminals of all the flip-flops. After the power is turned on, the flip-flops are resent through the master-clear operation. Note that it is not necessary to use an all-zero vector for the starting state. If any other vector is used, the master-clear circuit is designed to set and reset the flip-flops accordingly, through their preset and clear inputs.

Figure 8.34 shows a power-on-reset circuit. As the power is turned on, a step (rapidly increasing) signal is seen at V_p. The resistor-capacitor combination provides a slowly increasing signal V_i across the capacitor, as the capacitor charges. In response to V_i as the input, the Schmitt trigger produces an immediate high signal at its output, which switches to a low value as the capacitor is charged. The output of the Schmitt trigger can be used to reset or set the flip-flops in the circuit. The duration t of the reset signal is roughly proportional to $R \times C$, where R is in ohms, C is in farads, and t is in seconds.

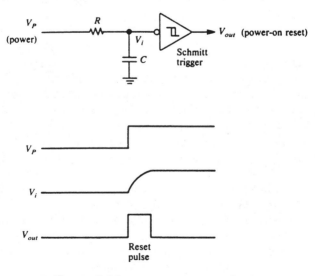

Figure 8.34   A power-on reset circuit.

8.9 Summary

This chapter introduced two types of asynchronous circuits. The analysis and synthesis of the first type (pulse-mode circuits) are very similar to that of synchronous circuits, except for some restrictions on the inputs. The second type (fundamental-mode circuits) are more difficult to analyze and design. Several examples were given to illustrate the analysis and design of such circuits. Hazards and race conditions common in such circuits were described and state assignment procedures to avoid critical race conditions were presented. Finally, the use of one-shots to generate timing signals and switch bounce avoidance techniques were presented. In practice, most digital systems are synchronous. However, there are occasions in which one encounters asynchronous behavior. The procedures provided in this chapter should be helpful in such occasions.

References

Caldwell, S. H. *Switching Circuits and Logical Design.* New York: John Wiley & Sons, 1958.

Greenfield, J. D. *Practical Digital Design Using ICs.* New York: John Wiley & Sons, 1983.

Mano, M. M. *Digital Design.* Englewood Cliffs, N.J.: Prentice-Hall, 1994.

McCluskey, E. J. *Introduction to the Theory of Switching Circuits.* New York: McGraw-Hill, 1965.

Nagle, H. T., Carroll, B. D., and Irwin, J. D. *An Introduction to Computer Logic.* Englewood Cliffs, N.J.: Prentice-Hall, 1975.

National Advanced Bipolar Logic Data Book, Santa Clara, Calif: *National Semiconductor*, 1995.

Roth, C. H. *Fundamentals of Logic Design.* St. Paul, Minn.: West Publishing, 1985.

TTL Data Manual. Sunnyvale, Calif.: Signetics, 1986.

Unger, S. H. *Asynchronous Sequential Switching Circuits.* New York: Wiley-Interscience, 1969.

Problems

8.1 Analyze the pulse-mode circuit shown in Figure 8.35. Determine the state table. Draw a timing diagram showing all input, output, excitation, and secondary-state values if the input sequence is x_1-x_3-x_2-x_1-x_3-x_2, starting from $x_1 = x_2 = x_3 = y_1 = y_2 = 0$.

8.2 Analyze the pulse-mode circuit shown in Figure 8.36 to determine the state table. Draw a timing diagram showing the inputs and output corresponding to the input sequence x_1-x_1-x_1-x_2-x_1-x_2, starting with state 00.

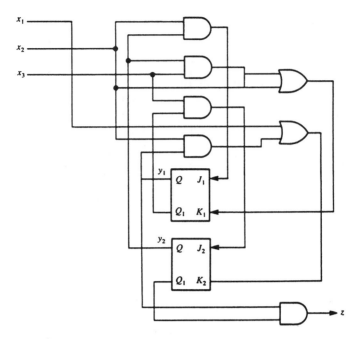

Figure 8.35 See Problem 8.1.

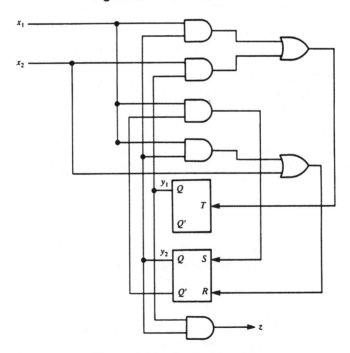

Figure 8.36 See Problem 8.2.

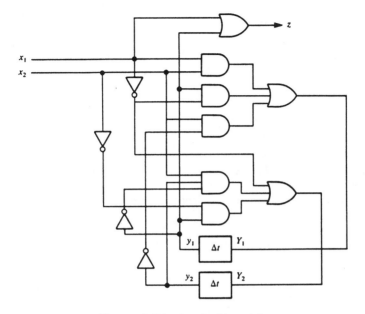

Figure 8.37 See Problem 8.3.

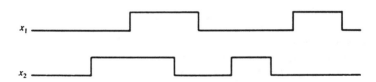

Figure 8.38 See Problem 8.3.

8.3 Analyze the fundamental-mode circuit shown in Figure 8.37 to determine the excitation and output tables. Complete the timing diagram for the input sequence shown in Figure 8.38.

8.4 Draw the circuit diagram for a pulse-mode circuit described by the following equations:

$$Z = xy_1y_2$$
$$J_1 = xy_2 \quad J_2 = x'$$
$$K_1 = x' \quad K_2 = x' + y_1'$$

Can you derive a state table for this circuit using K-maps for the above equations? Justify your answer.

8.5   For the flow table shown below:

| | x_1x_2 | | | |
	00	01	11	10
a	$d/0$	$c/0$	$b/0$	$a/0$
b	$c/0$	$c/0$	$b/1$	$d/1$
c	$c/0$	$c/0$	$b/1$	$d/1$
d	$d/0$	$d/0$	$c/0$	$a/0$

(a) Indicate all the stable states.
(b) If the machine is in the total (stable) state (00, d), find the output and
    next-state sequence for the input sequence

$$x_1x_2 = 00\text{-}01\text{-}11\text{-}10\text{-}00$$

8.6   For the following flow table:

| y_1y_2 | x_1x_2 | | | |
	00	01	11	10
00	00	11	00	11
01	11	01	11	11
10	00	10	11	11
11	11	11	00	11

(a) Find all the races.
(b) Indicate critical and noncritical races.
(c) Find a critical race-free state assignment.

8.7   For the following flow table:

| | x_1x_2 | | | |
	00	01	11	10
a	$\textcircled{a}/1$	$b/\!\!-$	$d/\!\!-$	$\textcircled{a}/1$
b	$c/\!\!-$	$\textcircled{b}/0$	$d/\!\!-$	$\textcircled{b}/\!\!-$
c	$\textcircled{c}/1$	$b/\!\!-$	$\textcircled{c}/0$	$a/\!\!-$
d	$c/\!\!-$	$\textcircled{d}/1$	$\textcircled{d}/0$	$a/\!\!-$

(a) Specify all the outputs.
(b) Find a critical race-free state assignment.

8.8 Find a critical race-free assignment for the following reduced table by each of the three state assignment methods:

	x_1x_2 00	01	11	10
a	fl—	(a)/0	(a)/1	fl—
b	(b)/1	al—	(b)/1	dl—
c	el—	(c)/1	al—	(c)/1
d	fl—	(d)/0	bl—	(d)/0
e	(e)/0	cl—	(e)/0	fl—
f	(f)/1	dl—	el—	(f)/1

8.9 Design a pulse-mode circuit using the following state table:

	x_1	x_2	x_3
a	a/0	b/0	c/1
b	b/0	c/0	d/0
c	c/0	d/0	a/1
d	d/0	a/0	b/1

Use the following assignments: $a = 00$, $b = 01$, $c = 11$, and $d = 10$. Use *JK* flip-flops.

8.10 Reduce the following primitive flow table:

	x_1x_2 00	01	11	10
1	(1)/1	3/—	—/—	5/—
2	(2)/0	3/—	—/—	4/—
3	1/—	(3)/1	6/—	—/—
4	2/—	3/—	—/—	(4)/—
5	2/—	—/—	6/—	(5)/1
6	—/—	3/—	(6)/1	5/—

8.11 Reduce the following primitive flow table:

	x_1x_2			
	00	01	11	10
1	8/—	—/—	6/—	①/0
2	—/—	3/—	②/1	1/—
3	4/—	③/1	2/—	—/—
4	④/1	3/—	—/—	5/—
5	4/—	—/—	2/—	⑤/1
6	—/—	7/—	⑥/0	1/—
7	8/—	⑦/0	6/—	—/—
8	⑧/0	7/—	—/—	5/—

8.12 A fundamental-mode circuit with two inputs, x_1 and x_2, and two outputs, z_1 and z_2, is required. The circuit operates so that z_i ($i = 1$ or 2) takes on the value of 1 if and only if x_i was the input that changed last. Assume that $z_1 = z_2 = 0$ to start with at $x_1 = x_2 = 0$. Draw the primitive flow table and reduce it.

8.13 A fundamental-mode circuit with two inputs (x_1 and x_2) and one output, z, is required. Whenever x_1 is 0, z is 0. The output z goes to 1 on the first 0-to-1 transition of x_2 when x_1 is 1. The output remains at 1 until x_1 returns to 0. No change in output occurs for any other conditions.
(a) Draw a primitive flow table and reduce it.
(b) Specify all the outputs in the reduced flow table.
(c) Find a race-free state assignment.
(d) Derive the circuit diagram using delay lines.
(e) Derive a circuit diagram using unclocked SR flip-flops.

8.14 A pulse-mode circuit with two inputs, x_1 and x_2, and one output, z, is needed. The output changes from 0 to 1 only on the occurrence of the last x_2 pulse in the sequence x_1-x_2-x_1-x_2. The output is reset by the first x_1 pulse occurring while z is at 1. Draw a state diagram.

8.15 Study the timing characteristics of Schmitt triggers (7413 and 74132) from the TTL data books.

8.16 Compare the operation and timing characteristics of other one-shots (e.g., 74123 and 74221) with those of 74121.

8.17 Determine the resistor and capacitor values required to generate pulses of the following width using 74121s:
(a) 40 ns.
(b) 1 second.
(c) 22 seconds.

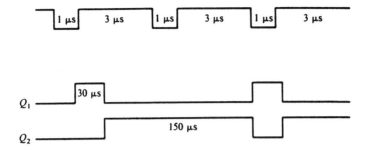

Figure 8.39 See Problem 8.20.

8.18 Extend the design in problem 8.17 to generate clocks with the appropriate pulse width. That is, the pulse should repeat. What is the time between pulses in each case?

8.19 Determine the resistor and capacitor values in the debounce circuit of Figure 8.33, if the switch bounce time is 9 ms and a pulse of 1 μs is needed.

8.20 Use two 74121s to derive the clock signals shown in Figure 8.39.

9

Programmable Logic

9.1 Introduction

The logic implementations discussed in previous chapters required the interconnection of selected SSI, MSI, and LSI components on printed circuit boards (PCBs). With the current hardware technology, the cost of the PCB, connectors, and the wiring is about four times that of the ICs in the circuit. Yet this implementation mode is cost-effective for circuits that are built in small quantities. With the progress in IC technology leading to the current VLSI era, it is now possible to fabricate a very complex digital system on a chip. But, for such implementation to be cost effective, large quantities of the circuit or the system would be needed, since the IC fabrication is a costly process. To manage costs while exploiting the capabilities of the technology, three implementation approaches are currently employed, based on the quantities of the circuits needed: custom, semicustom, and programmable logic. Custom implementations are for circuits needed in large quantities; semicustom for medium quantities, and the programmable logic mode is used for small quantities. In this chapter, we will first provide a brief description of the IC fabrication process and examine the relative merits of each of these approaches. We will then provide details on programmable logic design.

9.2 IC Fabrication

Figure 9.1 shows the steps in the IC manufacturing process. The process starts off with a thin (10 mil thick, with 1 mil equaling 1/1000th of an inch) slice of *p*-type semiconductor material, about two to five inches in diameter, called a *wafer*. Hundreds of identical circuits are fabricated on the wafer

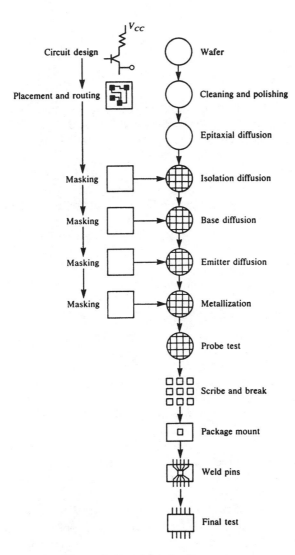

Figure 9.1 IC fabrication process.

using a multistep process. The wafer is then cut into individual dice, each
die corresponding to an IC.

The fabrication process consists of the following steps:

1. Wafer surface preparation.
2. Epitaxial growth.
3. Diffusion.
4. Metallization.
5. Probe test.
6. Scribe and break.
7. Packaging.
8. Final test.

The circuit designer first develops the circuit diagram. Before fabrication, the circuit must be at the transistor, diode, and resistor level of detail.
Today, however, the circuit designer does not need to deal with this level
of detail, since there are Computer-Aided Design (CAD) tools that translate
a gate-level design to the circuit level. Once the circuit is designed, it is
usually simulated to verify its functional, timing, and loading characteristics.
If these characteristics are acceptable, the circuit is brought to the placement
and routing stage.

Placement is the process of placing the circuit components at the appropriate positions so that the interconnections can be routed properly. Automatic placement and routing tools are available. For a very complex circuit, this step is the most time-consuming one, even with CAD tools. At the
end of this step, the circuit layout reflects its structure on the silicon.

The wafer is made of a high-resistivity, single-crystal, p-type silicon
material. The wafer is first cleaned, and both sides of it are polished, with
one side finished to mirror smooth.

The epitaxial diffusion is the process of adding minute amounts of n-
or p-type impurities (dopants) to achieve the desired low resistivity. The
epitaxial layer forms the collector of the transistors. This layer is then covered with a layer of silicon dioxide formed by exposing the wafer to an
oxygen atmosphere around 1000°C.

The most critical part of the fabrication process is the preparation of
masks to transfer the circuit layout into silicon. The mask plates are first
drawn according to the circuit layout. These plates are then reduced by
photographic techniques to the size of the final chip. The masks are then
placed on the wafer, one mask for each die, and a photoresist coating on
the wafer is exposed to ultraviolet light. The unexposed surface is etched
chemically, leaving the desired pattern on the wafer. The surface pattern is
then subjected to diffusion.

Although the steps vary depending on the process and the technology used to manufacture the IC, the diffusion can be classified into *isolation, base,* and *emitter diffusion* stages. These are the stages in which the corresponding terminals of the transistors are fabricated; each diffusion stage corresponds to one or more mask and etch operation.

By now, the wafer contains several identical dice with all the circuit components formed on each die. The wafer is then subjected to photo etching to open the windows to provide the connections between the components. The interconnections are made (i.e., through metallization) by vacuum deposition of a thin film of aluminum over the entire wafer, followed by another mask and etch operation to remove unnecessary interconnections.

Among the dice now on the wafer, some may be defective as a result of imperfections in the wafer or in the fabrication process. Selected dice are now tested to mark the failing ones. The percentage of good dies obtained is called the ''yield'' of the fabrication process.

The wafer is now scribed by a diamond-tipped tool along the boundaries of the dice to separate them into individual chips.

Each die is then tested and mounted on a leader, and pins are attached and packaged either in an inline or flat package.

Circuit layout and mask preparation are the most time-consuming and error-prone stages in the fabrication process and hence contribute most to the design cost. When the circuit is very complex, the circuit layout requires thousands of labor-hours, even with the use of CAD tools.

9.3 Circuit Implementation Modes and Devices

In the so-called custom-design mode of circuit implementation, all the steps in the IC fabrication process (described in the previous section) are unique to the application at hand. Although this mode offers the smallest chip size and highest speed, the design costs do not justify this mode for low-volume applications. Typically, the annual sales volume of the IC should be around five to ten times the nonrecurring engineering (NRE) costs of the design. Typical NRE costs of $100,000 to $250,000 indicate that the number of ICs produced should be in the order of 50,000 to 100,000 to be cost-effective. This makes the custom-design mode beyond the reach of most applications.

In the semicustom-design mode, the initial steps in the fabrication process remain standard for all applications. Only the last step (metallization) is unique to each application. This is accomplished using fixed arrays of gates predesigned on the chip real estate. These gates are interconnected to form the application specific IC (ASIC). The NRE costs for these ICs is of the order of $10,000 to $40,000, thus making them cost-effective for low-volume applications. Because of the use of standard gate patterns on the IC and simpler design rules employed, these ICs cannot use the chip area as efficiently and their speeds are also lower compared to custom-designed ICs. The ICs used in semicustom-design mode are *mask programmable*. That means, to make these ICs application specific, the user supplies the interconnection pattern (i.e., the program) to the IC manufacturer, who in turn prepares masks to program the IC. Obviously, once programmed, the function of these ICs cannot be altered.

In the programmable design mode, ICs known as programmable logic devices (PLDs) are used in the design. There are several types of PLDs each with their own pattern of gates and interconnection paths pre-fabricated on the chip. These ICs are programmed by the user with special programming equipment (called PLD programmers). That is, they are *field programmable*. Some PLDs allow erasing the program to reprogram them and some do not.

In the early development stages of the digital system, field-programmable devices are typically used to allow the flexibility of design alterations. Once the design is tested and the circuit's performance is deemed acceptable, mask-programmable ICs can be used to implement the system.

Because PLDs are available off-the-shelf, no fabrication expenses are involved. The designers simply program the IC to create the ASIC. PLDs typically replace several components from the typical SSI/MSI-based design, thereby reducing the cost through reduced number of external interconnections, PCBs, and connectors.

In this chapter, we will concentrate on design techniques utilizing the four popular PLDs:

1. Read only memory (ROM).
2. Programmable logic array (PLA).
3. Programmable array logic (PAL).
4. Gate arrays (GA).

The first three types are based on the two-level AND-OR circuit structure, while the last uses a more general gate structure to implement circuits. All these devices are available in both field- and mask-programmable versions. We will use the circuit in Figure 9.2 to illustrate the difference between the ROM, PLA and PAL structures, and defer the description of GA structure to Section 9.7.

Example 9.1

Consider the two-level implementation of the functions

$$F1 = AB + A'B'$$

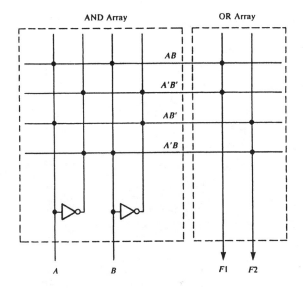

Figure 9.2 AND-OR array implementation of functions.

and

$$F2 = AB' + A'B$$

shown in Figure 9.2. The first level is implemented by an AND array. Each column corresponds to an input variable or its complement, and each row corresponds to a minterm of the input variables. The second level is implemented by an OR array. Each column corresponds to the OR combination of selected minterms generated by the AND array. For the purposes of this example, assume that each row-column intersection of these arrays can be programmed. That is, the electronic devices at these intersections can be used to either connect or disconnect the row line to the column line, and when connected, the intersection realizes either an AND or an OR operation (depending on the array in which the intersection is located). A dot at the intersections indicates that the two lines are connected.

In a ROM, the AND array is fabricated so that all the minterms corresponding to the input variables are available, and hence it is not programmable. The OR array is programmable to realize the circuit. With PAL, the OR array is completely fabricated and is not programmable while the AND-array is programmable. In a PLA, both arrays are programmable. We will describe each of these devices in more detail next.

9.4 Read Only Memory

A ROM is built out of an array of semiconductor devices such as diodes, bipolar transistors, and field effect transistors, interconnected to form the AND and OR arrays. The AND array is essentially a decoder circuit that generates all the minterms of the input variables. Each row in the OR array can store one or more bits of data. The ROM device derives its name from the fact that once the data are stored, they can only be read out but cannot be changed (or written into) under usual operating conditions. Special devices are needed to write data into a ROM. We will examine various types of ROMs later in this section. We will first provide a model for the ROM and illustrate the use of ROMs in the implementation of multiple-output combinational logic circuits. While ROM is a memory device, in the applications to be described here it will be used to construct purely combinational circuits.

Figure 9.3 shows a ROM with n inputs and m outputs. Since with n inputs we can generate 2^n minterms—whose decimal value ranges from 0 to $2^n - 1$—the data storage array in the ROM is organized into 2^n rows. There are m columns of data storage, each corresponding to one output. Each m-bit row of the array is called a *word* of the ROM. Thus, there are 2^n words, and each word has an *address* associated with it. The address is simply the row number and ranges from 0 to $2^n - 1$. This is a $2^n \times m$ ROM, meaning that the ROM contains 2^n words with m bits per word. The number of words in a memory device is usually a power of 2. In general, it is expressed in terms of Kilowords, where a kilo corresponds to 2^{10}, or 1,024. Thus, a $1K \times 4$ ROM contains 1,024 words with 4 bits per word.

The input lines to the ROM are called the *address* lines, and the output lines are the *data* lines. Thus, the function of the ROM is to produce an

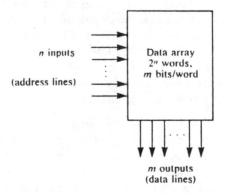

Figure 9.3 Read only memory.

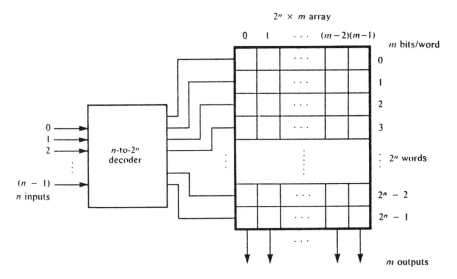

Figure 9.4 A ROM model.

output that is the bit pattern stored in the word addressed by the input lines. Figure 9.4 shows the internal structure of a ROM, consisting of the decoder and the storage array. The data stored in the word selected by the address appears on the output lines and remains on the output lines as long as the address is held on the input lines.

To implement an n-input, m-output combinational circuit, a $2^n \times m$ ROM is needed. The output portion (m columns) of the truth table is stored in the ROM words. Corresponding to each input combination (e.g., an address), the ROM produces the appropriate output on its output lines. The following examples illustrate the design.

Example 9.2

Figure 9.5 shows the implementation of the multiple-out function:

$$F1 \ (A, B, C) = \Sigma m \ (2, 4, 6)$$
$$F2 \ (A, B, C) = \Sigma m \ (2, 5, 6, 7)$$

Since there are three inputs, a ROM with three input lines and 2^3, or 8, words of storage is needed. Each word should be two bits wide, one bit for each output of the function being implemented.

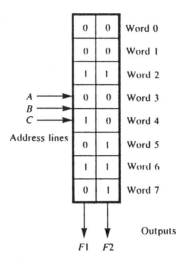

Figure 9.5   ROM implementation of a two-output function.

The following example further shows the structure of the ROM.

Example 9.3

Figure 9.6 shows an 8 × 4 ROM used to implement the function

$$F1 \ (A, B, C) = \Sigma m \ (0, 1, 5)$$
$$F2 \ (A, B, C) = \Sigma m \ (2, 3, 4, 6)$$
$$F3 \ (A, B, C) = \Sigma m \ (1, 5, 7)$$
$$F4 \ (A, B, C) = \Sigma m \ (2, 5, 6, 7)$$

A ROM with 3 input lines is needed to implement the 3-input function. A 4-bit ROM word is needed, since there are four outputs from the circuit. The 3-to-8 decoder generates the 8 minterms of the 3-variable function. Each minterm corresponds to a word line in the ROM. Each vertical line in the memory array corresponds to an output. Only those word lines that correspond to the minterms of the output (i.e., the output is 1) are connected to the vertical output line by a switching element. The switching element could be a diode or a transistor.

The decoder produces active-low outputs. When the selected word line goes low, the switching elements connected to that line conduct, driving the corresponding vertical (bit) line low. Thus, each vertical line implements a wired-AND function of all the word lines connected to it. The outputs produced by the circuit are thus active-low.

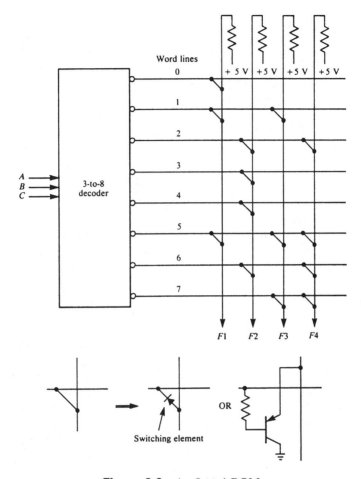

Figure 9.6 An 8 × 4 ROM.

Programming the ROM corresponds to the insertion of the switching elements where they are needed. In practice, a switching element is fabricated at each row column of the memory array. When programmed, only the desired switching elements are connected, leaving the others unconnected.

These examples are for purposes of illustration only. In practice, ROMs are available in various configurations with a storage capacity of 1 k-bit or above. As such, more complex circuits can be implemented using them. Implementing simple circuits using ROMs is not cost-effective.

9.4.1 Types of ROMs

There are three types of ROMs: (1) mask-programmable ROMs, (2) field-programmable ROMs (Programmable ROM or PROM), and (3) Erasable-Programmable ROM (EPROM). Both MOS and bipolar technologies offer ROMs with varying speeds. The data *access speed* of the ROM is the time it takes the ROM to produce an output once the address is input to it. Typical speeds are around 50 ns for bipolar and 400 ns for MOS ROMs. The ROM types available are shown below:

Bipolar	MOS
Mask programmable	Mask programmable
PROM	PROM
Fusible link	Fusible link
	Ultraviolet
	Electrically alterable

In a mask-programmable ROM, the data array is permanently stored during fabrication. This is done by selectively including a switching element where a 1 is desired in the data array. The designer of the circuit should provide the ROM program, which is simply the content of the storage array to the IC manufacturer. Once the ROM is fabricated, the data array cannot be changed. Mask-programmable ROMs are used when the ROM contents are not expected to change during the lifetime of the ROM. Fusible-link PROMs are manufactured with the switching elements in place at all the row-column intersections of the data array, with the connections made by "fusible" links. To program this ROM, the selected links are "blown" by sending the appropriate voltage pulse generated by a special device, known as a *PROM programmer*. The data once stored cannot be changed. This type of ROM is appropriate when the limited quantities needed do not justify a mask-programmable ROM.

EPROMs use a special charge-storage mechanism to enable or disable the switching elements in the data array. A PROM programmer is used to store the charge at the selected switching elements while the EPROM is programmed. The charge is retained by the EPROM, thereby retaining the program until the EPROM is erased by using an ultraviolet light. Once erased, the EPROM can be reprogrammed. This type of ROM is useful in the early development phases of digital circuit design, when it is often necessary to modify the data array.

Electrically alterable ROMs (EAROMs) are similar to ultraviolet PROMs, except that data can be erased and written while still in the circuit.

There is a limit to the number of times the data can be read from these ROMs (on the order of 2×10^{10} times).

9.4.2 Off-the-Shelf ROMs

Both bipolar and MOS ROMs are available in various configurations ranging in capacity from 256 bits to 64K bits. Typical word lengths are 4, 8, and 16 bits. The access time ranges from 15 to 100 ns for bipolar ROMs and 100 to 450 ns for MOS ROMs. ROMs of greater storage capacity and shorter access time are continually being announced. The periodicals listed in the reference section of this chapter carry listings of new IC announcements.

Figure 9.7 shows the details of a 32×8 bipolar PROM (82S23/82S123) with 5 address lines and 8 data lines. Internally, the data array is organized as a 32×8 array. The ROM is enabled by the active-low Chip Enable (CE). The outputs of 82S23 are open collector and those of 82S123 are tristate. These devices are supplied with all outputs at logic low. They are programmed to logic high at any specified address by fusing a nickel-chromium (Ni-Cr) link matrix. The programming sequence is also shown in Figure 9.7. The PROM programmer does these operations. This PROM is available in three versions with the access times of 50, 25, and 15 ns, respectively. The timing diagram shown in Figure 9.7 shows the relations between the address, enable, and output lines.

Figure 9.8 shows the details of a more complex PROM (82HS641). The storage array of this $8K \times 8$ ROM is organized as a 256×256 matrix. An 8-to-256 decoder selects one of the 256 rows based on the address lines $A_0 - A_7$. A group of 8 bits of the selected row is further chosen by the 1-of-32 multiplexer array based on the remaining address lines. The CE, when low, activates the tristate outputs. The access time is either 35, 45, or 55 ns, depending on which of the three versions of this device is used.

ROMs are useful in the implementation of complex multiple-output logic circuits directly from the truth table. They are slower than two-level gate implementations. They typically are used to implement code converters and arithmetic functions.

9.4.3 Multiple-Module Implementations

ROM capacity depends on the available technology. If the complexity of the circuit to be implemented using a ROM exceeds this capacity, multiple ROM modules are used. Assume that an n-input, m-output function is to be implemented and the largest ROM available has k address lines and r bits/word and the input variables are $(X_{n-1}, X_{n-2}, \ldots, X_1, X_0)$. There are three possibilities:

DESCRIPTION

The 82S23 and 82S123 are field programmable, which means that custom patterns are immediately available by following the Signetics Generic I fusing procedure. The 82S23 and 82S123 devices are supplied with all outputs at logical Low. Outputs are programmed to a logic High level at any specified address by fusing a Ni-Cr link matrix.

These devices include on-chip decoding and 1 chip enable input for memory expansion. They feature either Open collector or Three-state outputs for optimization of word expansion in bused organizations.

Ordering information can be found on the following pages.

The 82S23 and 82S2123 devices are also processed to military requirements for operation over the military temperature range. For specifications and ordering information consult the Signetics Military Data Book.

FEATURES

• **Address access time: 50ns max**
• **Power dissipation: 1.3mW/bit typ**
• **Input loading: −100μA max**
• **On-chip address decoding**
• **One chip enable input**
• **Output options:**
  − **N82S23: Open collector**
  − **N82S123: Three-state**
• **No separate fusing pins**
• **Unprogrammed outputs are Low level**
• **Fully TTL compatible**

APPLICATIONS

• **Prototyping/volume production**
• **Sequential controllers**
• **Format conversion**
• **Hardwired algorithms**
• **Random logic**
• **Code conversion**

PIN CONFIGURATION

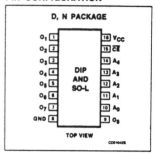

BLOCK DIAGRAM

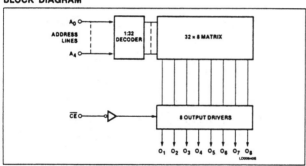

AC ELECTRICAL CHARACTERISTICS $R_1 = 470\Omega$, $R_2 = 1k\Omega$, $C_L = 30pF$, $0°C \leq T_A \leq +75°C$, $4.75V \leq V_{CC} \leq 5.25V$

PARAMETER	TO	FROM	LIMITS			UNIT
			Min	Typ[5]	Max	
Access time[4]						
T_{AA}	Output	Address		45	50	ns
T_{CE}	Output	Chip enable			35	
Disable time[6]						
T_{CD}	Output	Chip enable			35	ns

NOTES:
1. Positive current is defined as into the terminal referenced.
2. All voltages with respect to network ground terminal.
3. Duration of short circuit should not exceed 1 second.
4. Tested at an address cycle time of 1μsec.
5. Typical values are at $V_{CC} = 5V$, $T_A = 25°C$.
6. Measured at a delta of 0.5V from Logic Level with $R_1 = 750\Omega$, $R_2 = 750\Omega$ and $C_L = 5pF$.

Figure 9.7 A 32 × 8 Bipolar PROM (82S23/82S123) (Courtesy of Signetics Corporation).

VOLTAGE WAVEFORM

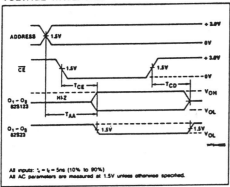

GENERIC I PROGRAMMING

The Signetics family of Advanced Junction Isolated Schottky PROMs are high performance bipolar devices which use a nickel/chromium (NiCr) alloy fuse to provide the many benefits of field programming. Programming is accomplished by application of voltages above those used for normal operation, therefore, no special pins are required for programming (except the 82S115 which has two fusing pins: FE1 and FE2). The programming voltages and timing requirements make unintentional programming virtually impossible. Arrays of devices may be programmed in the user's circuit, if desirable, as long as proper application of programming voltages is provided.

GENERIC I PROCEDURE

The Generic I family of Schottky PROMs uses no special pins for programming. The address pins remain TTL compatible during the programming procedure and are used to select the unique word to be programmed. The outputs are used to supply fusing current during the programming mode as well as selection of the bit to be programmed. Programming is performed one bit at a time. The programming mode is evoked by raising the V_{cc} pin to 8.75 ± .25V. This voltage is referred to as V_{ccp}. After the proper

delay the output corresponding to the bit selected is raised to 17.5 ± .5V. This voltage is known as V_{opt} and must be supplied by a voltage source with a low impedance and very fast transient response. Reliable programming depends on the V_{opt} power supply and circuitry. I_{opt} is the current which will be drawn by the part during the programming sequence. Again, after the proper delay the chip enable CE is pulsed to a TTL "0" level for 10 to 25 µs. It is during this time that the actual fusing of the NiCr link occurs. The actual time for fusing of a Signetics NiCr fuse link has been determined to be between .6 to 1.2µs. The shorter the fusing pulse (CE), within the recommended limits, the sooner the total programming sequence is completed. Note that unprogrammed Generic I (Junction Isolated) parts are supplied with all bits at a logic "0" level. Only the bits intended to be "ones" will be programmed. Verification of programming can be performed after each bit or after the entire device has been programmed.

A fuse which does not blow during the first programming cycle should be considered a defective device and should be discarded.

Figure 9.7 *Continued*

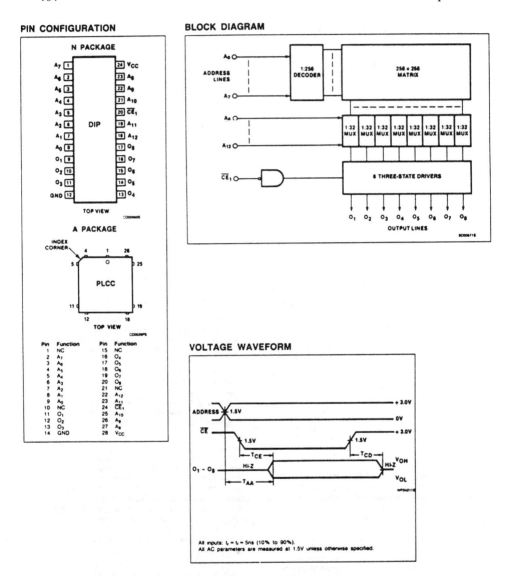

Figure 9.8 An 8K × 8 PROM (82HS641) (Courtesy of Signetics Corporation).

1. $n > k$ and $m < r$: The truth table for the function containing 2^n rows must now be partitioned into 2^{n-k} partitions, each containing 2^k rows. Each partition is implemented by a ROM. One of the partitions (i.e., ROMs) is selected by using either a multiplexer or a decoder. These implementations are shown in Figure 9.9.
2. $n < k$ and $m > r$: In this case, the output portion of the truth table is partitioned into m/r partitions of r columns each. Each partition is implemented by a ROM, and the address lines of all the ROMs are connected to the input lines, as shown in Figure 9.10.

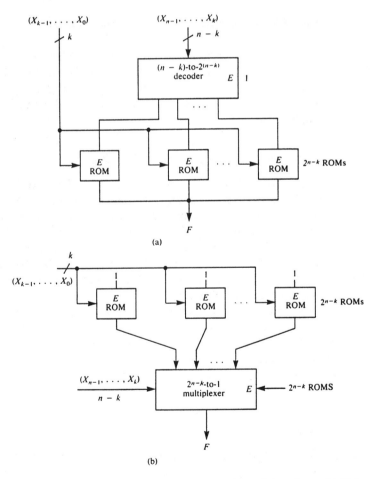

Figure 9.9 Multiple-module ROM implementation of functions. (a) Using a decoder; (b) using a multiplexer.

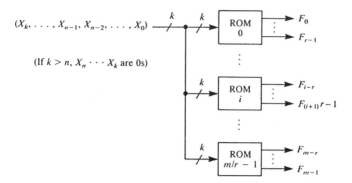

Figure 9.10   Multiple-ROM implementation of functions.

3. $n > k$ and $m > r$: In this case, the truth table is partitioned both along the rows (as in the first possibility) and along the columns (as in the second). The implementation is the combination of the implementations shown above.

9.5   Programmable Logic Arrays (PLAs)

In the implementation of logic circuits using ROMs, one ROM word is used for each minterm in the function. For an n-variable function, a ROM with 2^n words is thus required. Only few of these words are utilized when the function either has a large number of don't-care conditions or has only a few 1s in the output column of the truth table. Thus, it is wasteful to use a ROM in such cases. PLAs are LSI devices that are functionally equivalent to ROMs, except that they do not generate all the minterms. A SOP form of the function to be implemented is used in designing with PLAs. An AND gate is used to implement each product term and an OR gate for each sum term in the function. The AND and OR gates are realized using wired-logic rather than discrete gates. Since only the required sum and product terms are generated, the PLA implementation is more economical than the corresponding ROM implementation. Example 9.4 illustrates the structure and operation of PLAs.

Example 9.4

It is required to implement the following function using a PLA:

$$F1 \ (A, B, C) = \Sigma m \ (2, 3, 7)$$
$$F2 \ (A, B, C) = \Sigma m \ (1, 3, 7)$$

Unlike the ROM-based implementation, the minimized SOP form of the function is first obtained, for a PLA-based implementation. The K-maps of Figure 9.11(a) minimize this function. The multiple-output minimization procedures discussed in Chapter 3 can also be used to reduce the number of product terms. Note that the minimization procedure should reduce the number of product terms rather than just the number of literals in order to simplify the PLA.

Only three product terms are needed to implement this two-output function, since BC is common to both. A modified truth table is shown in (b). The first column lists the product terms. The second column lists the circuit inputs. For each product term, the inputs are coded "0," "1," or "-." A 0 indicates that the variable appears complemented; a 1 indicates that the variable appears uncomplemented; and a "-" indicates that the variable does not appear in the product term. This column provides the programming information for the AND array of the PLA. There are six inputs (A, A', B, B', C, C') to each AND gate in the array, as shown in (c), each input having a fusible link. Thus, a 0 implies that the link is retained corresponding to the complemented variable, a 1 implies that the link is retained corresponding to the truth variable and "-" implies "blowing" the link. This retaining/blowing operation on the links is the "programming" of the AND array.

The last column in (b) shows the circuit outputs. A 1 in this column in any row indicates that the product term corresponding to that row is a product term for the output corresponding to the column; a "-" indicates no connection. Thus, this column is useful in programming the OR array, as shown in (c).

In (c), the AND and OR gates are shown dotted, to indicate the wired logic. Each AND gate has six inputs to receive the three circuit inputs and their complements. But only the required connections, as indicated by the second column of (b), are made. There is one AND gate for each required product term. Similarly, the OR gates are shown with three inputs each, one for each product term in the circuit. Again, only the required product terms, as indicated by the third column of (b), are connected.

The actual PLA structure is shown in (d). Each dot in this figure corresponds to a connection using a switching device (a diode or a transistor, as in a ROM), and the absence of a dot at a row-column intersection indicates no connection. The arrow shown in the connection detail at the bottom of the figure can be considered the diode positive direction to verify that this structure implements the AND and OR logic.

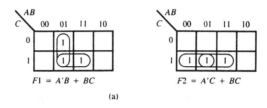

$$F1 = A'B + BC \qquad F2 = A'C + BC$$

(a)

Product Term	A B C	F1 F2
A'B	0 1 –	1 –
BC	– 1 1	1 1
A'C	0 – 1	– 1

(b)

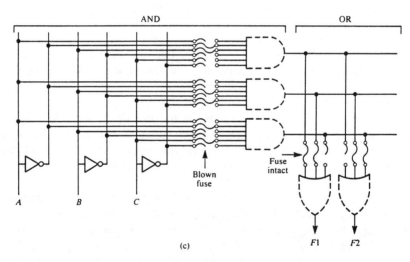

(c)

Figure 9.11 Programmable logic array. (a) K-maps; (b) PLA table; (c) PLA structure.

9.5.1 Types of PLAs

Two types of PLAs are available: mask and field programmable. They are identical to ROM types. A PLA program is provided by the designer so that the IC manufacturer can fabricate a mask-programmed PLA, which can never be altered once programmed. Special devices are used to program the field-programmable PLA (FPLA). For this type of PLA, switching devices

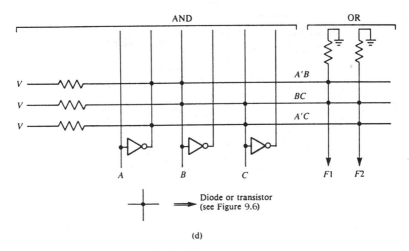

(d)

Figure 9.11 Continued

are fabricated at each row-column intersection and the connections are established by blowing the fusible links, as in ROM programming.

Figure 9.12 shows the general model for a $n \times k \times m$ PLA (e.g., with n inputs, m outputs, and k product terms). The complexity of the circuit implementable by a PLA device thus is constrained by the values of n, m, and k, which are in turn functions of the technology. When the complexity of the circuit exceeds these limits, multiple module implementations (similar to the ones using ROMs) are used.

Example 9.5

Figure 9.13 shows the implementation of the serial twos complementer controller of Example 7.15 using a PLA.

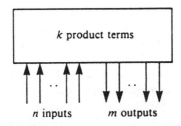

Figure 9.12 General PLA model.

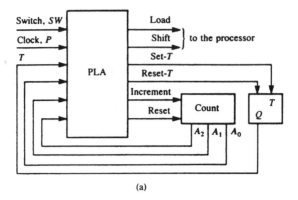

(a)

State	SW	T(t)	A₂	A₁	A₀	T(t+1)	Load	Shift	Reset-T	Set-T	Reset count	Increment count

Inputs/present state — Outputs/next state

State	SW T(t) A₂ A₁ A₀	T(t+1)	Load	Shift	Reset-T	Set-T	Reset count	Increment count
S_1	× 0 × × ×	0	0	0	1	0	0	0
	1 0 × × ×	1	1	0	0	1	1	0
S_2	× 1 0 0 0	1	0	1	0	1	0	1
	× 1 0 0 1	1	0	1	0	1	0	1
	× 1 0 1 0	1	0	1	0	1	0	1
	× 1 0 1 1	1	0	1	0	1	0	1
	× 1 1 0 0	1	0	1	0	1	0	1
	× 1 1 0 1	0	0	0	1	0	0	0

(b)

Figure 9.13 PLA implementation of the serial twos complementer controller. (a) Circuit; (b) operation.

Example 9.6

Figure 9.14 shows the details of the Signetics PLC105, a field-programmable logic sequencer. This is essentially a PLA along with some flip-flops on the chip to enable the implementation of Mealy-type sequential circuits. This CMOS chip contains an AND-OR gate array with user programmable connections that control the inputs of the on-chip state and output registers. The state register is made up of six *SR* flip-flops, and the output register is made up of eight *SR* flip-flops. These flip-flops are positive edge-triggered with a direct set (*preset*) input. During the power-on, all the flip-flops are set to 1.

PIN CONFIGURATION

FUNCTIONAL DIAGRAM

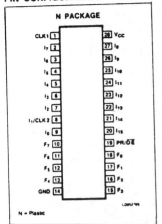

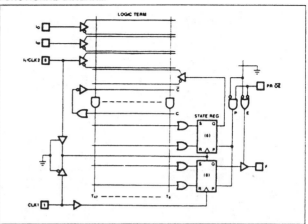

Figure 9.14 PLC105 field-programmable logic sequencer (16 × 48 × 8) (Courtesy of Signetics Corporation).

The AND array combines the 16 external inputs ($I_0 - I_{15}$) and the seven internal inputs ($P_0 - P_6$), which are fed back from the state registers to form up to 48 AND terms. These AND terms are then merged by the OR array to realize the output and the next-state circuits.

An internal input variable C is available that enables the generation of either true or complement AND terms, if needed. As a user programmable option, the PRESET input can also be used as an Output Enable. Figure 9.15 shows the realization of the circuit in Example 6.2 using this PLS.

9.5.2 Off-the-Shelf PLAs

PLAs with 12 to 20 inputs, 20 to 50 product terms, and 6 to 12 outputs are available off-the-shelf. In addition, Programmable Logic Sequencers (PLSs) that are PLAs with storage capabilities that are useful in sequential circuit implementations are also available. In recent PLAs the AND-OR structure is augmented with additional logic capabilities. We will discuss the details of one PLS device next.

Figure 9.16 shows the details of a 12 × 48 × 8 bipolar FPLA (Signetics PLS161). This device has 12 inputs, 8 outputs, and 48 product terms. The outputs can be programmed as active-high or active-low. There is an enable input (CE). In the unprogrammed state, this device is characterized by the following:

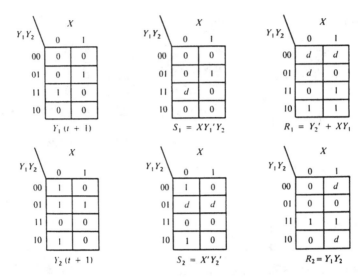

Figure 9.15 PLS implementation of the circuit in Example 6.2. *Notes*: (1) Tables for Y_1 and Y_2 are from Figure 6.24. (2) Also, from Figure 6.24 $Z = XY_1Y_2'$.

1. All Ni-Cr links intact.
2. Each product term contains the true and complement values of every input variable.
3. The OR array contains all 48 product terms.
4. The polarity of each output set to active-high.
5. All outputs are at a low logic level.

The typical propagation delay of this device is 35 ns.

Several CAD tools are available to enable designing with PLAs. These tools generate PLA programs, optimize PLA layout for custom fabrication, and simulate PLA designs. (Appendix D at the end of this book describes a PLA CAD software system.)

9.6 Programmable Array Logic (PAL)

The PAL is a version of PLA in which only the AND array is programmable and the OR array is fixed. PAL is thus easier to program and is less expensive than a PLA. But PAL is less flexible than a PLA in terms of circuit design, since the OR-array configuration is fixed.

Figure 9.17 shows PAL14H4, a PAL IC with 14 inputs (Pin numbers 1 through 9, 11, 12, 13, 18 and 19) and 4 outputs (Pin numbers 14 through 17). Each input is buffered, and the buffer produces the true and comple-

FPLS LOGIC DIAGRAM

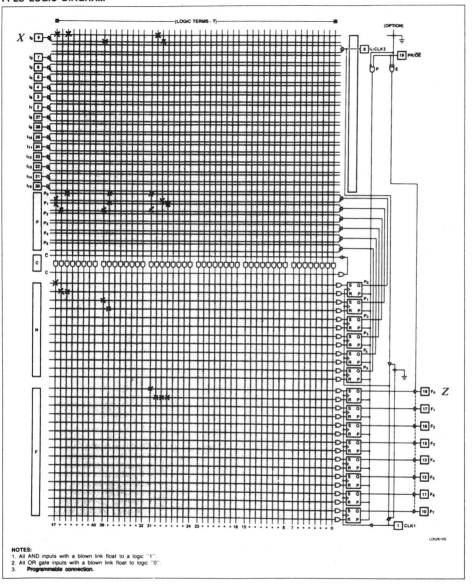

Figure 9.15 Continued

PIN CONFIGURATION   **FUNCTIONAL DIAGRAM**

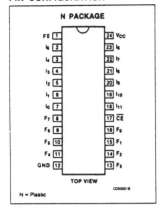

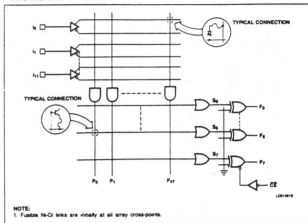

NOTE:
1. Fusible Ni-Cr links are initially at all array cross-points.

LOGIC FUNCTION

TYPICAL PRODUCT TERM
$P_0 = I_0 \cdot I_1 \cdot \overline{I_3} \cdot I_6 \cdot \overline{I_{11}}$

TYPICAL OUTPUT FUNCTIONS:
● $\overline{CE} = 0$:
$F_0 = (P_0 + P_1 + P_3) ● L = CLOSED$
$F_0 = (\overline{P_0} \cdot \overline{P_1} \cdot \overline{P_2}) ● L = OPEN$

NOTE:
For each of the 8 outputs either the function F_0 (active-high) or $\overline{F_0}$ (active-low) is available, but not both. The required function polarity is programmed via link (L).

Figure 9.16 A $(12 \times 48 \times 8)$ PLA (PLS161) (Courtesy of Signetics Corporation).

mented values of the input. The device layout consists of a 63-row, 32-column array. Each column corresponds to an input or its complement. Since there are only 14 inputs, only 28 of the 32 columns are utilized. Each row corresponds to a product term. Each output in this IC is realized by ORing four product terms. The product terms are realized by wired logic, as in ROMs. The AND gates shown at the inputs of OR gates thus are symbolic representations only. Only 16 of the 64 rows are used by this device.

Figure 9.18 shows the symbology used in PALs. In (a), the product term (ABC) is realized by retaining the fusible links at the row-column intersections of the AND array, as shown by an X. Absence of an X at any intersection indicates a blown link and hence no connection. Realization of the function $AB' + A'B$ is shown in (b). The unprogrammed PALs will have all the fuses intact. Fuses are blown during programming the PAL to realize the required function. PROM programmers with PAL personality modules are used in programming PALs.

A shorthand notation to indicate that all the fuses along a row are intact is shown in (c). This simply implies that a particular row is not utilized. A sample implementation using this convention is shown in (d).

FPLA LOGIC DIAGRAM

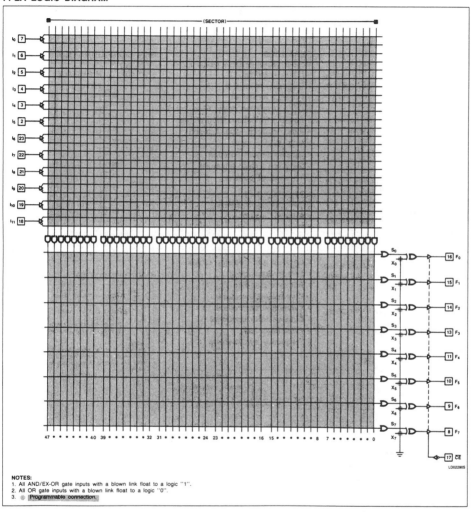

NOTES:
1. All AND/EX-OR gate inputs with a blown link float to a logic "1".
2. All OR gate inputs with a blown link float to a logic "0".
3. ◉ Programmable connection.

AC ELECTRICAL CHARACTERISTICS $R_1 = 470\Omega$, $R_2 = 1k\Omega$, $C_L = 30pF$, $0°C \leqslant T_A \leqslant +75°C$, $4.75V \leqslant V_{CC} \leqslant 5.25V$

PARAMETER		TO	FROM	LIMITS			UNIT
				Min	Typ[2]	Max	
Propagation delay							
T_{PD}	Input	Output	Input		35	50	ns
T_{CE}	Chip enable	Output	Chip enable		15	30	
Disable time							
T_{CD}	Chip disable	Output	Chip enable		15	30	ns

Figure 9.16   Continued

14H4

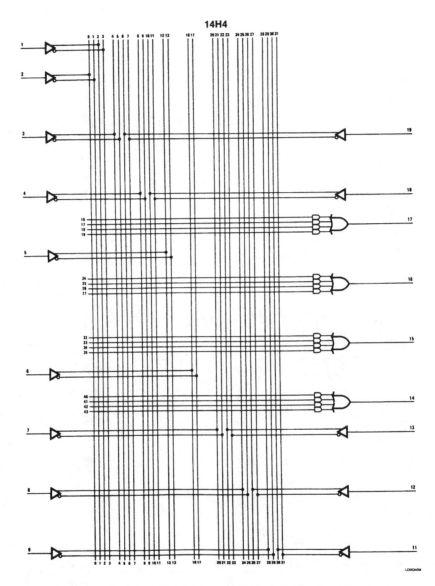

Figure 9.17   PAL14H4 (Courtesy of Monolithic Memories, Inc.).

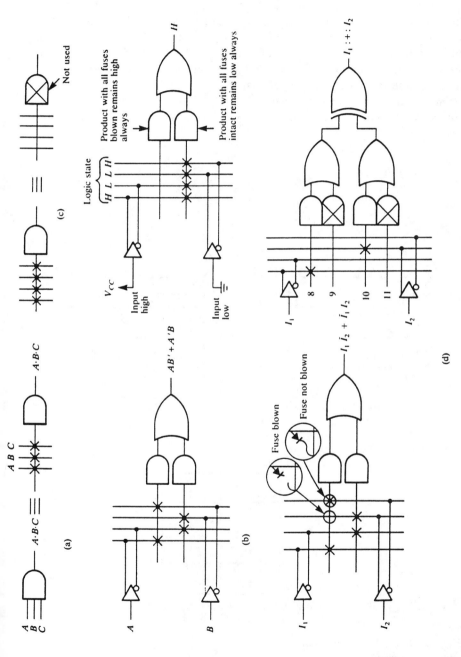

Figure 9.18 PAL symbology (Courtesy of Monolithic Memories, Inc.). (a) Product term realization; (b) realization of $A \oplus B$; (c) all fuses intact; (d) examples.

In realizing multiple-output circuits using a PLA, common product terms can be shared among the various outputs. With PAL, since the OR array cannot be programmed, such sharing is not possible. As such, each output function must be minimized separately before the implementation. Further, since the number of inputs to the OR gates at the output of the PAL is fixed, alternate ways of realizing the circuit may need to be explored when the number of product terms in the minimized function exceeds the number of inputs available. Example 9.7 shows a circuit implementation using PAL.

Example 9.7

Implement the following function using PAL:

$$F1\ (A, B, C, D) = \Sigma m\ (0, 1, 5, 7, 10, 11, 13, 15)$$

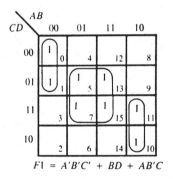

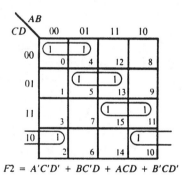

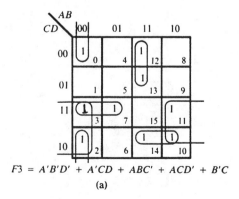

(a)

Figure 9.19 Implementation of functions in PAL. (a) Simplification; (b) implementation.

$$F2\ (A, B, C, D) = \Sigma m\ (0, 2, 4, 5, 10, 11, 13, 15)$$
$$F3\ (A, B, C, D) = \Sigma m\ (0, 2, 3, 7, 10, 11, 12, 13, 14)$$

Figure 9.19(a) shows the K-maps and the corresponding simplified functions, and (b) shows the implementation using the PAL14H4. The implementation of $F1$ and $F2$ does not present any problem, since the number of product terms in these functions is less than or equal to four. Since $F3$ has five product terms, one of the PAL outputs is used to realize the first four product terms (Z). Z is then fed into the PAL as an input and combined with the remaining product term of $F3$, to realize $F3$. Obviously, the implementation of this simple circuit using the PAL14H4 is not economical, since we did not use most the inputs and the product terms possible. Nevertheless, the example illustrates the design procedure.

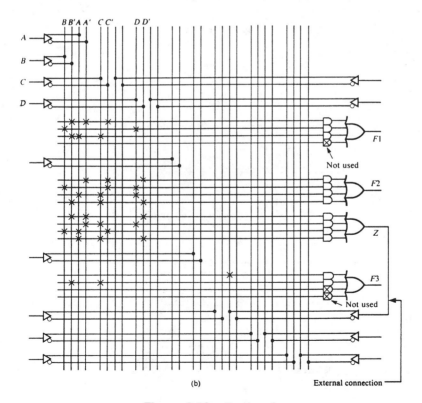

(b)

Figure 9.19 Continued.

Example 9.8

Figure 9.20 shows the PAL implementation of the modulo-4 up/down counter of Example 6.3. Only a portion of the PAL 12H6 is shown. The PAL is used to implement the combinational logic portion only and the flip-flops required for the design are supplied externally to the device.

9.6.1 Types of PALs

PALs are available in several output configurations, either in a 20-pin or a 24-pin package. The common output configurations are:

1. Active-high outputs.
2. Active-low outputs.
3. Programmable I/O (Input/Output).
3. Registered outputs with feedback.
5. EXCLUSIVE-OR outputs.
6. Arithmetic-gated feedback outputs.

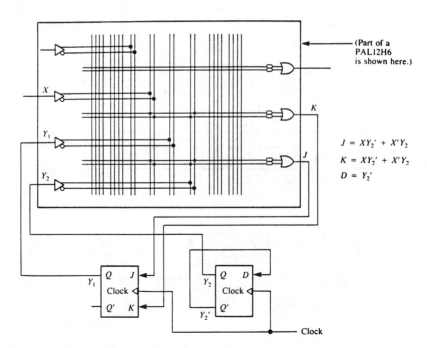

(Part of a PAL12H6 is shown here.)

$$J = XY_2' + X'Y_2$$
$$K = XY_2' + X'Y_2$$
$$D = Y_2'$$

Figure 9.20 PAL implementation of an up-down counter.

Three of these output configurations are shown in Figure 9.21. A PAL segment with programmable I/O is shown in (a). One of the product terms is used to enable the tristate buffer at the output, which in turn becomes an input to the PAL. When the tristate buffer is disabled, the I/O pin becomes an input. This feature is useful to allocate available pins for I/O functions and to form bidirectional outputs for shift and rotate operations.

A PAL with registered outputs is shown in (b). The output from the OR gate is stored in a *register*, which is a data storage device. A D flip-flop is shown as the one-bit storage device. This device simply stores the output of the OR gate and provides as its outputs the true (Q) and the complement (Q') of the data stored in it. (Flip-flops are described in Chapter 6, and registers in Chapter 7.) This type of PAL is useful in building sequential circuits (described in Chapter 6).

A PAL segment with an EXCLUSIVE-OR output with feedback capabilities is shown in Figure 9.21(c). This configuration is useful in building circuits for arithmetic operations. The PALs with arithmetic-gated feedback outputs provide a more elaborate arithmetic function at their outputs.

9.6.2 Off-the-Shelf PALs

Several PAL ICs are available with the following ranges: 10 to 20 inputs, 1 to 10 outputs (in the various configurations shown above), and 2 to 16 inputs per output-OR gate. The propagation delay of PALs ranges from 10 to 25 ns, though some ECL PALs with 6-ns propagation delays have been announced. PALs are manufactured using TTL Schottky bipolar technology. An *npn* emitter follower array forms the AND array. The *pnp* inputs provide high input impedance to the array. The outputs are standard TTL drivers with internal active pull-up transistors.

PAL technology received renewed interest during 1984 and 1985, resulting in PALs of very large sizes. For example, the AmPAL22V10 from Advanced Micro Devices has 22 inputs, 10 outputs, and 132 product terms. New programming features continue to be added to these devices.

Typically, PALs are programmed with PROM programmers that use a PAL personality card. During programming, half of the PAL outputs are selected for programming, while the other inputs and outputs are used for addressing. The outputs are then switched to program the unprogrammed locations.

One of the PAL design aids available is PALASM software (from Monolithic Memories Inc.). PALASM accepts the PAL design specification and verifies the design against an optional function table and generates the fuse plot required to program the PAL. The PALASM source program and typical outputs are listed in Appendix D at the end of this book.

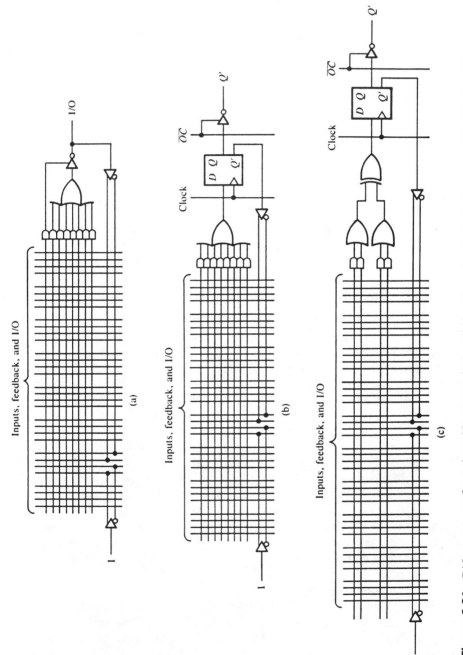

Figure 9.21 PAL output configurations (Courtesy of Monolithic Memories, Inc.). (a) Programmable I/O pin; (b) registered output with feedback; (c) EXCLUSIVE-OR output with feedback.

The mask-programmed version of PAL (called Hard Array Logic, or HAL) is also available. A HAL is to a PAL as a ROM is to a PROM. An MSI version of HAL technology yields the HMSI (i.e., HAL-MSI) family, which consists of devices to perform predetermined functions not available in TTL. (For further details, see the manufacturer's literature.)

9.7 Gate Arrays (GA)

Gate Arrays are LSI devices consisting of an array of gates fabricated over the chip area along with wire routing channels to facilitate their interconnection. GAs originated in the late 1970s as replacements for SSI- and MSI-based circuits built on PCBs. Figure 9.22 shows the structure of a typical GA. Each shaded rectangular area on the chip is an array of gates. The channels between these are the wire routing paths. The array of gates at the periphery of the chip are the input/output pads. In using this device, the designer specifies the interconnections within each rectangular area (i.e., cell) to form a function (equivalent to an SSI or MSI function). The intercell interconnections are generated using PCB routing software aids. The disadvantage of this structure is the increased propagation delays as a result of long path lengths and increased chip area to fabricate a given circuit, compared with a custom design chip for the same circuit or system.

In order to overcome the slow speeds caused by long path lengths and decreased density as a result of large areas dedicated for routing, devices evolved that allowed the interconnection over the GA area, rather than through dedicated channels. Figure 9.23 shows one such GA (Signetics 8A1260). This device uses Integrated Schottky Logic (ISL) NAND gates, arranged in two arrays of 26 rows × 22 columns. There are 52 Schottky buffers driving multiload enable signals. The 60 LSTTL input/output buffers can be programmed as inputs or bidirectional paths or as totem pole, tristate, or open collector outputs.

The ISL inverter circuit is shown in Figure 9.24. A 700-mV high level at the input produces a 500-mV low level at the output of Schottky diodes. The passive current source, consisting of V_{BB} and R_B, also acts as pull-up resistor for the output of the previous gate. The *pnp* transistor is used to control the base charge of the *npn* transistor, keeping it out of saturation and thereby providing fast switching. The gate has a fan-out of 4 (a fifth output is provided to allow improved wire routing). Two input points are provided. Input *A* is used when the previous stage is another ISL device. Input *B* is used if the previous stage is an I/O cell or a Schottky buffer. R_H is used to limit the current when the inputs of two gates are tied together.

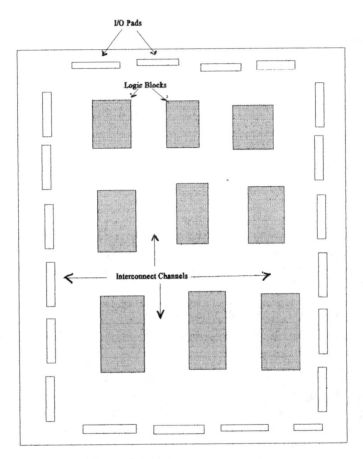

Figure 9.22   A gate array structure.

The inverter circuit discussed here can be used as a NAND circuit by wire-ANDing several inputs onto the common input line. Thus, only one input wire is used to realize multiple inputs. Any of the five outputs that are accessible can be used, and they allow for metallization runs over the unused outputs during routing. Hence, no dedicated routing channels are provided, thereby increasing the chip density.

By using a combination of appropriately configured NAND gates, any function can be realized. In fact, SSI and MSI functions from the TTL manuals can be copied. Unnecessary functionalities (such as multiple chip

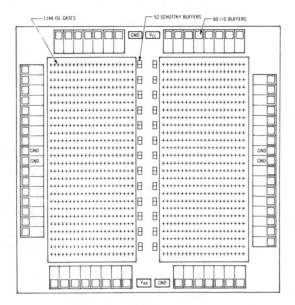

Figure 9.23 Signetics 8A1260 ISL gate array [*Source: Computer Design, 20* (August 1981)].

enables, for example) provided in TTL ICs can be eliminated during the copying to make those circuits more efficient.

In designing with GAs, the designer (user) interacts with the IC manufacturer (supplier) extensively through the CAD tools (see Figure 9.25). The user generates the logic circuit description and verifies the design through logic and timing simulation. A set of tests to uncover faults is also generated by the user and provided to the supplier, along with the design specifications (e.g., wire lists, schematics). The supplier performs the automatic placement and routing, mask generation, and IC prototype fabrication. Once the performance of the prototype is accepted by the user, the production run of the ICs is initiated.

It is no longer necessary to design at gate level of detail, while designing with GAs. The GA manufacturers provide a set of standard cells (functions) as part of the CAD environment. These standard cells are utilized in configuring the GA very much like using the SSI/MSI components in the design of the system. For instance, the Signetics' composite cell logic (CCL) GAs come with two standard cell libraries: the Extended Performance Library for high-speed applications and the ISL library for high density applications. The standard cells in each library correspond to gates of various

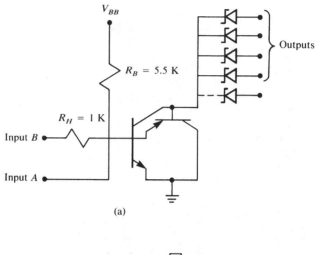

(a)

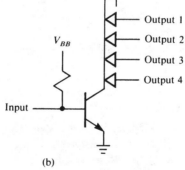

(b)

Figure 9.24 ISL inverter. (a) ISL gate structure; (b) symbol for ISL gate.

types and configurations and MSI-type cells such as multiplexers, counters, adders and arithmetic/logic units.

9.7.1 Off-the-Shelf GAs

As can be seen from the discussion above, GAs offer much more design flexibility compared to other three PLDs. But their disadvantage is that the turnaround time from design to prototype is several weeks, although GA manufacturers have started offering one- to two-day turnarounds. The disadvantage of the other three PLDs is their design inflexibility, since only two-level AND-OR realizations are possible.

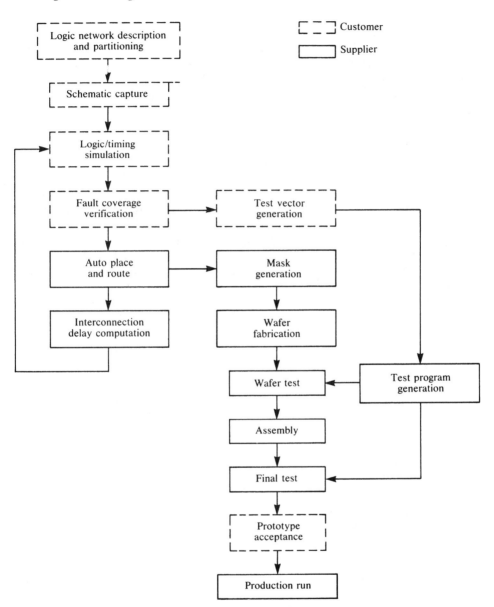

Figure 9.25 Semicustom logic design procedure (Courtesy of Signetics Corporation).

Several devices that combine the features of GA flexibility and pro-grammability are now being offered. These devices are essentially PLDs with enhanced architectures, some departing completely from the AND-OR structure and some enhancing the AND-OR structure with additional pro-grammable logic blocks (macro cells). We will collectively call all such devices field programmable GAs (FPGA). A brief description of two such devices follows.

Figure 9.26 shows a Logic Cell Array (LCS)—XC2064—manufac-tured by Xilinx Inc. This CMOS 1200 gate device uses logic elements in-terconnected through user-defined routing paths. Unlike the NAND elements of a GA, the logic elements of the LCA are configurable logic blocks (CLBs). Each CLB includes combinational logic elements and a storage element. The designer can program the combinational logic either as two three-input logic functions or as one four- or five-input function. The storage element can be programmed as a transparent latch or an edge-triggered flip-flop (see Chapter 6). The switching (toggle) frequencies of these flip-flops are on the order of 20 to 33 MHz.

There are 56 programmable I/O Blocks (IOBs) on the device surround-ing the CLB array. Each IOB can be configured as an input with an optional flip-flop, an output with an optional tristate control, or a bidirectional path with any of these options. Since device configuration takes only 12 ms, LCAs are being used in applications requiring dynamically reconfigurable logic. Thus, for example, the same device can be used to interface more than one peripheral device to a digital system by reconfiguring the LCA dynamically. An LCA design system consisting of a 113 SSI/MSI standard cell library, a graphics editor, and a design rule checker is available along with an optional simulator.

Excel Microelectronics offers a multilevel single-plane (MLSP) device (Erasic) as a departure from the two-plane AND-OR architecture. The Erasic uses a single NOR plane with programmable input and output polarity. Mul-tilevel logic is realized by feeding the outputs back into the plane as inputs. The NOR array can be programmed as either AND, NAND, OR, or NOR by selecting the proper polarity. In addition to the single NOR plane, this device contains macrocells in the feedback path for maximum design flexibility.

Figure 9.27(a) shows the structure of the Erasic XL78C800, and (b) shows the macrocell. The device includes 66 independent logic terms, un-limited term sharing at any level, two latchable four-input ports, 10 JK flip-flops, and 10 programmable macrocells. With 600 to 800 equivalent gates, this 24-pin CMOS device offers a 35-ns delay from input to output across the NOR plane and an internal delay of 20 ns per logic level. Each macrocell includes a JK flip-flop, two programmable multiplexers, a pro-

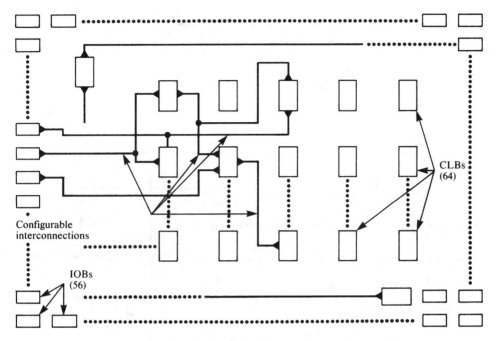

Figure 9.26 Logic cell array (XC2064).

grammable output driver, and three polarity control elements and is programmed by means of six fuses. Programming the Erasic takes less than 1 second.

Signetics offers MLSP devices called Programmable Macro Logic (PML). These are bipolar devices (PLHS501 and PLHS502) based on a single NAND plane that can be configured into multilevel logic circuits using feedback, as shown in Figure 9.28, with gate densities of 3,000 and 3,600.

PLHS501 is a 52-pin combinational device (or a gate bucket), aimed at replacing PALs and PLAs. It has 72 NAND terms, 24 dedicated inputs, eight bidirectional outputs, and eight dedicated outputs. The propagation delay across the NAND array is 17 ns, with an internal delay of 8 ns per logic level.

PLHS502 offers a propagation delay of 12 ns across the NAND plane and a 4-ns internal delay per logic level. It also offers sequential operation by way of 16 flip-flops that the designers can bury into internal logic or connect to I/O ports. Both these devices typically draw a current of 250 mA, compared with 20 mA drawn by an Erasic.

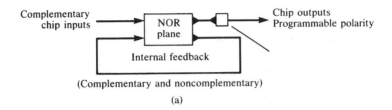

(a)

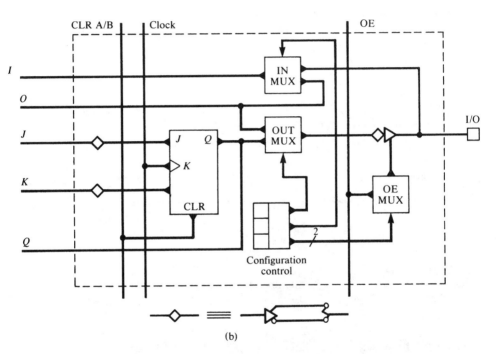

(b)

Figure 9.27   Erasic XL78C800. (a) Structure; (b) macrocell.

Various IC manufacturers offer an array of other PLDs. Lattice Semiconductors offers Generic Array Logic (GAL), which consists of CMOS devices and uses EPROM links rather than fuses. Since fuses are bulkier than EPROM devices, these PLDs (called EPROM-PLDs or EPLDs) offer higher densities and are designed to replace PALs. GALs have a programmable AND array, a fixed OR array, and programmable macrocells at the outputs. In addition to general PLDs, several IC manufacturers are now offering Application-Specific PLDs (ASPLDs).

It is very common to include memory cells as part of FPGAs just as the case with programmable logic sequencers (PLS) discussed earlier in

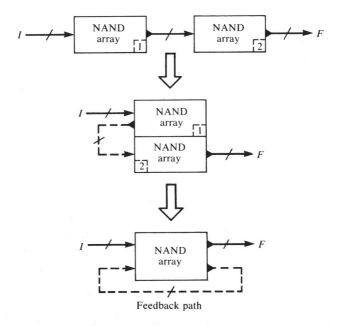

Figure 9.28 Programmable macro logic (PML) [*Source: Electronic Systems Design, 16*, No. 8 (July 1986): 34].

this chapter. These static random access memory (SRAM) based FPGAs are offered by several manufacturers, including Xilinx, Altera, Atmel, Motorola and AT&T microelectronics. In these FPGAs, logic functions and interconnections are determined by configuration data stored in internal static memory cells. The configuration information is loaded into these memory cells at system power-up. These configurations can be altered during system operation, thereby providing a dynamic reconfigurability. Newer characteristics and architectures are offered by FPGA manufacturers almost daily.

9.8 Summary

The relative merits of the three common modes of circuit implementation utilizing LSI and VLSI devices were discussed. Design techniques for combinational and sequential circuits utilizing PLDs and GAs were detailed.

It should be noted that the circuits designed using PLDs are in general slower than custom-designed ICs. They also do not use the silicon area on the chip very efficiently and hence tend to be less dense than their custom-designed counterparts. Nevertheless, these designs are more cost-effective for low-volume applications.

As with any IC technology, newer PLDs are continually being introduced by IC manufacturers. Their architecture and characteristics also vary. Because of the rapid rate of introduction of new devices, any description of these devices becomes obsolete by the time it is published in a book. As such, it is imperative that the designer consult the manufacturer's manuals for the most up-to-date information. The periodicals listed in the references section should also be consulted for new IC announcements.

References

Bipolar Memory Data Manual. Sunnyvale, Calif.: Signetics, 1986.

Blakslee, T. R. *Digital Design with Standard MSI and LSI.* New York: John Wiley & Sons, 1975.

Collett, R. "Programmable Logic Declares War on Gate Arrays." *Digital Design* 16, no. 8 (July 1986): 32–39.

Composite Cell Logic Data Sheets. Sunnyvale, Calif.: Signetics, 1981.

Computer. New York: IEEE Computer Society (monthly).

Computer Design. Littleton, Mass.: PennWell (semimonthly).

Davis, G. R. "ISL Gate Arrays Operate at Low Power Schottky TTL Speeds." *Computer Design* 20 (August 1981): 183–186.

Digital Design. Boston: Morgan-Grampian, April 1985 (monthly).

Greenfield, J. D. *Practical Digital Design Using ICs.* New York: John Wiley & Sons, 1983.

Integrated Circuits Magazine. Garden City, N.Y.: Hearst Publications (monthly).

Lau, S. Y. "High Performance 1200-Gate ISL Array." *Western Conference Proceedings,* September 1979.

Mano, M. M. *Digital Design.* Englewood Cliffs, N.J.: Prentice-Hall, 1994.

Nashelsky, L. *Introduction to Digital Technology.* New York: John Wiley & Sons, 1983.

PAL Data Sheets. Sunnyvale, Calif.: Monolithic Memories, 1984.

PAL Device Data Book, Sunnyvale, Calif.: AMD and MMI, 1988.

Pitts, R. C. "Gate Arrays—Cost-Slashing Replacements for SSI, MSI." *Electronic Design* 29, no. 30 (December 24, 1981).

Programmable Logic Data Manual. Sunnyvale, Calif.: Signetics, 1986.

Programmable Logic Data Book, San Jose, Calif.: Cypress, 1996.

Programmable Logic Devices Data Handbook, Sunnyvale, Calif.: Philips Semiconductors, 1992.

Roth, C. H. *Fundamentals of Logic Design.* St. Paul, Minn.: West Publishing, 1985.

Schewel, J. (ed.) "Field Programmable Gate Arrays for Fast Board Development and Reconfigurable Computing," *Proceedings of SPIE,* vol. 2607, Philadelphia, PA: October 1995.

Technical staff of Monolithic Memories. *Designing with Programmable Array Logic*. New York: McGraw-Hill, 1981.

TTL Data Manual. Sunnyvale, Calif.: Signetics, 1986.

Problems

9.1   Implement the circuits in problems 4-10, 4-14, 6-2 and 6-9 using appropriate (a) ROM, (b) PLA and (c) PAL.

9.2   Use a PLA and SR flip-flops to implement the sequential circuits in problems 6-10 and 6-11.

9.3   Use the PLS PLC105 to implement the circuits of problems 6-10 and 6-11.

9.4   Implement a converter to convert six-bit binary numbers into BCD, using a ROM of the appropriate size. (Note that the LSB of the binary input and BCD output will be the same.)

9.5   Implement the following using (a) a ROM, (b) a PLA, and (c) a PAL:

1.   A 4 × 4 multiplier.
2.   A circuit that squares the three-bit number input to it.

9.6   A decimal adder needs to be implemented using a ROM. Draw the block diagram of the circuit with the two BCD input digits, a CARRY-IN bit, the SUM digit, and the CARRY-OUT. Determine the size of the ROM required. Show the contents of the ROM words for typical inputs.

9.7   A divide-by-three circuit is needed to divide the input four-bit binary number to generate the quotient and the remainder. Draw the truth table for the circuit and implement the circuit with the PLA of minimum size.

9.8   Realize a triple 3-to-1 multiplexer using the PAL14H4. The selection signals are common to all three multiplexers.

9.9   Implement the following multiple-output function using the minimum PLA:

$$F1(A, B, C, D) = \Sigma m \ (0, 2, 8, 10) \ + d(1, 3, 9, 15)$$

$$F2(A, B, C, D) = \Sigma m \ (3, 6, 7, 14) \ + d(0, 8, 11)$$

$$F3(A, B, C, D) = \Sigma m \ (1, 5, 11, 13) + d(2, 7, 12)$$

10

Hardware Technologies

10.1 Introduction

So far we have treated gates as the basic components that are interconnected to form logic circuits. We have concentrated only on the functional aspects of gates and logic circuits in terms of manipulating binary signals. As mentioned earlier, a gate is an electronic circuit consisting of transistors, diodes, resistors, capacitors, and other components interconnected to realize a particular function. In this chapter we will examine the basic circuit configurations and characteristics of various IC technologies. In addition to the electronic IC technology, one area of intense research and development as an alternative technology is that of optics. We will introduce optical technology as applied to computing. An in-depth knowledge of electronics and solid-state physics is a prerequisite for fully understanding this material. Refer to Appendix B at the end of the book for a review of the fundamentals of electrical circuits.

There are two broad categories of silicon-based IC technology, one based on bipolar transistors (i.e., pnp and npn junctions of semiconductors) and the other based on unipolar metal oxide semiconductor field effect transistor (MOSFET). Within each of these technologies, several logic families of ICs are available. The popular bipolar logic families are transistor-transistor logic (TTL) and emitter-coupled logic (ECL). Section 10.2 provides details on these families. P-channel MOS (PMOS), N-channel MOS (NMOS) and complementary MOS (CMOS) are popular MOSFET logic families. These are covered in Section 10.3. Some believe that silicon technology has almost reached its physical limits of performance and newer technologies are needed. A new family of ICs based on gallium arsenide has recently been introduced. This technology, described in Section 10.4, has the potential of providing ICs that are faster than the ICs in silicon technology. Section 10.5 provides a brief introduction to optical technology, a nonelectronic technology.

10.2 Bipolar IC Technologies

In this section, we will examine the basic circuit forms in the popular bipolar IC technologies. We will first introduce the operating characteristics of a diode and demonstrate how logic gates can be built using diodes. Characteristics of bipolar transistors will then be examined, followed by an introduction to resistor-transistor logic (RTL) and diode-transistor logic (DTL) circuits. Although these two technologies are now obsolete, they are included here to supplement the description of TTL and ECL, which are currently the most popular bipolar technologies.

10.2.1 Diode Logic Gates

Consider the diode characteristics shown in Figure 10.1. The diode can pass the current from its *anode* to its *cathode* and block the current in the opposite direction. The diode symbol in (a) shows the direction of current flow. The diode is constructed of a tiny chip of silicon with a single *p-n* junction, as shown in (b). If V_a and V_b are voltages at the anode and cathode, respectively, with reference to the Ground, the diode is said to be "forward biased" when $V_a > V_b$ and acts as a short circuit, as shown in (c). When $V_a < V_b$, the diode is "reverse biased" and acts as an open circuit, as shown in (d). The volt-ampere characteristic of an ideal diode is shown in (e), wherein the diode current I is 0 when reverse biased and the diode voltage drop V is 0 when forward biased. In practice, the silicon diodes exhibit the characteristic shown in (f). When the diode is forward biased, a voltage drop of 0.7 V typically results. The diode also has a dc resistance of a few to tens of ohms. When forward biased, the current flows from anode to cathode. When the diode is reverse biased, a leakage current of the order of a few nanoamperes (rather than the zero current of an ideal diode) flows in the reverse direction.

Figure 10.2(a) shows the diode positive logic AND gate. *A* and *B* are inputs and *C* is the output. The supply voltage (*V*) is 5 V, and the reference voltage is the Ground (0 V). In (b), the *B* input is connected to the Ground.

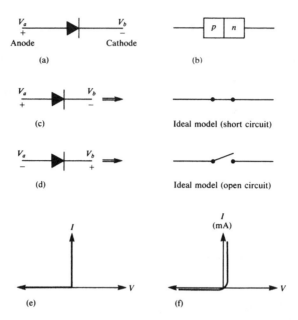

Figure 10.1 Diode. (a) Diode symbol; (b) construction; (c) forward biased; (d) reverse biased; (e) ideal characteristics; (f) actual characteristics.

Input A is at 5 V. The diode connected to B is now forward biased and hence it conducts, clamping the output to C to 0.7 V, which is the drop across the diode. The resistor R limits the current (I) through the diode to

$$I = (V - 0.7)/R \text{ (amperes)}$$

When both inputs A and B are at 5 V or higher, as shown in (c), both diodes are reverse biased; a negligible leakage current flows through the diodes and hence through the resistor R. Thus, C will be approximately at 5 V. The voltage table of (d) shows the function of the circuit, where the high-level voltage H is between 4.3 and 5 V and the low-level voltage L is in the range of 0 to 0.7 V. Using the positive logic convention, the truth table of (e) can be derived from (d). This is the truth table of an AND gate.

Figure 10.3 shows the positive logic OR gate circuit. If either of the inputs is at 5 V, the corresponding diode conducts, clamping the output C to $5 - 0.7 = 4.3$ V or ideally to 5 V. Only when both the inputs are at 0 V is the output at 0 V. The truth table in (d) for this circuit is that of an OR gate.

$$I = (V - 0.7)/R \quad \text{(amperes)}$$

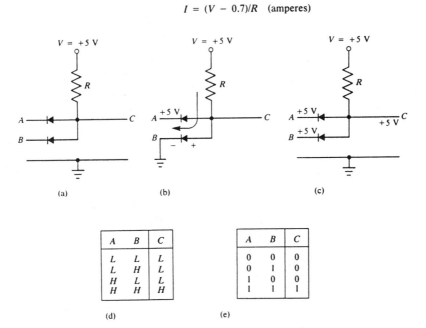

Figure 10.2 Diode positive logic AND gate. (a) Circuit; (b) $A = 5$ V, $B = 0$ V, C = 0.7 V (ideally 0 V); (c) $A = B = 5$ V, $C = 5$ V; (d) voltage table; (e) positive logic truth table.

These circuits can be extended to circuits with more than two inputs by including additional diodes at the input side. These diode-resistor circuits are not suitable for implementing ICs for the following reasons:

1.   It is not possible to realize the NOT operation with them.
2.   Since the difference between H and L at the output of the gate (0.7 to 4.3 V) is smaller than at its input (0 to 5 V), as the gates are cascaded, the voltage levels enter the forbidden region (transistor amplifiers are often used in conjunction with diode gates to maintain the operating voltage levels), and hence a low fan-out results.
3.   As the number of inputs connected to the output increases, the current drawn from the supply voltage through R increases, which causes the output voltage to rapidly enter the forbidden region, resulting in a low fan-out.
4.   The switching speeds of the diodes are limited.

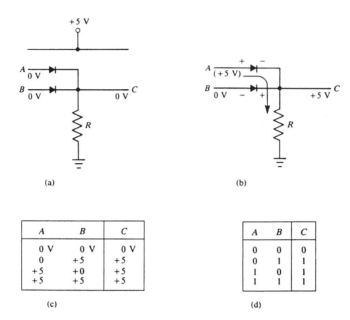

Figure 10.3 Diode positive logic OR gate. (a) $A = B = 0$ V $\therefore C = 0$ V; (b) $A = 5$ V, $B = 0$ V $\therefore C \cong 5$ V; (c) voltage table; (d) truth table.

10.2.2 Transistor Logic Gates

There are two types of *bipolar* transistors: *pnp* and *npn*. The *pnp* transistor has an *n* layer sandwiched between two *p* layers, and the *npn* has the reverse arrangement, as shown in Figure 10.4(a).

These transistors have three terminals: the Emitter (E), the Base (B), and the Collector (C). In the circuit symbols shown in (b), the arrow shows the positive direction of the collector current (I_C). A resistor R_C connected to the collector terminal as shown in (c) limits I_C whenever the transistor is conducting. The base voltage V_{BE} controls the emitter-collector current. The supply voltage (V_{CC}, V_{EE}) polarities required for the two types of transistors are also shown in (c).

Most bipolar ICs are made up of *npn* transistors. Therefore, we will use *npn* transistors in all the circuits in this section. The operation of circuits with *pnp* transistors is similar to those with *npn*, except that the current is drawn out of the base.

The bipolar *npn* transistor inverter gate is shown in Figure 10.5(a). V_i, V_{CC}, and V_o represent the input, supply, and output voltage, respectively, all

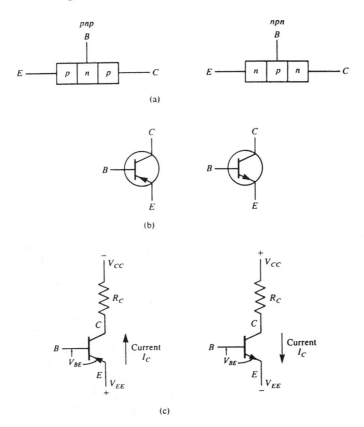

Figure 10.4 Bipolar transistors. (a) Construction; (b) circuit symbols; (c) notation.

with respect to the Ground. I_B, I_C, and I_E represent the base, collector, and emitter currents. Note that

$$I_E = I_B + I_C$$

In addition, V_{BE} and V_{CE} represent the base and collector voltages, respectively, with reference to the emitter.

The base-emitter characteristic of the *npn* transistor is shown in (b). When $V_{BE} < 0.6$ V, the base-emitter diode is reverse biased, and hence the transistor is said to be *cut off* (or simply *off*) and $I_{BE} = 0$. When $V_{BE} > 0.6$ V, the base-emitter junction is forward biased, and hence the transistor *conducts* (or is simply *on*). I_B increases very quickly as V_{BE} increases. V_{BE} never exceeds 0.8 V.

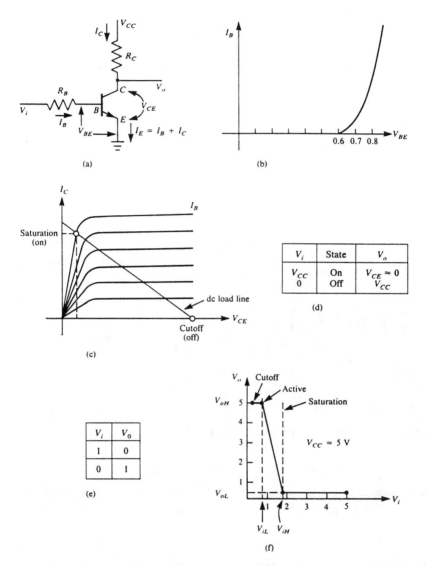

Figure 10.5 *npn* transistor inverter. (a) Circuit; (b) base-emitter characteristics; (c) transfer characteristic; (d) voltage table; (e) truth table; (f) operating voltages.

The transfer characteristics of the transistor in (c) show the relation between V_{CE} and I_C for various values of I_B. The transistor acts as a current amplifier in that a small change in base current (ΔI_B) results in a much larger change (ΔI_C) in I_C. The current gain of the transistor, beta (β), is defined as:

$$\beta = \Delta I_C / \Delta I_B$$

and is of the order of 20 to 100.

The *load line* shown in (c) represents the collector resistor R_C, which limits the collector current I_C. When V_i is 0, I_B is 0 and the transistor operates at the *cutoff* point on the load line with $V_o = V_{CC}$. The collector-to-emitter circuit then behaves as an open circuit and $I_C = 0$. When $V_i = V_{CC}$, the base-emitter junction is forward biased, the transistor is on (or in *saturation*), and $V_o = V_{CEsat}$, which is very close to 0 (about 0.2 V). Hence, $I_C = V_{CC}/R_C$. For operating as a binary device, the cutoff (V_{CC}) and saturation (V_{CEsat}) voltages can be selected as H and L levels, respectively. Circuits that use these levels are called *saturating circuits*. The input and output voltage values and the corresponding on and off states of the transistor circuit of (a) are shown in (d). As can be seen in (e), the truth table for the circuit using the positive logic notation, the circuit in (a) is an inverter. The ranges of input (V_{iL}, V_{iH}) and the output (V_{oL}, V_{oH}) voltages are shown in (f).

For the circuit in (a), in $R_C = 1$ kΩ and $R_B = 4$ KΩ, then

$$I_B = (V_{in} - V_{BE\,on})/R_B$$
$$= (5 - 0.7)/4 \qquad (V_{BE} = \text{1-diode drop})$$
$$= 1.075 \text{ mA}$$
$$I_{C\,sat} = (V_{CC} - V_{CE\,sat})/R_C$$
$$= (5 - 0.2)/1$$
$$= 4.8 \text{ mA}$$

Hence,

$$\beta = I_{C\,sat}/I_B = 4.8/1.075 = 4.47$$

That is, a transistor with a current gain of 4.47 is required to achieve a saturation voltage $V_{CE\,sat}$ of 0.2 V.

The operating characteristics of the transistor circuit are summarized below:

Transistor state	V_{BE}	V_{CE}	Current
Cutoff	0.6	Open circuit	$I_B = I_C = 0$
Saturation	0.7	0.2	$I_B > I_C/\beta$

The region between the cutoff and saturation is called the *active* or *linear* region of the transistor. Saturating circuits (i.e., circuits that use the cutoff and saturation points as the two states) are easy to build and are reliable, because the operating points are independent of the transistor current gain. It is sufficient to make the base current large enough to drive the transistor into saturation; and to ensure cutoff, it is sufficient to reverse-bias the base-emitter junction. To build *nonsaturating* circuits, the operating voltages are selected to be in the active region. More complicated circuits are needed to achieve the two stable points of operation in the active region of the transistor. However, nonsaturating circuits are faster than saturating circuits since the time it takes to switch the transistor from saturation is relatively longer.

Circuits using the cutoff point as an operating level have lower power dissipation compared with nonsaturating circuits, since I_C is 0 at cutoff and hence the circuit does not consume any power. The TTL family of logic circuits are examples of saturating circuits, and the ECL circuits are the nonsaturating type.

Because of the current amplification property of the transistor, H and L voltages at the input and the output of the circuit can be made similar. Furthermore, because of the current amplification characteristic of the transistor, these circuits provide larger fan-out and shorter switching time compared with diode circuits.

10.2.3 RTL Gates

Figure 10.6(a) shows a NOR circuit using transistors and resistors (and hence the resistor-transistor logic—RTL). When either of the inputs is at V_{CC}, the corresponding transistor conducts, thereby bringing Z to near 0 V (actually, $V_{CE\,sat}$, or 0.2 V); Z is equal to V_{CC} only when both transistors are off (that is, when both inputs are at 0). The voltage table in (b) summarizes the operating characteristics, and the positive logic truth table shown in (c) confirms the operation as a NOR circuit.

Figure 10.7 shows an RTL NAND gate. The output Z is 0 (actually, 0.4 V, the sum of the two $V_{CE\,sat}$) only if both inputs are V_{CC}; otherwise, Z is at V_{CC}. The operation of the circuit is shown in the voltage table in (b) and the truth table in (c) confirms circuit's operation as a positive logic NAND gate.

RTL gates have high power consumption (on the order of 12 mW), a fan-out of around 5, and a propagation delay of about 25 nanoseconds. Although easier to build, these circuits are sensitive to noise. When the output transistor is turned on, the sudden low impedance to the Ground attained by the circuit results in a rapid discharge of the stray capacitance.

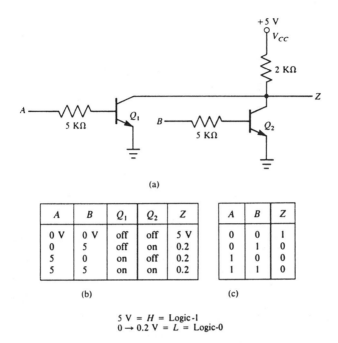

(a)

A	B	Q_1	Q_2	Z
0 V	0 V	off	off	5 V
0	5	off	on	0.2
5	0	on	off	0.2
5	5	on	on	0.2

A	B	Z
0	0	1
0	1	0
1	0	0
1	1	0

(b)   (c)

5 V = H = Logic-1
0 → 0.2 V = L = Logic-0

Figure 10.6 RTL NOR gate. (a) Circuit; (b) voltage table; (c) truth table.

Hence, H-to-L transitions are fast. When the transistor is switched off, R_C must charge up to the stray capacitance, slowing down the L-to-H transitions. Thus, the propagation delays are unequal during the two output transitions. Later in this section, these dynamic characteristics will be described further with respect to TTL circuits.

10.2.4 DTL Gates

DTL gates exhibit performance characteristics that are superior to RTL gates. Figure 10.8 shows a DTL NAND gate. When either of the inputs is at 0 V, the corresponding diode conducts, driving the voltage of P to 0.6 V (i.e., one diode drop above the Ground). For the transistor Q to be on, diodes D_1 and D_2 must be forward biased, which requires a voltage equivalent to two diode drops (0.6 V each) plus V_{BE} of Q (0.7 V), which is equal to about 1.9 V, at P. Since P is at 0.6 V, Q is off. Hence, $Z = V_{CC}$. When both inputs A and B are at V_{CC}, the input diodes do not conduct; D_1 and D_2 conduct, and the base current of the transistor turns it on, thereby clamping Z to 0.2 V. Assuming $H = V_{CC}$ and $L = 0$ to 0.2 V, the operation of the circuit is shown in the voltage table (b) and corresponds to the positive logic NAND circuit.

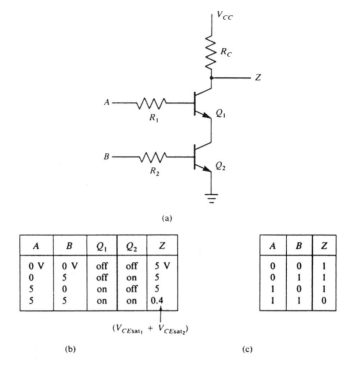

(a)

A	B	Q_1	Q_2	Z
0 V	0 V	off	off	5 V
0	5	off	on	5
5	0	on	off	5
5	5	on	on	0.4

$(V_{CEsat_1} + V_{CEsat_2})$

A	B	Z
0	0	1
0	1	1
1	0	1
1	1	0

(b)   (c)

Figure 10.7 RTL NAND gate. (a) Circuit; (b) voltage table; (c) positive logic truth table.

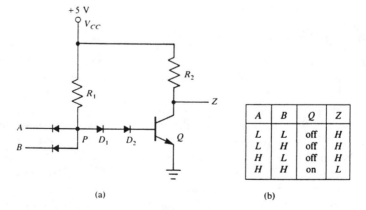

A	B	Q	Z
L	L	off	H
L	H	off	H
H	L	off	H
H	H	on	L

(a)   (b)

Figure 10.8 DTL NAND gate. (a) Circuit; (b) voltage table.

DTL circuits provide a fan-out of about 8 and a propagation delay of around 30 nanoseconds, and consume about 12 mW of power.

RTL and DTL gates are no longer used. The most popular bipolar technology is now TTL.

10.2.5 TTL Logic

Figure 10.9 shows the standard TTL NAND circuit. The circuit consists of three stages: the AND stage formed by the multiple-emitter transistor Q_1 (which essentially replaces the diode AND circuit portion in the DTL NAND circuit of Figure 10.8); the current amplifier stage formed by the transistor Q_2, and the inverter stage formed by Q_3.

Only one voltage source (V_{CC} = 5 V) is required in all TTL circuits. (The H and L voltage ranges were shown in Figure 2.2.) When any of the inputs is low (e.g., 0.2 V), transistor Q_1 is turned on. For Q_3 to be on, A must be at 2.1 V, which corresponds to the sum of the V_{BE} of Q_3 and Q_2 and the V_{BC} of Q_1, which are 0.7 V each. Since Q_1 is on, the voltage at A is 0.9 V (input voltage of 0.2 V plus a V_{BE} drop of 0.7 V). Hence, Q_3 is off; no current flows through R_3, and the output F will be at V_{CC} (or high). The

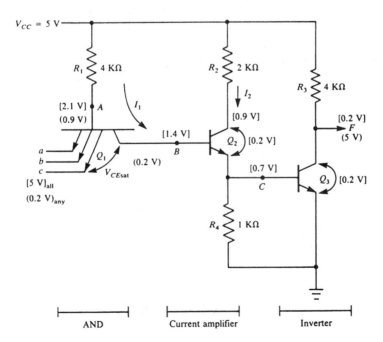

Figure 10.9 Standard TTL NAND gate.

voltage values corresponding to this operation are shown enclosed in parentheses in Figure 10.9.

When all the inputs are high, Q_1 is reverse biased, and hence Q_1 base
current flows through the base of Q_2, driving Q_2 on, which in turn drives
Q_3 on, thereby driving the output F low (or 0.2 V), which is V_{CEsat} of Q_3.
The voltage values corresponding to this operation are shown in brackets in
Figure 10.9. The voltage at A is now equivalent to three V_{BE} drops, or 2.1
V.

Thus, the circuit of Figure 10.9 is a positive logic NAND circuit. Note
also that in a TTL circuit, floating an input (i.e., not connecting it to any
voltage) is equivalent to connecting it to V_{CC}.

The NAND gate circuit in Figure 10.9 is the basic circuit from which
circuits with various types of outputs are derived. Three of the most common
outputs with which TTL circuits are available (*totem pole, open collector,*
and *tristate*) are described later in this section.

Propagation Delay   Figure 10.10 shows the dynamic characteristics of the
TTL NAND circuit. When the input changes from L to H, the output changes
from H to L after a propagation delay of t_{PHL}; when the input changes from
H to L, the output changes from L to H after a delay of t_{PLH}. The output
signal rise time (t_R) and the fall time (t_F) are also shown in the figure.

The propagation delays and rise and fall times are functions of such
transistor characteristics as carrier transit time, parasitic capacitance and resistance of the circuit, and the load connected to the circuit's output. We
will now briefly examine the effect of load on the rise and fall times.

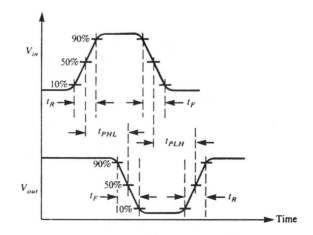

Figure 10.10   Dynamic characteristics.

The load on the circuit output can be modeled in terms of an equivalent resistance R and an equivalent capacitance C, as shown in Figure 10.11(a). All the elements (e.g., wires, connectors, other circuits) that are connected to the output contribute to the values of R and C. The capacitance is charged when the output changes from low to high. The charging time constant that determines t_R is approximately R_3C, since $R_3 >> R$. If $R_3 = 3$ kΩ and $C = 12$ pico farads (pF), the rise time would be 99 nanoseconds (ns). The rise time reduces to 66 ns as the capacitance drops to 9 pF. For large capacitance loads, the rise time can be reduced by lowering the value of R_3.

The capacitance discharges as the output changes from H to L. Since the resistance of transistor (Q_3) in saturation is low compared with R_3, the discharge time is much shorter, and hence $t_F < t_R$. These characteristics are shown in Figure 10.11(b).

Thus, in order to improve the rise time, R_3 will need to be reduced. R_3 cannot be reduced arbitrarily, however, since reducing the value of R_3 results in increased current through Q_3 when it is on. The current through Q_3 is limited by its power dissipation. Furthermore, increasing R_3 reduces the current output at the gate output (when Q_3 is off) and hence decreases the fan-out. To overcome these problems, a fourth transistor (Q_4) is added to the output portion of the standard TTL NAND circuit, as shown in Figure 10.12. The value of R_3 is reduced to around a few hundred ohms.

Totem Pole Output In the circuit of Figure 10.12, when Q_3 conducts, Q_4 is cut off, and hence the current through Q_3 is low. When Q_3 is off, Q_4 will be on, providing a small resistance for the charging of the capacitance and thereby improving the rise time. This circuit is called *active pull-up* or *totem pole* output circuit. Q_4 is the "pull-up" transistor, and Q_3 is the "pull-down"

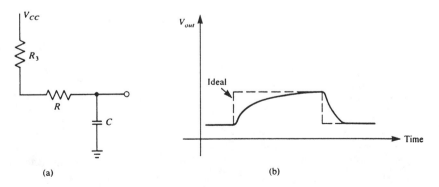

Figure 10.11 Effect of load on the output signal. (a) Load model; (b) rise/fall time.

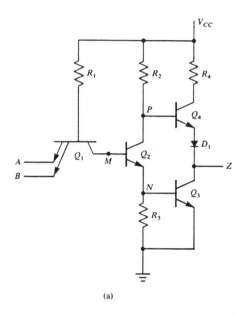

(a)

A	B	Q_1	M	Q_2	N	P	Q_3	Q_4	Z
0 V	0 V	on (Forw)	0.4 V	off	0.2 V	5	off	on	3.6 V
0	5	on	0.4	off	0.2	5	off	on	3.6
5	0	on	0.4	off	0.2	5	off	on	3.6
5	5	on (rev)	1.4	on	0.7	0.7	on	off	0

(b)

Figure 10.12   TTL NAND with totem pole outlet. (a) Circuit; (b) voltage table.

transistor. The standard circuit of Figure 10.9 is called a *passive pull-up* outside circuit since R_3 is a passive component.

When Q_4 is on, the output Z is clamped to V_{CC}. When Q_3 is on, the output Z is clamped to L. The circuit is designed so that both Q_3 and Q_4 can never be on at the same time.

When either or both the inputs A and B are at L, transistor Q_1 is driven on. The voltage at M (the base of Q_2) is V_{CE} saturated of Q_1, which is not enough to turn Q_2 on. Since Q_2 is off, the small current flowing through R_3 is not sufficient to raise the voltage at N enough to turn on Q_3; but the bias current through R_2 turns on Q_4, thereby clamping Z to the level of V_{CC}. The actual voltage at Z is

$$V_{CC} - V_{BE} \text{ of } Q_4 - \text{ drop at } D_1$$
$$= \quad 5 \quad - \quad 0.7 \quad - \quad 0.7$$
$$= 3.6 \text{ V}$$

When all the inputs are at V_{CC}, Q_1 operates in its reverse on condition. The bias current through R_1 then raises the voltage at M enough to turn on Q_2. The emitter current of Q_2 raises the voltage at N high enough to turn on Q_3. Since Q_2 is on, the voltage at P is

$$V_{BE} \text{ of } Q_3 + V_{CE\,sat} \text{ of } Q_2$$
$$= \quad 0.7 \quad + \quad 0.2$$
$$= 0.9 \text{ V}$$

This is not sufficient to turn on Q_4, since when Q_3 is on, it takes a voltage of

$$V_{BE} \text{ of } Q_3 + \text{ diode drop}$$
$$= \quad 0.7 \quad + \quad 0.7$$
$$= 1.4 \text{ V}$$

to turn on Q_4. The function of the diode D_1 is thus to keep Q_4 off when Q_3 is on. The voltage at Z is actually $V_{CE\,sat}$ of Q_3, which is 0.2 V. Thus, the totem pole circuit behaves as a NAND gate.

The gates with the totem pole output provide a shorter propagation delay compared with those with other types of outputs.

Open Collector Outputs   The outputs of two totem-pole gates cannot be connected, since the current through the conducting pull-down transistor Q_3 of one gate and the conducting pull-up transistor Q_4 of the other gate (see Figure 10.13) is too large and can end up damaging the gate. However, when the outputs of two gates with passive pull-up are tied together (Figure 10.14), a positive logic AND of the output signals results, since any Q_3 that is conducting pulls the output to L.

Figure 10.15 shows a NAND gate with an open collector output. This circuit is identical to that in Figure 10.11, except that Q_4, R_4, and D_1 are removed. When any of the inputs is at L, Q_1 is on, Q_2 is off, and hence Q_3, making Z an open circuit. An external resistor is provided to pull Z up to V_{CC}.

When all the inputs are at H, Q_1 is reverse biased and is driving Q_2 on, which in turn drives Q_3 on, and hence Z will be at L.

There is a limit to the number of open collector gates that can be tied together, a limit imposed by the load resistor value and the operating voltage levels. Because a passive pull-up is used, rise time is high and hence the open collector gates can be used only for low capacitance loads. Gates with tristate outputs are used for larger loads.

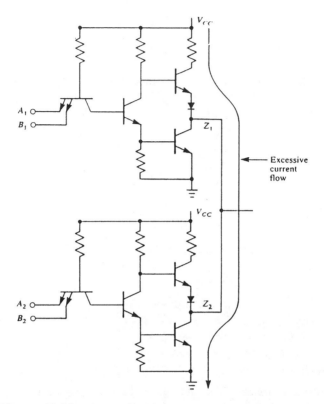

Figure 10.13   Connecting the outputs of totem pole circuits.

Tristate Output   The NAND circuit shown in Figure 10.16 incorporates the advantages of both the totem pole and open collector outputs. This circuit provides a third output state of "high impedance" when the control signal C (enable) is low. The circuit behaves as a NAND gate, providing either a 0 or a 1 output based on the other inputs when it is enabled (e.g., $C = 1$).

When C is 1, if either of the inputs A or B is low, Q_1 will be on and both Q_2 and Q_3 will be off. Q_4 is then driven on, as is Q_5, clamping the output Z to V_{CC}. When C is 1, if both A and B are high (V_{CC}), Q_1 will be reverse on, driving Q_2 and Q_3 on; Q_4 and Q_5 are both off and Z goes low (0 V).

When C is 0, the diode D_1 keeps Q_4 off, which in turn keeps Q_5 off; Q_1 is on and both Q_2 and Q_3 are off. The output Z is then an open circuit.

The circuit thus has a high impedance state in addition to the H and L output states. It is as though the gate is not connected to the rest of the circuit, when it is in the high impedance state.

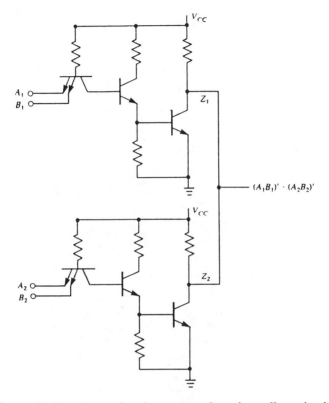

$(A_1B_1)' \cdot (A_2B_2)'$

Figure 10.14 Connecting the outputs of passive pull-up circuits.

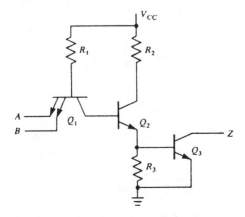

Figure 10.15 TTL NAND with open collector output.

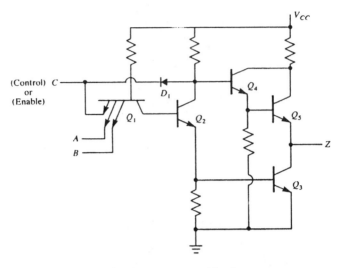

Figure 10.16   TTL NAND with tristate output.

Loading and Fan-out   The limit on the current flowing through the output terminal of the gate determines its fan-out. Figure 10.17 shows the two driving configurations of a gate with passive output. When the output is at H, the driving gate acts as a current source. The current I_3 through the pull-up resistor R_3 is the sum of currents drawn by each input connected to the output. The output voltage V_O then is

$$V_O = V_{CC} - I_3R_3$$

For proper operation, V_O should not be less than V_{Hmin}. That is,

$$V_{CC} - I_{3\,max}\, R_3 > V_{H\,min}$$

If $R_3 = 3$ KΩ, the maximum output current I_{3max} is

$$(5 - 2.4)\ V/3K\Omega = 0.867\ mA$$

Assuming a typical gate input current of 40 μA, the number of gates that can be driven is about 21.

When the driving gate output is low, Q_3 is in saturation and acts as a current sink. The current through Q_3 is

$$I_o = I_3 + I_{in}$$

where I_{in} is the sum of input currents and

$$I_3 = (V_{CC} - V_{CE\,sat})/R_3$$

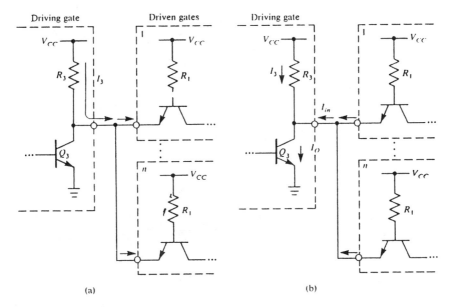

Figure 10.17 Fan-out computation. (a) High output; (b) low output.

The maximum value of the current through Q_3 ($I_{o\,max}$) is determined by the *heat dissipation* limit of the transistor. Thus, the fan-out is limited by ($I_{o\,max}$ − I_3). A standard TTL gate can sink at most 16 mA. Assuming an input current of 1.6 mA per gate, a maximum of 10 gates can be driven.

Thus, the fan-out is the lower of the above two values, which is 10. *Buffer gates* are used for higher driving capability.

Buffer Gate Figure 10.18 shows a buffer gate. These gates have a smaller pull-up resistor and large transistors Q_3 and Q_4 compared with a standard gate. The transistor Q_5 provides the additional amplification to drive Q_4. A standard TTL buffer can drive 30 standard TTL gates.

Note that the effect of load on the gate is to increase the value of V_{OL} and decrease the value of V_{OH}, thus reducing noise margins. Thus, lightly loaded gates have higher noise margins compared with fully loaded gates. The *guaranteed noise margins* specified by the gate manufacturers correspond to the noise margins of fully loaded gates.

Variations of TTL As mentioned earlier, TTL ICs are available in two series: 7400 series for commercial applications and 5400 series for military

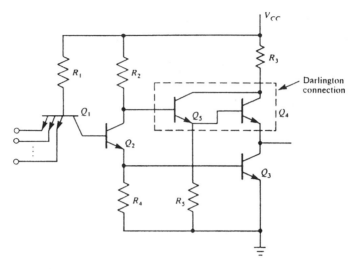

Figure 10.18   TTL buffer circuit.

applications where temperature ranges are extreme. The IC manufacturers use standard designations to denote the various characteristics and functionality of ICs within the family. Figure 10.19 shows the details of the standard designation, which consists of five parts. The first part consists of either one or two characters identifying the manufacturer; the second part identifies the

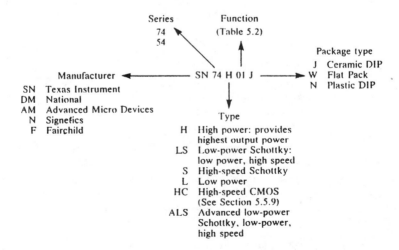

Figure 10.19   TTL IC designation.

series (74 or 54); the third part identifies the type or variation (e.g., standard, low-power Schottky, and so on); the fourth part designates the function performed by the IC, and the last part designates the package type. For example, SN74LS00N designates an IC manufactured by Texas Instruments (SN), 7400 series, low-power Schottky (LS), NAND gate (00), packaged in plastic DIP (N), whereas SN7400N designates the same NAND package but with standard TTL. Table 10.1 lists a small set of available TTL ICs. For a complete listing, see the IC manufacturers' catalogues listed at the end of this chapter.

The variants of the standard TTL family offer a choice of power dissipation, speed, density, and cost characteristics. Some of these characteristics are listed below and summarized in Table 10.2.

The high-power TTL (H type) provides the highest output power and also dissipates the most power. It is obtained by decreasing the resistance values in the standard TTL circuit and hence is faster than standard TTL.

The low-power TTL (L type) is obtained by increasing the resistance values. This results in lower speed than standard TTL. Since the power reduction is higher than the speed reduction, the speed-power product of this type is lower than that of standard TTL.

The Schottky TTL (S type) uses a Schottky transistor that consists of the regular transistor along with an additional diode in its base-collector junction. The diode prevents the transistor from going into saturation, thus reducing the switching time and hence increasing speed. The power dissipation is not increased. Variations of Schottky TTL with better speed-power products are now available and include the Texas Instruments Advanced Schottky (AS) and Advanced Low-Power Schottky (ALS) and Fairchild Advanced Schottky (FAST). ALS is currently the most popular because of its high speed and low power dissipation.

High-speed CMOSs (HC type) are CMOS circuits that are identical in function to TTL ICs of similar numeric designation.

In general, when different types of TTL gates (e.g., standard, low-power, Schottky) are used in the circuit, it is desirable to express the driving capability in terms of the number of standard TTL loads that a gate can drive. Similarly, the load of an input can be expressed in terms of the number of standard TTL loads. These standard values can be used to identify any loading problems. Actual current values can then be used to calculate loads accurately.

The current from a standard TTL load is 1.6 mA, while the load current of an LS circuit is about 0.4 mA. Thus, using the standard TTL load as the unit (UL), the LS circuit load is 0.25 UL. A standard TTL gate can sink 10 standard loads of 1.6 mA. The following table lists the loads of the representative TTL variants.

TABLE 10.1   Typical TTL ICs

7400	Quad two-input NAND gates (totem pole)
7401	Quad two-input NAND gates (open collector)
7402	Quad two-input NOR gates (totem pole)
7404	Hex inverters
7415	Triple three-input AND gates (totem pole)
7420	Dual four-input NAND gates (totem pole)
7428	Quad two-input NOR buffer
7486	Quad two-input EXCUSIVE-OR
74125	Quad tristate buffer (low enable)
74126	Quad tristate buffer (high enable)

Note: Variants are not shown here. For example, variants of 7400 are 74H00, 74S00, 74LS00, and so on.

Type	Input load (UL) High	Input load (UL) Low	Output load (UL) High	Output load (UL) Low
74	1	1	20	10
74S	1.25	1.25	25	12.5
74LS	0.5	0.25	10	5
74H	1.25	1.25	25	12.5

Example 10.1 shows the fan-out calculations using unit loads.

Example 10.1

The following table shows the fan-out calculations for various driving configurations:

TABLE 10.2   Characteristics of TTL Variants

TTL type	Delay (ns)	Power dissipation (mW)	Fan-out
Stanford	10	10	10
High power (H)	6	22	10
Lower power (L)	33	1	20
Schottky (S)	3	19	10
Low-power Schottky (LS)	9.5	2	20
Advanced low-power Schottky (ALS)	4	1	20

Driving gate type	Load gate type	Low-level fan-out	High-level fan-out	Overall fan-out
Standard	Standard	10/1 = 10	20/1 = 20	10
LS	LS	5/0.25 = 20	10/0.5 = 20	20
H	LS	12.5/0.25 = 50	10/0.5 = 20	20

where the fan-out at low or high level is the ratio of the UL of driving gate to the UL of the driven gate at that level. The overall fan-out limit is the minimum of the high- and low-level fan-outs.

10.2.6 ECL Gates

As mentioned earlier, ECL gates have the lowest propagation delay of any IC family and as such are used in applications requiring very high speeds. They have a very low noise immunity and a high power dissipation. The low propagation delay (of the order of 1 nanosecond) is achieved by operating the transistors in the nonsaturated region. In these circuits, the change of logic state is brought about by switching the current path rather than changing its magnitude.

Figure 10.20(a) shows the basic ECL inverter circuit. V_{BB} establishes a bias (or reference) voltage level for the operation of the gate. When the input voltage V_{in} is less than V_{BB}, Q_2 is on and Q_1 is off. Hence, the current I flows through R_2. Since no current flows through R_1, point P is at 0 V. When $V_{in} > V_{BB}$, Q_1 conducts and Q_2 is off. Hence, the current I flows through R_1, clamping the voltage at P to $(- IR_1)$. Q_3 and R_4 form a level shifter network to transform the voltage levels at point P into the desired V_{OL} and V_{OH} values.

As we can see by the voltage transfer characteristics of Figure 5.21(b), the input voltage swing for ECL circuits is of the order of 0.2 V, compared with 1.2 V for TTL circuits. In addition, the current steering mode of operation eliminates the power spikes introduced by fast TTL circuits during switching, thereby facilitating power distribution. But the power consumption of ECL gates is of the order of four times that of a standard TTL circuit (10 mW for a 10-ns TTL compared with 40 mW for a 1-ns ECL gate).

In this circuit, any noise at point S is passed by the input stage, while the noise at R is rejected. Since the Ground is less susceptible to noise, V_{CC} is connected to Ground and a negative power supply $V_{EE} = -5.2$ V is used to achieve better noise immunity.

Figure 10.21 shows the Motorola ECL OR-NOR circuit. The voltage levels used by the circuit are $H = -0.75$ V and $L = -1.55$ V. In (a), one of the inputs (R) is high, and the other inputs are low. Hence, Q_3 is on and

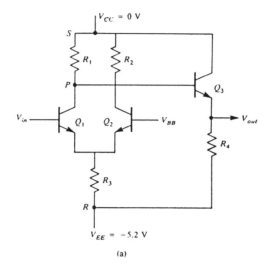

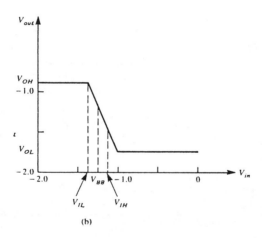

Figure 10.20   ECL inverter (Courtesy of Motorola Inc.). (a) Circuit; (b) transfer characteristics.

point A is at $-0.75 - 0.75$ (V_{BE} of Q_3) $= -1.5$ V. This decreases the base-emitter drop at Q_4, and thus Q_4 turns off. The current I through R_{c1} now results in a voltage of -0.8 V at B, as set by the value of R_{c1}. Hence, the Y output will be at $-0.8 - 0.75 = -1.55$ V, or at L. The Y' output is at -0.75 V, or at H.

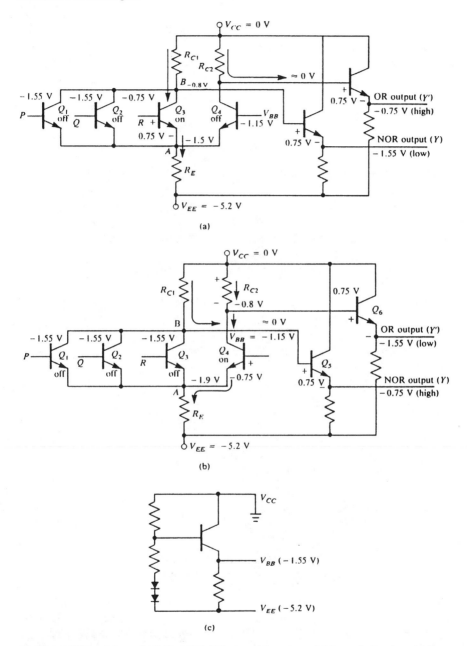

Figure 10.21 Motorola ECL NOR/OR gate (Courtesy of Motorola Inc.). (a) Low output; (b) high output; (c) bias voltage.

If all the inputs are low, as shown in (b), all the input transistors are off and Q_4 conducts. The voltage at A is $(-1.15 - 0.75) = -1.9$ V. The base voltages of input transistors (-1.55 V) are not sufficiently higher than the voltage at A and hence all the input transistors are off. The current I now flows through R_{c2}, Q_4, and R_E and results in a voltage drop of -0.8 V as set by the value of R_{c2}. Then, the Y' output will be at $(-0.8 - 0.75) =$ -155 V, or at L. If we neglect the small voltage drop across R_{c1} as a result of the very small base current of Q_5, the voltage at B is 0 V, and hence the Y output will be at -0.75 V or H. Thus, the circuit functions as a positive logic OR/NOR circuit. The calculation of resistance values in this circuit is left as an exercise.

The circuit used to obtain the bias voltage V_{BB} is shown in (c). This circuit is a temperature-compensated voltage divider network that sets V_{BB} to -1.15 V using the supply voltages V_{CC} and V_{EE}. Thus, only one voltage, V_{EE}, is required for the ECL circuit.

ECL circuits have a typical fan-out of 20. Because of their high operating speed, care must be taken to reduce the line reflections in external connections when two or more ECL ICs are used in a circuit, especially when the line length exceeds a few centimeters. Special packaging and housing of ECL circuits is needed because of their high power dissipation.

Several ECL families of logic units are available, the 10K and the 100K series being the most popular ones. Typical characteristics of the 100K family include the following:

1. Voltage levels:

$$V_{OL} = -1.7 \text{ V} \qquad V_{OH} = -0.9 \text{ V}$$

$$V_{IL} = -1.4 \text{ V} \qquad V_{IH} = -1.2 \text{V}$$

$$V_{EE} = -5.2 \text{ V}$$

2. Noise margin = 0.3 V; fan-out = 10
3. Power dissipation: 40 mW per gate
4. Propagation delay: 0.75 ns

10.3 Metal Oxide Semiconductor (MOS) IC Technology

MOS technology uses MOSFETs as switching elements. These are *unipolar* devices and hence can be fabricated in a smaller silicon area than can the bipolar devices used in TTL and ECL technologies. MOS circuits thus provide higher integration levels and hence are suitable for LSI and VLSI levels of integration. The power dissipation of these circuits is also less than that of bipolar circuits. Because the fabrication process is less complex—resulting in fewer defects per unit area of silicon—fabrication of larger chips is cost-effective. MOS circuits are slower than bipolar circuits and cannot drive large capacitance loads. Thus, MOS circuits are used extensively in slow-speed devices requiring dense circuits. The speed of MOS circuits depends on the transistor size. Advances in fabrication technology are rapidly reducing MOS transistor size, thereby increasing speed.

There are two types of MOSFETs: *P-channel* and *N-channel*, shown in Figure 10.22. The *substrate* is a thin slice of silicon doped with either *n*-type or *p*-type impurities. *Source* and *drain* are formed by a heavier doping of the opposite type of impurity as that of the substrate. The region between the source and the drain is the channel. The *gate* is a metal plate placed over the channel and separated from it by insulating material of silicon dioxide.

Figure 10.22(c) shows the symbols used to represent MOSFETs, and (d) shows their transfer characteristics. A voltage applied to the gate produces an electric field that controls the conduction between the source and the drain. When there is no electric field, the path between the source and the drain is equivalent to two diodes connected back to back, and thus the equivalent resistance is high. When the gate voltage exceeds a *threshold voltage* V_T, the transistor conducts and the direction of the current flow is indicated by the arrow in the transistor symbols in (c). We will use the simplified symbols shown in (e) to represent these transistors.

The *n*-channel MOS (NMOS) device requires a positive voltage exceeding V_T for it to conduct, while the *p*-channel MOS (PMOS) device requires a negative voltage value exceeding V_T. The threshold voltage is determined by the doping concentration of the channel and can be positive, zero, or negative. If the threshold voltage is positive (corresponding to a *p*-type implant) the NMOS transistor is called an *enhancement-mode device*; if negative (corresponding to an *n*-type implantation), it is called a *depletion-mode device*.

Logic circuits can be built using either NMOS or PMOS devices. NMOS is the more popular technology, predominantly for LSI and VLSI circuits. The other popular MOS technology is Complementary MOS (CMOS), which uses both PMOS and NMOS devices and is suitable for LSI as well as MSI and SSI levels of integration. The primary characteristic of CMOS is its low power consumption. We will now examine the basic circuits in NMOS and CMOS technologies.

10.3.1 NMOS Logic Gates

Figure 10.23(a) shows an inverter circuit built out of an NMOS transistor Q_1 (pull-down) and a (pull-up) resistor R. When $V_{in} < V_T$, Q_1 is off, thereby

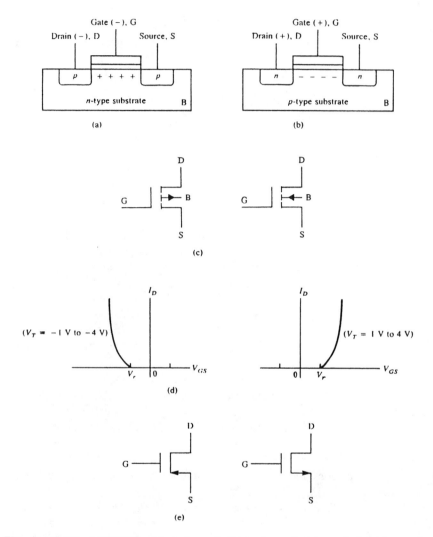

Figure 10.22 MOSFETs. (a) *p*-channel; (b) *n*-channel; (c) symbols; (d) transfer characteristics; (e) simplified symbols.

driving V_{out} high (i.e., V_{DD}). When $V_{in} > V_T$, Q_1 conducts and clamps V_{out} to the Ground, confirming the inverter action where logic-0 corresponds to 0 V and logic-1 corresponds to V_{DD}, which is usually $+10$ V.

The *V-I* characteristics of the inverter are shown in (b) for various values of V_{GS}. When Q_1 is off, $V_{out} = V_{DD}$, which should be V_{OH} for reliable

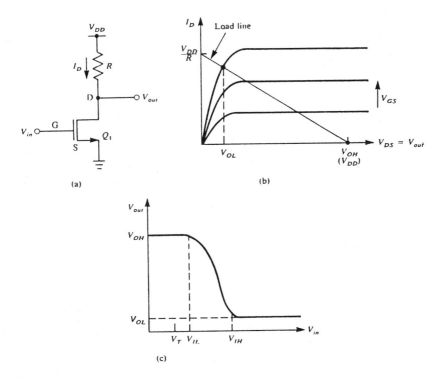

Figure 10.23 NMOS inverter. (a) Circuit; (b) *V-I* characteristics; (c) voltage transfer characteristics.

operation of the circuit. V_{OL} is selected by the intersection of the load line with the *V-I* curve corresponding to the selected value of V_{GS}. For reliable operation, V_{OL} should be less than V_T. The voltage transfer characteristics of the inverter circuit are shown in (c).

To bring the power consumption to values compatible with LSI and VLSI circuits, the resistance R should be around 100 K. Since fabrication of such resistance values requires a large silicon area, another MOS transistor is used as a pull-up transistor, just as in TTL totem pole circuits. The pull-up transistor can be of either the enhancement mode or the depletion mode, as shown in Figure 10.24. In (a), for Q_2 to conduct, its V_{GS} must be greater than V_T. Hence, when Q_2 is off, V_{out} at the maximum will be at $(V_{DD} - V_T)$, thereby reducing V_{OH} and hence the noise margin. This problem is solved by using the depletion-mode pull-up transistor, as shown in (b).

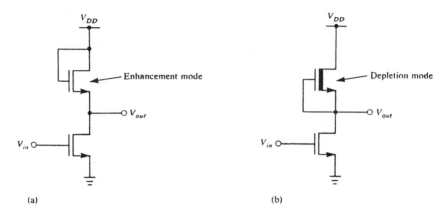

Figure 10.24 Totem pole NMOS inverter. (a) Enhancement mode pull-up transistor; (b) depletion mode pull-up transistor.

The physical characteristics of the transistor determine such performance aspects as noise margin and power consumption. For dense circuits to be achieved, the transistors must be as small as possible. Transistor size is determined by the channel width (W) and length (L) of the transistor. The transfer characteristics depend on the ratio of width to length of the channel. This ratio (W/L) is called the *aspect ratio*. For the transistor to be small, the aspect ratio should also be small. The smallest dimension possible is determined by the definition of features allowed by the fabrication process. In current technology, a feature size of 3 micrometers (μm) is common; ICs with 1-μm feature sizes are now possible, and the technology is progressing toward achieving the physical limit of feature size, which is about 0.2 μm.

A low aspect ratio of the pull-down transistor is desirable to allow a low effective resistance of the load transistor, thereby producing a fast rise time of the output voltage when the pull-down transistor is driven on. However, the greater the aspect ratio, the lower the output voltage will be below V_T, and hence the noise immunity at low output voltage is better. Thus, the aspect ratio is selected to compromise between these two aspects while observing the technology limits on feature sizes. Typical values for the aspect ratio for pull-up and pull-down transistors are $\frac{1}{2}$ and 2, respectively. Thus, one of the transistors must be about four times larger than the other.

Figure 10.25(a) shows a two-input NMOS NAND gate constructed by connecting two NMOS transistors in series. In general, for an n-input NAND gate, n transistors are connected in series. The aspect ratio of each pull-down transistor should be n times that of the basic inverter, since the com-

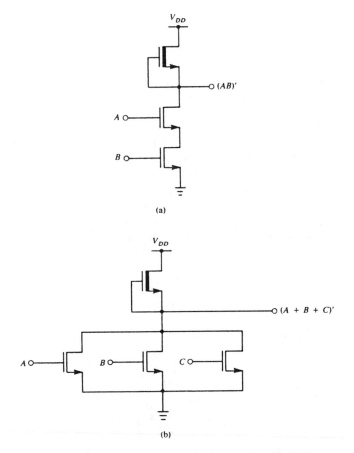

Figure 10.25 NMOS circuits. (a) NAND; (b) NOR.

bined aspect ratio should be such that the desired value of V_{OL} is retained when all the inputs are high. This means that the transistor size increases as the number of inputs increases.

Figure 10.25(b) shows the NMOS NOR circuit obtained by connecting transistors in parallel. In this circuit, since the output will be at V_{OL} when only one pull-down transistor is conducting, the aspect ratio of all the pull-down transistors is the same as in the basic inverter. Hence, NOR circuits are more attractive than NAND circuits in NMOS when the number of inputs is high.

It is possible to connect transistors in various series and parallel combinations to derive complex functions (see problem 10.12).

10.3.2 CMOS Logic

CMOS circuits use a combination of NMOS and PMOS enhancement-type transistors. Since two types of transistors are used, the fabrication of CMOS ICs is more complex than that of NMOS ICs. The silicon area occupied by these devices also increases, thereby reducing circuit density. The characteristics of CMOS are low power consumption, high noise immunity, and temperature stability.

Figure 10.26(a) shows a basic CMOS inverter circuit built out of an NMOS and a PMOS device. The input is formed by connecting the two devices together. When V_{in} is positive and greater than the threshold voltage V_{TN} of the NMOS transistor, it conducts and clamps the output V_{out} to the Ground (0 V). When V_{in} is negative and exceeds the threshold voltage V_{TP} of the PMOS transistor, it conducts, driving V_{out} to V_{DD}. Since only one of the transistors is on in either case, a negligible current flows from V_{DD} to Ground, and hence the power dissipation is almost zero at static operation.

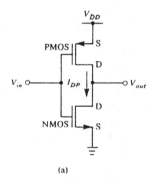

(a)

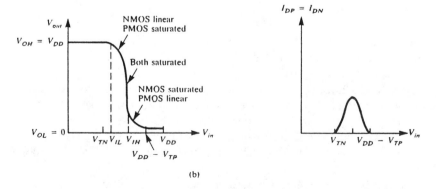

(b)

Figure 10.26 CMOS inverter. (a) Circuit; (b) transfer characteristics.

When the output changes, however, both transistors may momentarily be on during the switching, thereby contributing to the V_{DD}-to-Ground current and hence power consumption. Power consumption increases with the frequency of switching and approaches that of TTL circuits at high frequencies. CMOS circuits are thus useful when the lowest power consumption is needed.

The voltage transfer and current characteristics of the inverter circuit are shown in Figure 10.26(b). Note that the current flows when V_{in} is between both threshold voltages (i.e., when both the transistors are conducting). Since one of the transistors is conducting in both the states, CMOS circuits have good dynamic characteristics.

The noise immunity of CMOS circuits is also better than that of NMOS circuits since the voltage swing is between 0 and V_{DD}. Assuming a V_{DD} of +5 V, if V_{IH} = 3 V, then any input between 3 and 5 V produces a low output (e.g., a low-level noise margin of 2 V). Similarly, if V_{IL} = 2 V, then any input between 0 and 2 V will produce a high output (e.g., a high-level noise margin of 2 V). CMOS gates provide a typical noise immunity of 0.45 V_{DD} and a guaranteed noise immunity of 0.3 V_{DD}. V_{DD} ranges from +5 to +18 V.

CMOS circuits provide very high fan-outs of the order of 50 or higher. Ideally, the fan-out is infinite since no loading occurs when the output is connected to the gate of an enhancement MOSFET.

Propagation delay of CMOS gates ranges between 20 to 100 μs, but increases with greater load capacitance, thereby limiting fan-out.

Figure 10.27 shows a CMOS NOR gate and Figure 10.28 shows a CMOS NAND gate. Note that in these circuits for each input to the circuit, there is a PMOS and an NMOS transistor.

Table 10.3 lists some CMOS 4000 series of ICs available. Some manufacturers (e.g., Motorola) number this series starting at 14000.

Mixing Logic Families   As mentioned earlier, a special series of CMOS ICs that emulate TTL ICs are also available. The ICs in this series are designated 74C or 74HC, indicating their compatibility with corresponding TTL ICs. (For example, 74C00 is a CMOS quad two-input NAND IC similar to TTL 7400.) Although these ICs are functionally equivalent and are pin-compatible to the corresponding TTL circuits, their fan-out, propagation delay, and power dissipation characteristics correspond to CMOS.

It is not always possible to connect ICs of two different technologies directly, since the logic voltage levels and polarities may not agree and the current sourcing and sinking capabilities may differ. Special ICs (such as CMOS 4504-level shifter) are available that allow such interfacing.

The TTL totem pole or tristate output can be directly connected to the input of a CMOS circuit with V_{DD} = 5 V, since the logic levels of H near 3.6 V and L near 0 V are compatible. However, the CMOS low-level output

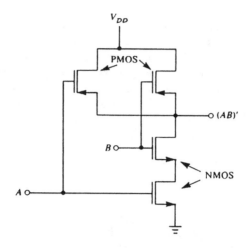

Figure 10.27   CMOS NOR.

cannot sink the standard TTL input. The TTL LS type of circuit provides the best match because of its low current levels. As the number of loads increases, this interface also will not work, since the CMOS output cannot sink the combined input currents.

Because the V_{DD} value is between +5 and +12 V for both CMOS and NMOS, a direct connection between these types of ICs is possible. In general, PMOS circuits have higher supply voltage levels than those of CMOS;

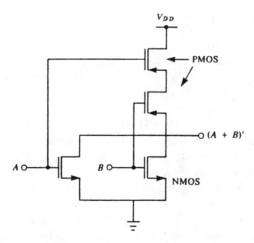

Figure 10.28   CMOS NAND.

TABLE 10.3 Some CMOS ICs

Part number	Function
14001	Quad two-input NOR
14002	Dual two-input NOR
14011	Quad two-input NAND
14012	Dual two-input NAND
14023	Triple three-input NAND
14025	Triple three-input NOR
14069	Hex inverter
14070	Quad EXCLUSIVE-OR
14071	Quad two-input OR
14072	Dual four-input OR
14073	Triple three-input AND
14075	Triple three-input OR
14077	Quad EXCLUSIVE-NOR
14503	Hex tristate buffer
14504	TTL-CMOS level shifter

hence, resistor networks to drop the voltage levels of PMOS are needed to interface them with CMOS circuits.

10.4 Gallium Arsenide IC Technology

Currently, the highest switching speeds are offered by the silicon ECL technology. GaAs technology is projected to compete with ECL and has been called "the technology of the 1990s" for fast ICs with low power consumption. Compared with silicon, GaAs offers superior electronic properties, such as high electron mobility, high saturation velocity, and semi-insulating substrates. As such, GaAs technology has the potential for high-speed circuits with low power dissipation and high current driving capabilities.

Despite considerable research into developing satisfactory oxides or dielectrics on GaAs, it has been difficult to achieve stable dielectric-semiconductor interfaces of sufficient quality. The high cost of GaAs material, lack of standard packaging, lack of general design rules and tools, and lack of appropriate test equipment are some of the problems with this technology—problems that are slowly being solved.

TABLE 10.4 Comparison of GaAs and Silicon technologies

Characteristic	GaAs	Silicon
Transistors/chip	20000–30000	300,000–400,000
Gate delay	50–150 ps	1–3 ns
Transistors/gate	1 + fan-in	1 + fan-in
Fan-in (typical transistor)	3–4	5
Fan-out (typical transistor)	3–4	5

Adapted from *Computer Design*, October 15, 1985.

Prototype GaAs ICs with an integration level of around 10,000 gates per chip are currently available in limited functionalities, though most commercially available ICs are at the integration level of 3,000 gates. Table 10.4 compares the performance of GaAs and silicon technologies. Figure 10.29 shows the speed-power products for various technologies. The diagonal lines represent equal speed-power products. Characteristics of three generations of GaAs devices and two generations of Advanced Digital Bipolar (ADB)

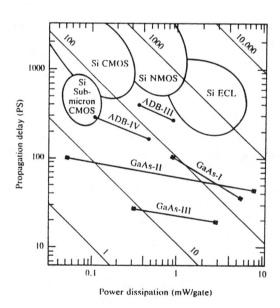

Figure 10.29 Comparison of IC technologies. (Reprinted with permission of *Computer Design*, October 15, 1985.) (*Note:* Diagonal lines represent equal speed-power products.)

ICs produced by Honeywell are shown. The GaAs technology shows clear superiority.

In addition to off-the-shelf logic parts, the commercial emphasis is currently on such semicustom GaAs devices as gate arrays and memory devices (see Chapter 9).

Most GaAs ICs now available are based on either the depletion- or the enhancement-mode MEtal Semiconductor Field Effect Transistor (MES-FET) shown in Figure 10.30. These transistors are fabricated by direct ion implantation onto a high-quality semi-insulating GaAs substrate, and their structure is very similar to that of a MOSFET.

Figure 10.31 shows a GaAs inverter circuit using a depletion mode MESFET. This MESFET uses a threshold voltage V_T in the range of -1 V to -2.5 V. A negative gate voltage exceeding V_T is required to turn this transistor off when the drain voltage is positive. The cascade of three diodes in the output stage is used to shift the output voltage levels to the operating range of $+0.5$ to -2 V. Figure 10.32 shows the positive logic NAND and NOR circuits that are constructed by connecting input MESFETs either in series or in parallel.

This logic approach is known as Buffered FET Logic (BFL). The limitation of this approach is its high power consumption as a result of the

Planar depletion-mode MESFET

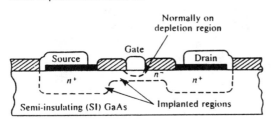

Enhancement-mode FET

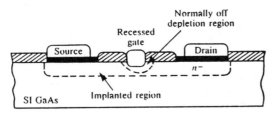

Figure 10.30 MESFET.

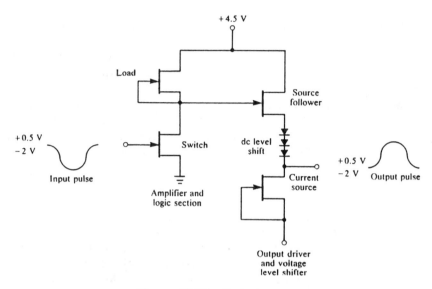

Figure 10.31   GaAs inverter.

diode cascade in the load path. BFL gates typically consume 4 mW of power per gate. These power levels limit this logic to MSI levels of integration.

Various other GaAs logic technologies based on more complex fabrication processes are in various stages of development. These technologies

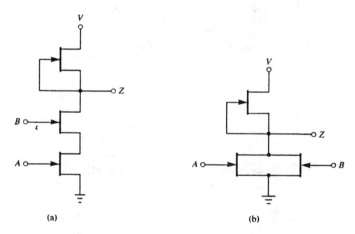

Figure 10.32   GaAs circuits. (a) NAND; (b) NOR.

are based on two transistor configurations known as the high-electron mobility transistor and the heterojunction bipolar transistor.

10.5 Optical Technology

Optical computing has been an active area of research over the last few years, as an alternative to electronic digital computing to design high speed computer systems. Data communication through optical fibers has already been proven to be more versatile and cost effective than that through the electronic media. Optical interconnection networks are widely being investigated [Brenner, 1988; Jahns, 1988]. Optical interconnections between the components on an integrated circuit chip would save the chip area occupied by the interconnect silicon and eliminate the signal interference problems. Several analog computer-like structures based on optical devices have also been in use in image processing applications. But the technology has not yielded any practical equivalent to current day electronic digital computers yet. The AT&T Bell laboratory has recently announced the invention of a practical optical transistor. This has led to the speculation that optical computers may be around the corner.

The speed characteristics of electronic computational structures are functions of device speeds and the architecture. There is a large disparity between the speed of the fastest electronic switching component and the speed of the fastest digital electronic computers. The switching speed of transistors are as high as 5 picoseconds, while the fastest computers operate at clock periods of the order of a few nanoseconds. The limitations of electronic technology that cause this speed disparity are [Jordan, 1988]: electromagnetic interference at high speeds, distorted edge transitions, complexity of metal connections, drive requirements for pins, large peak power levels and impedance matching effects.

Electromagnetic interference is the result of coupling of the inductances of two current carrying wires. Sharp edge transitions are a requirement for proper switching. But higher frequencies attenuate greater than lower frequencies, resulting in edge distortions at high speeds. The complexity of metal connections on chips, circuit boards and between system components introduce complex fields and unequal path delays. The signal skews introduced by unequal path delays are overcome by slowing the system clock. Large peak power levels are needed to overcome residual capacitances. Impedance matching effects at connections require high currents and in turn cause lower system speeds.

There are several advantages to using free-space optics for interconnections [Jordan, 1988]. By imaging a large array of light beams onto an array of optical logic devices, it is possible to achieve high connectivity.

Since physical interconnects are not needed (unless fibers or waveguides are used) connection complexity and drive requirements are reduced. Optical signals do not interact in free space (i.e., beams can pass through each other without any interference) and hence a high bandwidth can be achieved. There is no feedback to the power source as in electronic circuits and hence there are no data dependent loads. The inherently low signal dispersion of optical signals implies that the shape of a pulse as it leaves its source remains virtually unchanged until its destination. Another advantage of optics over electronics is communication. Optical devices can be oriented normal to the surface of an optical chip such that light beams travel in parallel between arrays of optical logic devices rather than through pins at the edges of chips as in electronic integrated circuits. Lenses, prisms, and mirrors can convey an image with millions of resolvable points in parallel.

It is important to note the differences in basic characteristics of electrons and photons. Electrons easily affect each other even at a distance, thus making it easy to perform switching. But, this ease of interaction complicates the task of communication, since the signals must be preserved. Since it is very difficult to get two photons to interact, it is very difficult to get two optical signals to interact. Thus, optics is bad for switching but good for communications. A solution may be to stay with hybrid technology where electronics performs all the computations and the optics performs all the communication.

There are several problems associated with the optical technology. Most of these problems stem from an inability of optical signals to interact and thus perform switching. Electronics technology is mature, cost effective and allows the fabrication of high density switching components. Photonics on the other hand, is less mature and requires tight imaging tolerances and constant power consumption for modulator-based optical devices. Optical devices can be spaced a few micron apart on optical chips but require several centimeters of interaction distance for lenses, gratings, and other imaging components. Microoptic techniques are being investigated as solutions to this problem.

The lack of a suitable optical memory has been another problem. The development of optical memories has followed the requirements imposed by the modified finite state machine which is a serial structure, rather than the classical parallel model. The parallelism offered by optics was thus ignored and the emphasis was to incorporate an addressing mechanism. This required beam deflectors, page composers, and detector arrays which were slow, awkward, and expensive. Also note that in a modified finite state machine the memory elements are required to preserve their contents indefinitely since they are addressed in random order. (In the classical finite state machine the storage elements need only preserve their information for one cycle). It has been difficult to fabricate such an optical device. An optical disk device provides this capability but is hardly satisfactory as a main memory element.

A number of research efforts were started in early 1960s to build digital logic devices utilizing the semiconductor laser diodes and the nonlinear phenomena of saturable absorption. The infancy of optoelectronic technology and a critical study of power dissipation vs speed for optical logic led to the conclusion that the inherently higher switching speeds of optical phenomena could not be exploited to build optical computers that can offer the performance of electronic technology. The early 1980s saw rapid improvements in optoelectronic technology. New materials such as multiple quantum wells (MQW) and device configurations such as self-electro-optic effect devices (SEED) and optical logic talons (OLE) were developed.

Many approaches have been proposed for forming general purpose optical computers, although none of them have been completely implemented. One among them is Huang's symbolic substitution [Huang, 1983], which is based on binary pattern substitution. The general idea is to search for a two-dimensional pattern in a binary grid and to replace that pattern with another pattern everywhere the search pattern is found.

Consider the example shown in Figure 10.33. The search pattern is the left-hand side (LHS) of the transformation rule and the pattern that replaces the LHS is the right-hand side (RHS) as shown in (a). In (b), the LHS of the rule is satisfied at two locations. The RHS is written at those locations and the cells that do not contribute to the LHS disappear after the rule is applied.

A number of rules can be applied either in series or in parallel over a number of iterations to realize complex functions. Transformation rules can be customized to perform specific functions such as addition or subtraction

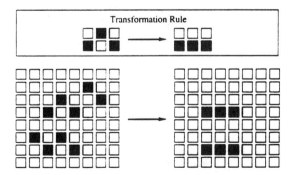

Figure 10.33 Symbolic substitution [Huang, 1983]. Reprinted with permission from IEEE.

or they may be made to perform Boolean logic primitives in which case the configuration of the grid is customized to implement specific functions.

A schematic diagram of an optical setup for a single transformation rule with four cells in the LHS and four cells in the RHS is shown in Figure 10.34. Here, a two-dimensional input pattern is combined with a two-dimensional control image (produced by imaging light through a mask) onto the optically nonlinear OR-array A. The feedback path from array A is split into four identical copies which are shifted and superimposed onto the AND-array B to implement the LHS of the rule. Array B performs a threshold operation and normalizes the signals. The output of array B is split into four copies which are shifted and superimposed onto array A to implement the RHS of the rule, normalize signals, and provide an output. Figure 10.35 shows a time sequence of symbolic substitution using this setup. Here, the input image (A) contains one binary pattern that matches the LHS of the rule. It is split into four identical copies (B), which are each shifted (C) and superimposed (D) according to the positions of the bits in the LHS. Each of the bits in the LHS is one position away from the center cell in the x and y directions, and hence each image is shifted according to its distance from the center cell. A threshold operation (E) sets all cells to 0 except those cells that have the original intensity after the images are superimposed. The array (F) is then split into four identical images (G) which are shifted (H)

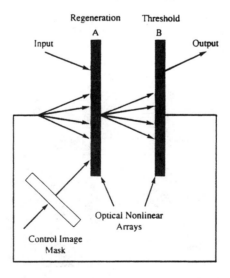

Figure 10.34 Optical implementation of symbolic substitution [Huang, 1983]. Reprinted with permission from IEEE.

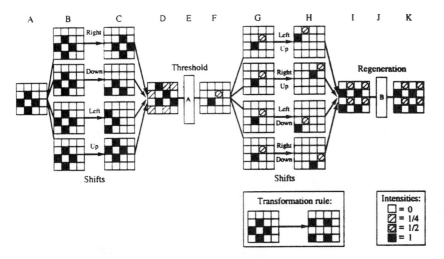

Figure 10.35 Time sequence of symbolic substitution for a 4 × 4 system [Huang, 1983]. Reprinted with permission from IEEE.

and superimposed (I) according to the RHS pattern. The intensity values in the final image are restored by regeneration element B (J). This mechanism will locate in parallel all such areas that contain such pattern.

It is possible to apply several rules in parallel. Figure 10.36 shows a setup for implementing four rules with two cells in the LHS and RHS of

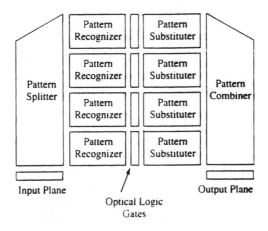

Figure 10.36 Optical implementation of four 2 × 2 symbolic substitution rules [Huang, 1983]. Reprinted with permission from IEEE.

each rule [Brenner et al., 1986]. The input image is passed through a cascade of beam-splitters to produce four copies. Each copy is passed to a pattern recognizer/substituter pair and the signals are regenerated by an array of optically nonlinear logic gates. The outputs are then combined on the output plane through a cascade of beam-combiner.

While quite different from the current approach for computing, the symbolic substitution technique is quite general and powerful. Refer to [Huang, 1983] for further details. For further details on optical computing refer to the book by Murdocca [1990], and excellent surveys on various aspects of optical computing in [Feitelson, 1988; Bell, 1986].

10.6 Summary

In this chapter, we have examined pertinent details of ICs that are useful in circuit design using them. Although the electronic circuit-level details provided here may not be needed by the average logic designer, familiarity with them enables a designer to better appreciate loading, timing, and compatibility problems.

We have examined representative IC technologies available for realizing logic circuits. The selection of technology is a compromise among such characteristics as speed, power dissipation, cost, and level of integration, among others.

ECL is used in such high-speed applications as real-time signal processors and supercomputers. Where the cost or power consumption of ECL is not suitable but high-speed circuits are needed, TTL and its variants are used. In the fabrication of microprocessors and memory chips where high densities are needed, MOS technologies are used. NMOS is the popular technology for such medium-speed applications. CMOS is suitable for those applications requiring the lowest power dissipation.

Several popular IC families such as Integrated Injection Logic (I^2L) were not included in this chapter. IC fabrication technology is probably the most rapidly changing technology in recent times. As such, any book on the technology is virtually obsolete by the time it is published. It is hoped that the basics provided in this chapter will make readers comfortable about exploring such technologies.

One of the technologies that has been of renewed interest is the area of "optical" logic elements. Although optical fibers have been used successfully as data transmission media, optical computing is still in the laboratory stage. Very high switching times, on the order of 10^{-14} s, are envisioned with this technology.

References

Bell, T. E., "Optical Computing: a field in Flux," *IEEE Spectrum*, vol. 34, Aug. 1986.

Brenner, K. H., A. Huang, and N. Streibl, "Digital Optical Computing with Symbolic substitution," *Applied Optics*, vol. 25, 1986.

Brenner, K. H. and A. Huang, "Optical Implementation of the Perfect Shuffle Interconnections," *Applied Optics*, Vol. 27, pp. 135, 1988.

Computer, New York: IEEE Computer Society, October 1986 (Special issue on GaAs technology).

Computer Design, Littleton, MA: Penn Well (semimonthly).

Feitelson, D., *Optical Computing*, The MIT Press, 1988.

Hnatek, E. R., User's Guidebook to Digital CMOS ICs, New York: McGraw-Hill, 1981.

Howes, M. J. and Morgan, D. V. ed., Gallium Arsenide Materials, Devices and Circuits, New York: Wiley, 1985.

Huang, A., "Parallel Algorithms for Optical Digital Computing," *IEEE 10th International Optical Computing Conference*, Cambridge, MA. 1983.

Jhans, J. and M. J. Murdocca, "Crossover Networks and Their Optical Implementation," *Applied Optics*, Vol. 27, No. 15, pp. 3155, 1988.

Jewell, J. L., M. J. Murdocca, S. L. McCall, Y. H. Lee, and A. Scherer, "Digital Optical Computing: Devices, Systems, Architectures," *Proceedings of The Seventh International Conference on Integrated Optics and Optical Fiber Communication*, Kobe, Japan, Jul. 18, 1989.

Jewell, J. L., A. Scherer, S. L. McCall, A. C. Gossard, and J. H. English, "GaAS-AIAs Monolithic Microresonator Arrays," *Applied Physics Letters*, Vol. 51, No. 2, pp. 94, July 13, 1987.

Jordan, H. F., "Report of the workshop on all-optical, stored program, digital computers," *Technical Report*, Department of Electrical and Computer Engineering, University of Colorado at Boulder, 1988.

Murdocca, M., *A Digital Design Methodology for Optical Computing*, The MIT Press, 1990.

Problems

10.1 Out of the 14 pins available in a standard TTL SSI circuit package, two pins are used for V_{CC} and Ground. The following gates are required:

    (a) Two-input NAND.
    (b) Three-input NOR.
    (c) Four-input OR.
    (d) Three-input tristate AND.
    (e) Eight-input OR.

Determine how many gates can be packed into an IC in each case.

10.2   It is required that a NOR gate drive 30 other NOR gates. The fan-out of the IC logic family is 10, and no buffer gates are available. Show the circuit diagram with the appropriate number of additional gates if needed to realize this load condition.

10.3   Repeat problem 10.2 with buffer gates with a fan-out of 20 available in the logic family.

10.4   Determine the operating current and voltage values and the function of the following diode circuit when:

(a)   Logic-0 is 0 V and logic-1 is -3 V.
(b)   Logic-0 is -3 V and logic-1 is 0 V.

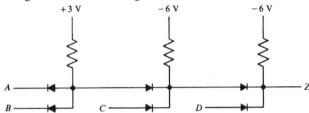

10.5   Determine the function realized by the following DTL circuit. Assume logic-1 is 5 V and logic-0 is 0 V.

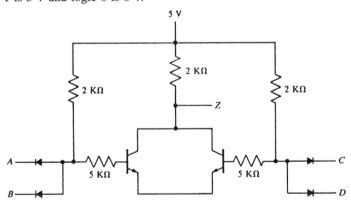

10.6   Determine the current gain (β) of the circuit in Figure 5.6(a) if $R_B = 8$ KΩ and $R_C = 500$ Ω.

10.7   Determine the values of all the resistors in the RTL NAND gate of Figure 5.8 assuming a $\beta = 10$ for each transistor and that the maximum current allowed through each transistor is 0.5 mA. What are the operating ranges of voltages with these resistor values?

10.8   It is required that the data from three sources (A, B, C) be transmitted to either of the two destinations D and E using a bus. Each data item is three bits wide. Draw the bus circuit using tristate gates.

10.9 Analyze the TTL tristate circuit of Figure 10.17 to determine the output voltage ranges for each input combination. What are the acceptable input voltage ranges for the proper operation of the circuit?

10.10 Show that two open collector TTL inverters when their outputs are connected together produce the NOR function.

10.11 Show that two open collector ECL NOR gates when wired together with external resistor and $-V_{EE}$ produce an OR function.

10.12 Show the circuit that implements the following function using (a) NMOS and (b) CMOS transistors:

$$Z = (AB + CD + CE)'$$

10.13 Draw the truth table for the following CMOS circuit:

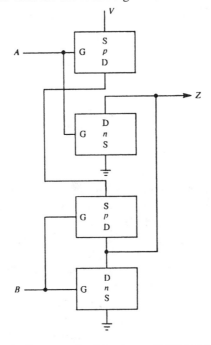

10.14 Draw the circuit diagram of a three-input CMOS NOR gate with one input at 0 V and the other inputs at V_{DD} showing all the voltage levels.

10.15 Determine the high- and low-level noise immunities of the TTL LS gate driving a standard TTL gate.

10.16 What is the average power dissipation of a standard TTL gate that provides high and low collector current values of 10 mA and 2.5 mA, respectively?

10.17 Determine the noise margins of an ECL gate.

11

Laboratory Experiments

Introduction

This chapter presents 21 experiments designed to provide hands-on experience, supplementing the concepts presented in this book. Although these experiments specifically use TTL ICs, they are general enough to be used with CMOS ICs. Several logic breadboards are commercially available that enable mounting ICs and interconnecting them with wires, without requiring any soldering. In addition to the ICs, wires, and a breadboard, the only other requirements are a TTL data book and a dual-trace oscilloscope. TTL data books are available from IC manufacturers, and some are listed in the references section at the end of this chapter. The references section also includes several books that introduce digital logic design through experiments. Those books can be consulted for further details on various aspects of the design, construction, and troubleshooting of digital circuits.

The logic breadboards available from various vendors (e.g., LD-1 and LD-2 from E&L Instruments, Digital Trainer from Heathkit, and Power Ace from AP Products) provide the basic facilities needed to run all the experiments in this chapter. Figure 11.1 shows the details of E&L Instruments' LD-2. It consists of:

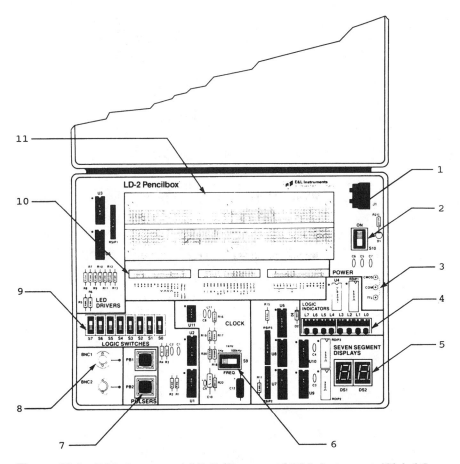

Figure 11.1 E&L Insruments' LD-2 (Courtesy of E&L Instruments/Global Spe-
cialties and Interplex Electronics Company).

1. A five-pin DIN type receptacle for power supply.
2. Power on/off switch along with the power-on indicator LED
    (D1).
3. A CMOS/TTL selection switch that allows experiments with
    CMOS and TTL components. In the CMOS mode, the operating
    voltage of the LD-2 circuits is +12 V, and in the TTL mode it
    is +5 V.
4. The eight LEDs L0–L7 can be used to monitor the logic states
    of circuit points. A logic 1 on the LED input will light it.

5. There are two independent seven-segment displays DS1 and DS2. The input to these displays should be a four-bit BCD value. The D, C, B, and A tie points on the row of connectors (10) are used for these inputs.

6. A three-position frequency switch S9 allows the selection of frequencies of 1, 10, or 100 KHz.

7. Pulsers PB1 and PB2 are two fully debounced pushbuttons with true and complementary outputs. The outputs of these pulsers are available on the row of connectors (10). The PB1 output, for instance, stays at logic 1 when the pulser PB1 is not depressed and goes to logic 0 wen PB1 is depressed.

8. Two additional connectors, BNC1 and BNC2, allow for input and output interfacing of signals from external devices to LD-2.

9. Eight logic switches S0–S7 are provided with outputs at the connector (10) tie points. These switches provide a logic 1 at their outputs when pushed up and a logic 0 when pushed down.

10. All functions are permanently tied to the three tie point connectors. Each tie point has two solderless connection points.

11. A solderless breadboard SK-10 accommodates up to eight 14-pin DIP ICs. Once the IC is inserted at the center of the breadboard, up to four wires can be connected to each pin of the IC. If the experiment requires a larger number of ICs than can be accommodated on this socket, additional sockets can be used external to LD-2 and connected to the circuit on it. This breadboard also contains eight power rails (or busses), each with 25 tie points.

We will now examine the experiments. Most of the experiments use the example circuits from the earlier chapters. No attempt is made here either to provide all the details of the required ICs or to repeat the theoretical aspects of logic design. Thus, frequent references to the earlier chapters of the book and to TTL data books are required while performing these experiments. Although 21 experiments are listed here, a laboratory session can cover more than one experiment, depending on the length of time available. After the first few experiments, students may be encouraged to build the circuits on the breadboard (or a separate socket) outside the laboratory and use the time in the laboratory simply to test the circuit.

This chapter provides broad guidelines for a set of experiments. It is not intended to serve as a laboratory manual. It is assumed that the laboratory instructor will enhance the experimental details provided here, while tailoring the laboratory to the equipment available. Furthermore, since the operating details of the laboratory equipment vary, such details are left to laboratory instructors.

Experiment 1

Objective: To familiarize the student with the logic breadboards and mechanical characteristics of ICs.

Procedure:

1. Connect the output of a logic switch to an LED and observe the output of the switch at logic-1 and logic-0 positions.
2. Connect a pulser to an LED and observe its output.
3. Observe the output of the clock generator at frequencies of 1 and 10 Hz on LEDs.
4. Connect the clock generator to one channel of the oscilloscope and observe the clock waveform at frequencies of 10 Hz and higher.
5. Select the two-input AND gate IC (7408) and note the function of all 14 pins. (If you hold the IC with the notch facing up, the pins are numbered from 1 through 14 in counterclockwise manner. This is the standard convention.)

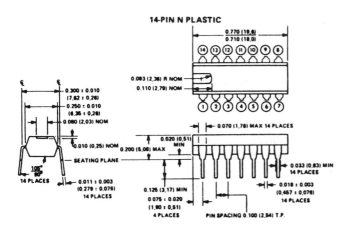

14-PIN N PLASTIC

Falls Within JEDEC TO-116 and MO-001AA Dimensions

QUADRUPLE 2-INPUT
POSITIVE-AND GATES

08

positive logic:
Y = AB

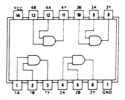

SN5408 (J, W)   SN7408 (J, N)
SN54LS08 (J, W)   SN74LS08 (J, N)
SN54S08 (J, W)   SN74S08 (J, N)

6. Turn the breadboard power off. Insert the 7408 into the socket. Connect the inputs of one of the gates (say, pins 1 and 2) to two logic switches and the output (pin 3) to an LED. Turn the power on. Apply all the combinations of inputs and observe the output and generate the truth table for the AND gate.

7. Disconnect one of the inputs from the logic switch. Toggle the switch connected to the other input and observe the output. Recall that an "open" intput on a TTL IC corresponds to a logic-1.

Experiment 2

Objective: To introduce the student to other gate ICs.

Procedure: Repeat Steps 5, 6, and 7 of Experiment 1 using the following ICs:

(a) Three-input AND (7411).
(b) Two-input OR (7432).
(c) Two-input NAND (7400).
(d) Two-input NOR (7402).
(e) Two-input Exclusive-or (7486).

Experiment 3

Objective: To illustrate the use of universal gates.

Procedure:

1. Using only 7400s, realize two-input AND and OR circuits and a NOT circuit (see Chapter 2). Test these circuits for all combinations of inputs.
2. Repeat Step 1 using a 7402.

Experiment 4

Objective: To compute propagation delay.

Procedure:

1. Select the Hex-Inverter IC (7404) and connect the output of one NOT gate to the input of the other to form a circuit with three gates in series. Connect the input of the series of gates to a logic switch and the output to an LED. Apply the two possible input values and observe the output. The output must be the complement of the input.
2. Disconnect the logic switch and connect the clock generator to the input of the series of gates in Step 1 and also to one channel of the oscilloscope. Connect the output to the other channel of the oscilloscope. Set the clock frequency to 10 kHz and observe the waveforms.

3. Compute the displacement of the output waveform from the input waveform by noting the time axis for corresponding edges. If you do not see much displacement, increase the number of gates in the chain to five or six. Now, the propagation delay of each NOT gate is given by the total displacement of the output signal from the corresponding edge of the input signal divided by the number of gates in the series.
4. Does your computation agree with the delay specified in the data book?

Experiment 5

Objective: To build two-level circuits.

Procedure:

1. Design a three-input circuit that realizes $F(A, B, C) = \Sigma m(0, 1, 3, 4, 6, 7)$. Use minimum AND-OR and NAND-NAND implementations. Select the appropriate ICs from the data book and wire both circuits on the breadboard.
2. Apply all the input combinations and verify the operation of both circuits.

Experiment 6

Objective: To understand the operation of a binary counter.

Procedure:

1. TTL 7493 is a *binary counter*. It has four outputs (Q_0, Q_1, Q_2, and Q_3), two clock inputs (CP_0 and CP_1), and two reset inputs (MR_1 and MR_2). Tie MR_1 and MR_2 to Ground to put the counter in COUNT mode. Connect a pulser to CP_0. Connect $\overline{CP}_1$ to Q_0 to make up the four-bit counter circuit that counts from $Q_0 Q_1 Q_2 Q_3 = 0000$ to 1111 and returns to 0000. The count is incremented at each pulse input from the pulser. Connect the four outputs to LEDs and observe the count sequence.
2. Note that the output of 7493 can be used to drive any circuit with up to four inputs. All 16 combinations of input values are generated by the counter. Just using the pulser, we can now change the input values and verify the operation of the circuit for all the input combinations. We can also connect the clock generator instead of the pulser to the $\overline{CP}_0$ input of 7493 and cycle through the 16 combinations automatically.
3. Connect $\overline{CP}_0$ to Q_0 to two channels of the oscilloscope. Observe that Q_0 changes at each negative (or 1-to-0) transition of $\overline{CP}_0$. That is, the frequency of Q_0 is half that of $\overline{CP}_0$.

4. Observe the relationship between Q_0 and Q_1, then between Q_1 and Q_2, and finally between Q_2 and Q_3 by connecting them to the two channels of the oscilloscope. In each case, the frequency of the latter should be half that of the former signal.
5. Plot Q_0, Q_1, Q_2, and Q_3. This should be the timing diagram showing all 16 combinations of four inputs, with Q_0 as the LSB that changes every clock pulse.
6. The 7493 can be converted into a BCD counter to count from 0000 through 1001. Note that both MR_1 and MR_2 must be 1 to reset the counter to 0000. At the count 1010 (i.e., after 1001, the last BCD value), Q_1 and Q_3 are at 1. By connecting Q_1 to R_1 and Q_3 to R_2, the counter is reset right after the count of 1001 and starts all over again from 0000. Note that R_1 and R_2 are asynchronous inputs and hence do not require a clock along with them to reset the flip-flops in 7493. Wire the 7493 to behave as a BCD counter, and observe the operation.
7. Design the wiring scheme to convert 7493 to count from 0000 to (a) 0101; (b) 1010; and (c) 0111. Wire each circuit and observe its operation.

Experiment 7

Objective: To observe and compensate for hazards in combinational circuits.

Procedure:
1. Build the AND-OR circuit of Experiment 5.
2. Connect the Q_0, Q_1, and Q_2 outputs of the 7493 (wired in binary counter mode as in the previous experiment) to C, B, and A inputs of the AND-OR circuit, respectively.
3. Connect Q_2 to one of the channels of the oscilloscope and the output of the circuit to the other channel. Adjust the oscilloscope and the clock frequency to see one cycle of Q_2 on the oscilloscope. This is equivalent to displaying the complete truth table.
4. Determine all the hazards (see Chapter 4).
5. Include additional gates to remove hazards.

Experiment 8

Objective: To build and test adder circuits.

Procedure:
1. Build a half-adder using NAND gates only and test it.
2. Build a full-adder using two half-adders and test its operation.
3. Connect the half- and full-adders to form a two-bit adder.

4. Use a 7493 to apply all the combinations of inputs to the circuit above and test it.
5. Convert the circuit above into an adder/subtractor. Include a control signal that is 0 for ADD and 1 for SUBTRACT. Include the minimum number of additional gates. Test the circuit.

Experiment 9

Objective: To understand the decoder operation.

Procedure:

1. Select the 3-to-8 decoder (74S138). Drive its inputs from a 7493 and observe its outputs.
2. Build a circuit to realize $F(A, B, C) = \Sigma m(0, 1, 6, 7)$ using 3-to-8 decoder and observe its operation. Note that the outputs of 74S138 are active-low.

Experiment 10

Objective: To examine a seven-segment display operation.

Procedure:

1. Select an available seven-segment display device and determine the inputs needed to display digits 0 through 9.
2. Design and build a BCD-to-seven-segment decoder using a minimum number of NAND gates.
3. Connect a 7493 to drive the decoder above that in turn is connected to the display device. Verify the operation.

Experiment 11

Objective: To design with a read-only memory.

Procedure:

1. Select an available ROM IC.
2. Determine the ROM pattern needed to implement the BCD-to-seven-segment decoder of Experiment 10 using a ROM.
3. Program the ROM and place the ROM pattern starting from location 0. Refer to the operating instructions of the ROM programmer for details on programming the ROM.
4. Use a 7493 to supply the addresses for the ROM.
5. Drive the seven-segment display from the outputs of the ROM.

Experiment 12

Objective: To design with PLAs and PALs.

Repeat Experiment 11 using an available (a) PLA and (b) PAL and the corresponding programmers.

Experiment 13

Objective: To understand the operation of flip-flops and latches.

Procedure:

1.   Wire the cross-coupled NOR circuit of Chapter 6.
     (a)   Using the S and R inputs, alternately set and reset the output and verify the operation of the latch.
     (b)   Set both inputs to 1 and observe the outputs.
     (c)   Return both inputs simultaneously to 0 and observe the outputs. Repeat (b) and (c) several times and observe the outputs. Do the outputs always reach the same state?
     (d)   Include additional gates to the above circuit to realize a clocked SR flip-flop.
     (e)   Use a pulser as the clock input. Apply various input combinations and observe the output after each clock pulse.
2.   Connect the J and K inputs of 7473 JK flip-flop to two logic switches. Connect a pulser to its clock input and flip-flop outputs to two LEDs.
     (a)   Connect a logic switch each to direct-set and direct-reset inputs and the outputs of the flip-flop to LEDs. Toggle the logic switches and observe the operation of the flip-flop. Leave the switches at the position in which the direct inputs are inactive.
     (b)   Apply each combination of JK inputs and observe the output after applying a clock pulse. (Observe that the flip-flop does not change its state until the clock is applied.)
     (c)   Disconnect the pulser from the flip-flop. Connect a logic switch to the clock input. To apply a clock pulse, the logic switch must be changed from 0 to 1 and then to 0. When does the flip-flop change its state: on the rising or the falling edge? Note that the switch may bounce. That is, when the switch position is changed from, say, 1 to 0, the output of the switch may oscillate between 0 and 1 several times before settling to 0 (and vice versa). Observe the effect of switch bouncing on the flip-flop state. If you observe a switch bounce, debounce the switch using another latch (see Chapter 8).
     (d)   Draw a timing diagram to reflect the operation of the flip-flop showing the clock, JK inputs, and the outputs of the flip-flop.

3. Repeat Step 2 using a 7474 *D* flip-flop.
4. Connect a logic switch to the Enable input of a two-bit latch in 7475 and the outputs of the latch to LEDs. Connect two other logic switches to *D* inputs.
    (a) Set the Enable input at 1 and observe the outputs by changing the *D* inputs.
    (b) Set the Enable input at 0 and observe the operation of the latch when inputs are changed.

Experiment 14

Objective: To build a synchronous sequential circuit.

Procedure:

1. Design a 0101 sequence detector using *JK* flip-flops. Assume that the sequences may overlap.
2. Wire the circuit. Set the circuit to initial state and apply the input sequence to verify its operation. Make sure you cover all the transitions in the state diagram.

Experiment 15

Objective: To build registers with parallel load and parallel outputs and to construct data transfer circuits.

Procedure:

1. Build a four-bit register using 7474 *D* flip-flops. Apply inputs through four logic switches and observe the output on four LEDs. Note that the input data enter the register only at the clock edge.
2. Include a Load control input to the register above. The input data should enter the register only if Load is 1; otherwise, the data in the register must remain intact. Verify the operation by applying several inputs and Load conditions.
3. Build another four-bit register, as in step 1, and interconnect the two registers so that a four-bit piece of data from the logic switches is loaded into the first register (at one edge of the clock) and is transferred to the second register at the other edge of the clock. This is essentially a master/slave configuration.
4. Change the circuit in Step 3 so that two clock pulses are needed for the data entry and transfer. That is, data enter the first register at the first clock pulse and are transferred to the second register at the subsequent pulse.
5. Repeat the above steps using 7476 *JK* flip-flops.

Experiment 16

Objective: To construct shift registers and data transfer circuits.

Procedure:

1.   Construct a four-bit serial-in, parallel-out right shift register using 7474
     D flip-flops. Connect the serial input to a logic switch, the outputs to
     LEDs, and the clock (shift) input to a pulser. Data are loaded one bit at
     a time by setting the switch to the appropriate value and pressing the
     pulser, thus loading the data into the MSB and shifting the register right.
     Load several four-bit data items into the register and observe the outputs
     at each pulse.
2.   Study the operation of the four-bit shift register 7495. Using the mode
     control and clock inputs, perform serial-in, parallel-out and parallel-in,
     serial-out operations.
3.   Connect two 7495s so that the four-bit data item enters the first register
     in parallel. The data are then shifted to the second register one bit at a
     time (i.e., in series). (Note that since both registers are triggered by the
     same clock edge, the pulser should be connected directly to the transmit
     register and inverted and connected to the receive register. Thus, the
     transmit register is loaded and shifted when the pulser is pushed, and
     the receive register is shifted when the pulser is released.) Load several
     four-bit data items and shift them to observe the circuit's operation.

Experiment 17

Objective: To design the control circuitry for the serial four-bit data transfer circuit in Experiment 16.

Procedure: Note that in the data transfer circuit of the previous experiment
there are five steps: the first one is to load the data into the transmit register
in parallel and the last four are to shift the data to the receive register. This
process can be automated.

1.   Wire a 7493 counter to count from 0000 through 0100, the five steps
     required in the circuit. Use a pulser to provide clock input to 7493.
2.   Connect the three least significant outputs of 7493 to the inputs of the
     3-to-8 decoder (74S138). Connect the Enable inputs appropriately to
     enable the decoder.
3.   Using the outputs of the decoder, design the circuit to control the mode
     control and clock inputs to the data transfer circuit in the previous ex-
     periment, at each of the five steps involved. (Refer to Chapter 7.)
4.   Verify the operation of the circuit.

Experiment 18

Objective: To construct a data transfer and arithmetic circuit and the necessary control circuitry.

Procedure: Here, we will extend the circuit in the previous experiment to include a four-bit addition capability. The circuit should perform the following operations:

Step	Function
0	Load the first four-bit operand into the transmit register.
1 to 4	Transmit the operand to the receive register.
5	Load the second operand into the transmit register.
6 to 9	Add the four-bit operands in these registers using a full-adder and a flip-flop, and send the results to the transmit register.

Extend the control circuit in Experiment 17 to control the above operations. Apply several inputs and observe the operation of the circuit.

Experiment 19

Objective: To observe the operation of a Random Access Memory (RAM).

Procedure:

1. Study the functions of all the input and output signals of the RAM IC 74S189.
2. Connect the outputs of a 7493 to the four address inputs of the RAM. Use a pulser to change the count and hence the address. Set the chip enable signal to 0. Connect the write-enable to another pulser, the RAM inputs to four logic switches, and the outputs to LEDs.
3. To write data into the memory, the address and the data lines are first set to the proper values and the write enable ($\overline{\text{WE}}$) is brought to 0. Once the data are written, the $\overline{\text{WE}}$ line should be set to 1. During the write operation, the address and data should not change.
4. Write 16 four-bit data values into the 16 words of the RAM, entering one word at a time.
5. Now leave the RAM in read mode. Change the addresses from 0000 through 1111 and observe the outputs.
6. Wire up the memories of the following organizations using the proper number of 74S189s and test the memory circuit:
   (a) 16 × 16.
   (b) 32 × 4.
   (c) 48 × 4.

Experiment 20

Objective: To build ripple counters and divide-by-*n* circuits.

Procedure:

1. Wire a 7476 *JK* flip-flop as a divide-by-two circuit. This is done by connecting the clock to the clock input of the flip-flop. The output of the flip-flop changes half as quickly as the clock input.
2. Extend the circuit above into a three-bit ripple counter by including two more flip-flops, and verify its operation as a divide-by-eight circuit.
3. Design the circuitry around the three-bit counter above to convert it into a divide-by-five circuit. Note that at the count of five (i.e., 101), the outputs of the first and last flip-flop are both at 1. Connect these outputs to a two-input NAND gate and connect its output to the direct-reset input of all the flip-flops, thereby resetting the counter at the count of 101.
4. Using the data book, look up the delay characteristics of the flip-flops and the gates used in the circuit, and draw a timing diagram to depict the operation of the divide-by-five circuit. How long does the counter stay in state 101 before it is reset to 000?
5. Concatenate a divide-by-two and a divide-by-five circuit to obtain a divide-by-ten circuit. Apply a 10 kHz clock to the input of the circuit. Observe the input and output on the oscilloscope to verify the circuit's operation.

Experiment 21

Objective: To design clocks using one-shots.

Procedure:

1. Determine the minimum pulse width needed to observe the pulse on an LED. (If the pulse width is too small, the human eye cannot observe it in an LED.)
2. Calculate the timing resistor and capacitor values needed to produce the pulse width determined above, and wire a 74121 circuit to produce the pulse.
3. Trigger the 74121 using either pulsers or logic switches and observe the output. Verify the operation of the 74121 for all combinations of input triggers. The output of the 74121 should be on only for the pulse duration.
4. Use two 74121s to produce a clock with a pulse width of 100 ns and a frequency of 500 kHz.

References

Bipolar Memory Data Manual. Sunnyvale, Calif.: Signetics, 1986.

Blakeslee, T. R. *Digital Design with Standard MSI and LSI*. New York: John Wiley & Sons, 1975.

LD-2 Pencilbox Logic Designer Instruction Manual. New Haven, Conn.: E&L Instruments.

Pasahow, E. J. *Learning Digital Electronics Through Experiments*. New York: Mc-Graw-Hill, 1982.

Passafiume, J. F., and Douglas, M. *Digital Logic Design—Tutorials and Laboratory Exercises*. New York: Harper & Row, 1985.

The TTL Data Book. Dallas: Texas Instruments, 1986.

TTL Data Manual. Sunnyvale, Calif.: Signetics, 1986.

Young, G. *Digital Electronics—A Hands-on Learning Approach*. Hasbrouck Heights, N.J.: Hayden, 1980.

Appendix A

IEEE Standard Logic Symbols

A.1 Introduction

In 1984, the Institute of Electrical and Electronic Engineers (IEEE) and the International Electrotechnical Commission (IEC) introduced a new standard for logic symbols. This standard provides for representation of devices ranging from logic gates to complex integrated circuits. It emphasizes the use of rectangular symbols for all components, though traditional symbols for AND, OR, and other gates are permitted. In this appendix, we will summarize the major characteristics of the standard. References listed at the end of this appendix give further details.

The standard symbol shown in Figure A.1 consists of an outline (or a combination of outlines) and one or more qualifying symbols. The general qualifying symbol indicates the operation performed by the element, represented by the rectangular line. In general, input lines are placed on the left and output lines are placed on the right of the rectangle. When this is not possible, an arrow shows the direction of signal flow. Input and output lines each can have a qualifying symbol (or label) both inside and outside the outline. Table A.1 shows the general qualifying symbols. Table A.2 summarizes the qualifying symbol for the input and output lines, and Table A.3 lists the symbols allowed inside the outline. Note that the half-arrow used to indicate polarity and the bubble used to indicate negation are valid symbols in the standard.

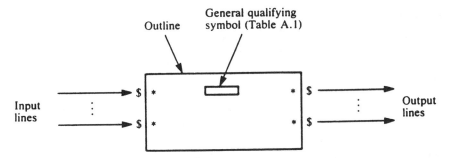

Figure A.1 General structure of a symbol. $: Qualifying symbols for input and output lines (see Table A.2). *: Qualifying symbols within the outline (see Table A.3).

When the outlines of elements in the logic diagram are abutted or embedded, the following conventions apply. When the line common to the outlines of two elements is parallel to the direction of the signal flow, there is no logical connection between the elements. If the line common to the two outlines is perpendicular to the signal flow direction, there is at least one logic connection between the elements. Qualifying symbols are then used to clarify the connection.

A.2 Representation of SSI Functions

Figure A.2 shows the standard symbol for an IC consisting of four NAND gates (TTL 7400). The logic symbol and the pin configuration are also shown for reference purposes. The ampersand symbol (&) represents the AND operation and, the half-arrow at the output represents the negation. Because all the lines common to the outlines of the four elements in the IC are parallel to the signal flow direction (from left to right), there is no logic connection between the elements. Figure A.3 shows the standard symbols for other representative gate ICs.

Figure A.4 shows an IC with two *JK* flip-flops. Each flip-flop has an active-low asynchronous clear input, *R*. The clock input is designated by $C1$. The letter C is used to designate control signals, and 1 indicates the control signal number. If there were more control signals, they would be designated $C2$, $C3$, $C4$, and so on. *J* and *K* inputs bear the designation of $1J$ and $1K$, indicating that $C1$ is the control signal that activates them. The flip-flops in (a) are pulse-triggered, and, as shown by the "⌐|" symbol at the output, the *JK* information is transferred to the output at the clock's high-to-low transition. The flip-flops in (b) are negative edge-triggered. The

TABLE A.1   General Qualifying Symbols

SYMBOL	DESCRIPTION
&	AND gate or function.
>1	OR gate or function. The symbol was chosen to indicate that at least one active input is needed to activate the output.
=1	Exclusive OR. One and only one input must be active to activate the output.
=	Logic identity. All inputs must stand at same state.
2k	An even number of inputs must be active.
2k+1	An odd number of inputs must be active.
1	The one input must be active.
▷ or ◁	A buffer or element with more than usual output capability (symbol is oriented in the direction of signal flow).
⎍	Schmitt trigger; element with hysteresis.
X/Y	Coder, code converter (DEC/BCD, BIN/OUT, BIN/7-SEG, etc.).
MUX	Multiplexer/data selector.
DMUX or DX	Demultiplexer.
Σ	Adder.
P−Q	Subtracter.
CPG	Look-ahead carry generator.
π	Multiplier.
COMP	Magnitude comparator.
ALU	Arithmetic logic unit.
⎍⎍	Retriggerable monostable.
1⎍⎍	Non-retriggerable monostable (one-shot).
G⎍⎍	Astable element. Showing waveform is optional.
!G⎍⎍	Synchronously starting astable.
G!⎍⎍	Astable element that stops with a completed pulse.
SRGm	Shift register. m = number of bits.
CTRm	Counter. m = number of bits; cycle length = 2^m.
CTR DIVm	Counter with cycle length = m.
RCTRm	Asynchronous (ripple-carry) counter; cycle length = 2^m.
ROM	Read-only memory.
RAM	Random-access read/write memory.
FIFO	First-in, first-out memory.
I=0	Element powers up cleared to 0 state.
Φ	Highly complex function; "gray box" symbol with limited detail shown under special rules.

*Not all of the general qualifying symbols have been used in this book, but they are included here for the sake of completeness.

Source: Courtesy of Texas Instruments Inc.

TABLE A.2 Qualifying Symbols for Inputs and Outputs

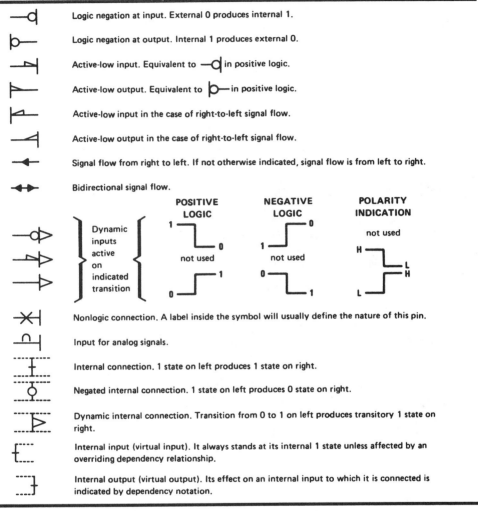

	Logic negation at input. External 0 produces internal 1.
	Logic negation at output. Internal 1 produces external 0.
	Active-low input. Equivalent to ⎯◁ in positive logic.
	Active-low output. Equivalent to ▷⎯ in positive logic.
	Active-low input in the case of right-to-left signal flow.
	Active-low output in the case of right-to-left signal flow.
	Signal flow from right to left. If not otherwise indicated, signal flow is from left to right.
	Bidirectional signal flow.

	POSITIVE LOGIC	NEGATIVE LOGIC	POLARITY INDICATION
Dynamic inputs active on indicated transition	1 ⎤⎦ 0 / not used / 0 ⎦⎤ 1	⎡⎣ 0 / not used / 0 ⎤⎦ 1	not used / H ⎤⎦ L , L ⎤⎦ H / L

	Nonlogic connection. A label inside the symbol will usually define the nature of this pin.
	Input for analog signals.
	Internal connection. 1 state on left produces 1 state on right.
	Negated internal connection. 1 state on left produces 0 state on right.
	Dynamic internal connection. Transition from 0 to 1 on left produces transitory 1 state on right.
	Internal input (virtual input). It always stands at its internal 1 state unless affected by an overriding dependency relationship.
	Internal output (virtual output). Its effect on an internal input to which it is connected is indicated by dependency notation.

The internal connections between logic elements abutted together in a symbol may be indicated by the symbols shown. Each logic connection may be shown by the presence of qualifying symbols at one or both sides of the common line and if confusion can arise about the numbers of connections, use can be made of one of the internal connection symbols.

The internal (virtual) input is an input originating somewhere else in the circuit and is not connected directly to a terminal. The internal (virtual) output is likewise not connected directly to a terminal.

Source: Courtesy of Texas Instruments Inc.

TABLE A.3 Qualifying Symbols Inside the Outline

Symbol	Description
⌐├	Postponed output (of a pulse-triggered flip-flop). The output changes when input initiating change (e.g., a C input) returns to its initial external state or level.
─┤ˌ	Bi-threshold input (input with hysteresis)
◇├	NPN open-collector or similar output that can supply a relatively low-impedance L level when not turned off. Requires external pull-up. Capable of positive-logic wired-AND connection.
◉├	Passive-pull-up output is similar to NPN open-collector output but is suplemented with a built-in passive pull-up.
◇├	NPN open-emitter or similar output that can supply a relatively low-impedance H level when not turned off. Requires external pull-down. Capable of positive-logic wired-OR connection.
◈├	Passive-pull-down output is similar to NPN open-emitter output but is supplemented with a built-in passive pull-down.
▽├	3-state output
▷├	Output with more than usual output capability (symbol is oriented in the direction of signal flow).
─┤EN	Enable input When at its internal 1-state, all outputs are enabled. When at its internal 0-state, open-collector and open-emitter outputs are off, three-state outputs are at normally defined internal logic states and at external high-impedance state, and all other outputs (e.g., totem-poles) are at the internal 0-state.
J, K, R, S, T	Usual meanings associated with flip-flops (e.g., R = reset, T = toggle)
─┤D	Data input to a storage element equivalent to:
─┤→m ─┤←m	Shift right (left) inputs, m = 1, 2, 3 etc. If m = 1, it is usually not shown.
─┤+m ─┤−m	Counting up (down) inputs, m = 1, 2, 3 etc. If m = 1, it is usually not shown.
(binary grouping symbol)	Binary grouping. m is highest power of 2.
─┤CT = 15	The contents-setting input, when active, causes the content of a register to take on the indicated value.
CT = 9├	The content output is active if the content of the register is as indicated.
(input line grouping symbol)	Input line grouping . . . . indicates two or more terminals used to implement a single logic input. e.g., The paired expander inputs of SN7450.
"1"├	Fixed-state output always stands at its internal 1 state. For example, see SN74185.

Source: Courtesy of Texas Instruments Inc.

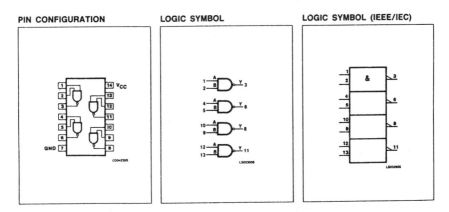

Figure A.2 TTL 7400 (Courtesy of Signetics Corporation).

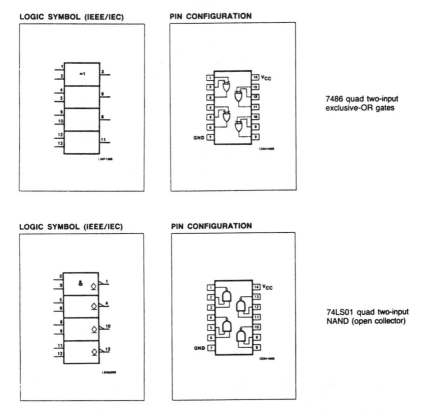

Figure A.3 Some TTL ICs (Courtesy of Signetics Corporation).

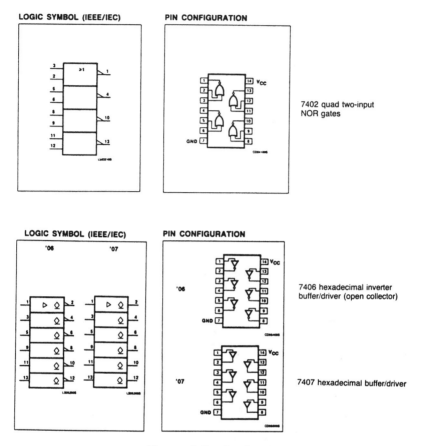

Figure A.3 Continued

triangle at the clock input inside the outline indicates edge triggering, and the half-arrow indicates the negative edge required to trigger.

Figure A.5 shows a dual *D* flip-flop IC. Each flip-flop has active-low asynchronous set and reset inputs and is triggered by the positive edge of the clock.

Figure A.6 shows the IC with four bistable latches. The upper two latches are controlled by *C*1 as indicated by 1*D*, and the lower two are controlled by *C*2, as indicated by 2*D*. *C*1 and *C*2 are the Enable inputs. The distinctively shaped element on the top is called a *common control block*. This shape is used to represent signals that are common inputs to more than one element.

PIN CONFIGURATION   LOGIC SYMBOL   LOGIC SYMBOL (IEE/IEC)

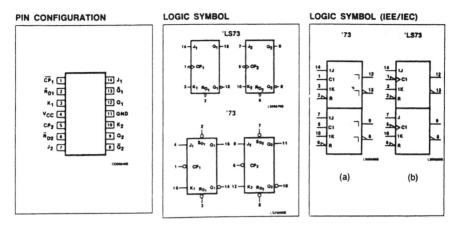

Figure A.4   TTL 7473, 74LS73 (Courtesy of Signetics Corporation).

PIN CONFIGURATION   LOGIC SYMBOL   LOGIC SYMBOL (IEEE/IEC)

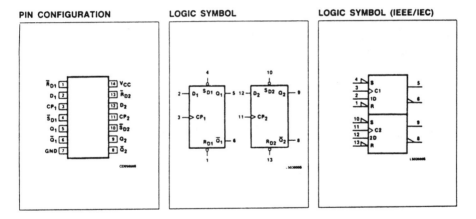

Figure A.5   TTL 7474 (Courtesy of Signetics Corporation).

A.3   Representation of MSI Functions

Figure A.7 shows a 3-to-8 decoder IC. The inputs are designated by

$$G \begin{array}{c} 0 \\ - \\ 7 \end{array}$$

and only the first (0) and the last (2) inputs are explicitly numbered. The IC is enabled by the AND of three enable inputs, two of which are active-low

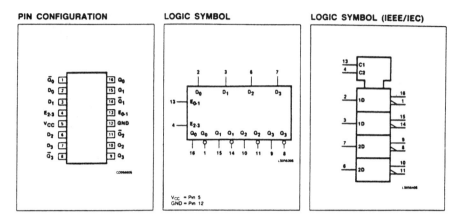

Figure A.6   TTL 7475 (Courtesy of Signetics Corporation).

and the third is active-high. Because there is only one element in this diagram, no common control block symbol is necessary.

Figure A.8(a) shows a four-bit shift register (7495). The standard symbol is shown in (b). The four-bit shift operation is indicated by the qualifying symbol SRG4. Control signals $M1$ and $M2$ are derived from the mode control signal S. CP_1 forms the control signal $C3$ and is activated by $M1$. In addition, as indicated by "/1→," $C3$ is the shift signal. CP_2 forms the control signal $C4$ activated by $M2$. The serial input to the leftmost flip-flop

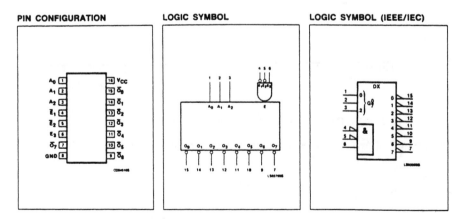

Figure A.7   TTL 74LS138 decoder/demultiplexer (Courtesy of Signetics Corporation).

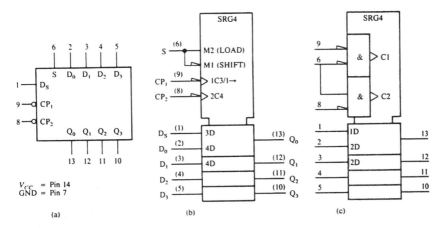

Figure A.8 TTL 7495 Four-bit shift register (Courtesy of Signetics Corporation [a and c] and Texas Instruments [b]). (a) Logic symbol; (b) IEEE standard symbols; (c) alternative symbol.

is controlled by $C3$, and the parallel inputs are controlled by $C4$. An alternative symbol is shown in (c).

Figure A.9 shows a counter IC. The logic symbol is shown in (a). The standard symbol for the divide-by-10 counter (74160) is shown in (b). Control signals R, $M1$, $G3$, and $G4$ correspond to MR, PE, CEP, and CET, respectively. The active-low clock input forms the control signal $C2$. The counter is incremented when $C2$ is active, along with $M1$, $G3$, and $G4$, as indicated by $/1,3,4 +$. A four-bit data item enters the counter through the D inputs when $M1 = 0$ and $C2$ is active, as indicated by $(1, 2D)$. The TC output (pin 15) is active when $G4 = 1$ and the count $= 9$, as indicated by $(4\ CT = 9)$. Note the double horizontal lines separating the bottom-most block from the others, indicating a logical connection between that block and the other blocks. The standard symbol for the four-stage (divide-by-16) counter 74163 shown in (c) is similar to that in (b), except that the control signal R is activated by $C2$.

Figure A.10 shows a 16×14 RAM device. It has an Enable input (EN), four address inputs designated by

$$A \genfrac{}{}{0pt}{}{0}{-}{} 15$$

and a clock input ($C1$). The memory is made up of 16 sections of four transparent latches with tristate outputs.

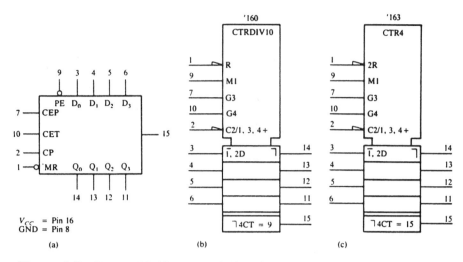

Figure A.9 Counter IC (Courtesy of Signetics Corporation). (a) Logic symbol; (b) standard symbol for 74160; (c) standard symbol for 74163.

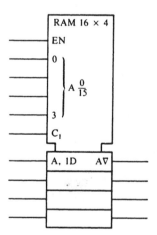

Figure A.10 A 16 × 4 random access memory.

A.4 Dependency

The compact and meaningful symbols used in the above standard are the result of the so-called *dependency notation*, which provides a means of denoting the relationship between the inputs and the outputs without showing the logic interconnection details within the element. So far in this appendix we have examined the following types of dependencies:

1. *G*—Essentially an AND type of dependency.
2. *C*—Control dependency.
3. EN—Enable dependency.
4. *M*—Mode dependency.
5. *A*—Address dependency.

References

IEEE Standard Graphic Symbols for Logic Functions, ANSI/IEEE Standard 91–1984. New York: Institute of Electrical and Electronic Engineers, 1984.

Mann, F. A. "Explanation of New Logic Symbols," in *The TTL Data Book* (Volume 1). Dallas: Texas Instruments, 1984.

Appendix B

Fundamentals of Electrical Circuits

B.1 Introduction

Gates and flip-flops are the basic components used in the design of digital systems. These components transform the binary signals consisting of two voltage levels (high and low). Because the analysis and design techniques described herein for digital systems do not use the more detailed electronic circuit levels of detail, very little knowledge of electronics is needed to learn these topics. However, when the electronic details of the above components are examined, as in Chapter 10, or when a detailed loading and timing analysis is needed, a knowledge of fundamental concepts of electrical and electronic circuits comes in handy. This appendix is a brief and informal introduction to circuit theory.

An electronic circuit is an interconnection of such electronic components as resistors, capacitors, diodes, and transistors. In addition to these components, a circuit might contain one or more sources of voltage (such as a battery or generator) and current (such as a transistor). In general, if the current flows out of a device, it is called a *current source* (i.e., the supplier of the current), and if the current flows into the device (i.e., the device is the receiver of the current), it is called a *current sink*.

Consider the circuit shown on the next page:

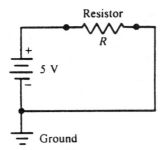

The circuit contains a voltage source with two terminals marked "+" and "−." The positive terminal marked "+" is at a higher potential than the negative terminal marked "−." The potential difference between the two terminals is equivalent to 5 V. The usual convention is to show the negative terminal connected to the ground, which forms the reference level for all the voltages in the circuit. That is, the positive terminal is at 5 V above the ground.

The circuit contains a resistor. In this circuit, the terminals of the voltage source are connected to the two terminals of the resistor. That is, a voltage of 5 V is "impressed" across the resistor. As a result, a current flows through the resistor. The transistor offers a resistance to the current flowing through it. As can be expected, the higher the value of the resistance, the lower the value of the current flowing through it for the same voltage impressed across the resistor. This property is governed by the so-called *Ohm's law*, which is stated below.

The current (I) through a circuit is inversely proportional to the resistance (R) of the circuit, and the voltage (V) across the terminals of the circuit is given by

$$V = IR \qquad (B.1)$$

Let us now include a switch in the circuit. If the switch is in the open position, as shown below, the continuity of the circuit is broken. The resistance offered by the circuit is infinity, and hence the current through it is 0. We say that the circuit is an *open* circuit.

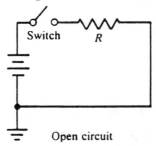

When the switch is in the other position, the circuit is said to be *closed*, and a current flows. The magnitude of the current (I) is given by

$$I = V/R \tag{B.2}$$

Switch R

I

Closed circuit

The current flowing through the resistor creates a voltage drop across its terminals. In the above circuit, the voltage drop is equivalent to IR, which is equal to V, since there is only one resistor in the circuit. *Kirchhoff's voltage law* states that the sum of all voltages around a closed circuit is 0. The polarity of the voltage across the resistor is thus shown to be opposite of that of the voltage source.

Thus, in the following circuit, the voltage across the resistor V_r is given by

$$V_a + V_r + V_b = 0 \tag{B.3}$$

$R = 50\Omega$ I

V_r

5 V V_a V_b 2 V

Following the direction of the current shown, $V_a = +5$ V and $V_b = 2$ V (since the two voltage sources are opposing each other), and hence

$$5 + V_r - 2 = 0 \tag{B.4}$$

Thus,

$$V_r = -3 \text{ V}$$

and the current I is given by

$$I = V_r/R = 3/50 = 0.06 \text{ A}$$

Figure B.1 shows several circuits. In (a) there are three resistors connected in *series* with a voltage source. The current flowing through the circuit is then determined by

$$V = IR \qquad \text{(B.5)}$$

where R is the equivalent resistance of the circuit. From Kirchhoff's voltage law, we see that

$$V = IR_1 + IR_2 + IR_3 \qquad \text{(B.6)}$$

since the same current I flows through each resistor. From (B.5) and (B.6), we derive

$$IR = IR_1 + IR_2 + IR_3 \qquad \text{(B.7)}$$

and hence,

$$R = R_1 + R_2 + R_3 \qquad \text{(B.8)}$$

Thus, the resistance of a series circuit is the sum of the resistances of its components.

In (b), the three resistors are connected in *parallel*. Since the same voltage is impressed on all the resistors, the currents through them are V/R_1, V/R_2, and V/R_3, respectively. If R is the equivalent resistance of the circuit, then the total current I is given by V/R.

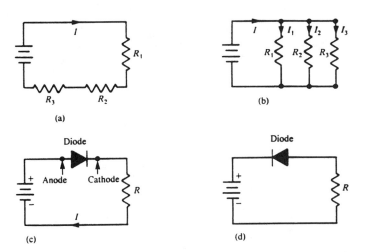

Figure B.1 Typical circuits. (a) Series; (b) parallel; (c) forward-biased diode; (d) reverse-biased diode.

Kirchhoff's current law states that the sum of all the currents flowing toward a junction is 0. Thus,

$$I - I_1 - I_2 - I_3 = 0 \tag{B.9}$$

(using the direction of I as the positive direction). That is,

$$I = I_1 + I_2 + I_3 \tag{B.10}$$

Hence,

$$V/R = V/R_1 + V/R_2 + V/R_3 \tag{B.11}$$

Thus,

$$1/R = 1/R_1 + 1/R_2 + 1/R_3 \tag{B.12}$$

That is, the reciprocal of the equivalent resistance of a parallel circuit is the sum of the reciprocals of individual resistances in the parallel circuit.

In (c) a *diode* is included in series with a resistor. The diode is said to be *forward-based*, since its *anode* is connected to a higher voltage than its *cathode*. When the diode is forward-based, it acts like a closed switch and offers a negligible resistance to the current flow through it (in the direction of the arrow of the diode symbol). The magnitude of the current and the voltage drop across each component are calculated as above for a series circuit.

In (d), the diode is *reverse-biased*, and hence the current I is 0.

A capacitor (usually formed of two metal sheets placed very close together) stores the electric charge. The amount of electric charge q stored on the capacitor is proportional to the voltage v across the two metal sheets forming the capacitor. That is,

$$q = Cv \tag{B.13}$$

where C is the *capacitance* of the capacitor. In the circuit

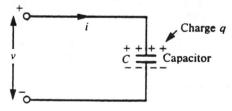

the current i is determined by the amount of charge flowing from one plate to the other. That is,

$$i = dq/dt \tag{B.14}$$
$$= d(Cv)/dt$$
$$= C\ dv/dt \tag{B.15}$$

When a voltage is applied across the capacitor, the voltage across the plates changes from its initial value v_0 (at time $t = 0$, say) to v_1 at time t_1. From (B.15) above,

$$C\ dv = i\ dt \tag{B.16}$$

Hence,

$$C \int_{v_0}^{v_1} dv = \int_{0}^{t_1} i\ dt$$

That is,

$$C(v_1 - v_0) = \int_{0}^{t} i\ dt$$

and hence,

$$v_1 = v_0 + 1/C \int_{0}^{t_1} i\ dt \tag{B.17}$$

Thus, the voltage across the capacitor at any time t is equal to the initial voltage at $t = 0$, plus a change that is proportional to the integral of the current after $t = 0$.

This property of the capacitor is used in electronic circuits. For instance, in the simple resistor-capacitor circuit shown below, the voltage at point A slowly rises to the magnitude V, t seconds after the voltage V is applied at the input. Time t is proportional to the product (RC).

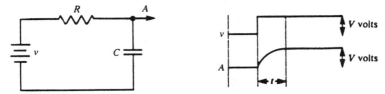

Thus, the effect of the capacitor is to introduce a time delay into the state changes of a signal. Circuit components will have inherent capacitance that contributes to slowly rising and falling signal characteristics.

This appendix has provided a very brief introduction to the fundamental concepts of electronic circuits. The books listed below may be consulted for further details.

References

Blackwell, W. A., and Grigsby, L. L. *Introduction to Network Theory*. Boston: PWS Engineering, 1985.

Director, S. W. *Circuit Theory—A Computational Approach*. New York: John Wiley & Sons, 1975.

Hostetter, G. H. *Engineering Network Analysis*. New York: Harper & Row, 1986.

Huelsman, L. P. *Basic Circuit Theory*. Englewood Cliffs, N.J.: Prentice-Hall, 1984.

Madhu, S. *Electronics: Circuits and Systems*. Indianapolis, Ind.: Howard W. Sams, 1985.

Appendix C

Simulation

Once a digital system is designed, it is imperative that its operation is validated before converting the design to silicon. The majority of design errors can be eliminated by using a systematic design practice and employing a top-down, modular design process. However, as the complexity of the system increases, additional verification aids would be needed. In the early days of hardware design where system conplexity was not very high, it was a common practice to build prototypes and verify their operation. In the VLSI era, prototyping is almost impossible since a proper prototype for a system is the system itself and fabrication of the system on the silicon is an expensive process to be used in a trial and error mode. Computer simulation of the system being designed is the most cost effective way of verifying the design.

In general, it is impractical to simulate the complete system to the level of verifying the operation of each transistor in the design, since the simulation complexity becomes astronomical for any reasonably complex system. As such, it becomes necessary to partition the system into several subsystems, verify the operation of each subsystem separately and then verify the operation of the integrated system. Typically, the subsystems are modeled at a very detailed level for simulation and the integrated system is modeled at a very coarse level of detail in terms of its operation.

A digital system can be modeled at the following levels of complexity (detail):

1.   Algorithmic level, which specifies only the algorithm used by the hardware for problem solution.
2.   Processor, Memory, Switch (PMS) level, which describes the system in terms of processing units, memory components and switching networks for their interconnection. This is the block diagram level.
3.   Instruction level (programming level) where instructions used by the system and their interpretation rules are of main concern.
4.   Register transfer level, where registers are system elements and the data transfer between them are verified.
5.   Switching circuit level, where the system structure consists of an interconnection of gates and flip-flops and the behavior is verified at the binary signals level.
6.   Circuit level, where the gates and flip-flops are replaced by circuit elements such as transistors, resistors, etc. and the operation is verified in terms of the current and voltage levels.
7.   Mask level, where the emphasis is on the fabrication details and the design rules for the particular IC technology used.

Note that at the topmost level in the above hierarchy the emphasis is on the behavior of the system. As we move down the heirarchy, the emphasis changes to the structure of the system.

Figure C.1 shows the utility of simulation in a typical design environment for digital systems. Here, the system is first described at a very high (behavior) level using a Hardware description Language (HDL). HDLs are similar to high-level programming languages except that their primitives allow the description of hardware characteristics and timing more readily. Several HDLs exist today. Verilog and VHDL (Very High Speed Integrated Circuit HDL) are the most popular ones in commercial and military use.

The HDL description of the system is translated into an intermediate database representation. The first simulation is performed at this level (LOOP 1) to refine the description. Typically, the functionality of the system is verified at this level leaving the timing and other performance checks to lower levels. The description is complete and correct once the simulation at this level does not reveal any deficiency.

The database is then translated (hardware compilation) into more structural form. The second level of simulation shown in Figure C.1 is at the logic diagram level. At this level, one can simulate for faults, timing problems and functionality.

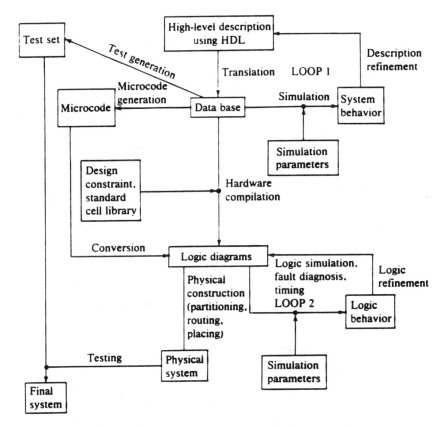

Figure C.1 Automatic hardware design process.

The logic diagrams are then converted into the silicon structures (physical system). The mask patterns are used in simulation at this level to verify design rules, to optimize silicon area, etc. This simulation at this level is not shown in Figure C.1.

Obviously, no single simulation system that allows the simulation of the design at all the above levels. In general, we use multiple simulators depending on the complexity of the system being designed. Some simulation systems allow simulation at multiple levels. That is, one subsystem can be simulated at gate level while the other subsystem could be at the register transfer level, so on.

There are several simulation systems now commercially available from companies such as Mentor Graphics, Zycad, IKOS, etc. There are also special processors called simulation accelerators that provide a fast simulation capability, available from several vendors. Refer to publications such as EDN, Computer Design, Integrated Systems Design, etc. for further details on the latest simulation systems. There are also several academic and low cost simulation systems available. We will provide a very brief description of one such system in the next section.

Logic Works

Logic works is an interactive circuit design software system available in PC Windows and Macintosh versions from Capilano Computing Systems, Ltd. Its major characteristics are:

1. Schematic development and editing from standard component libraries (SSI, MSI levels).
2. Simulation at various levels and corresponding simulation control facilities.
3. Capability to change device characteristics and parameters for simulation.
4. Facilities to create new library components.
5. Input/Output simulation through pseudo-devices such as switches, probes, keyboards, etc.
6. Creating PLDs and corresponding program files to serve as inputs to PLD programmers.
7. Creating text reports.

We will just concentrate on showing the simulation features of this software system through two simple examples and leave the details to the book on Logic Works listed in the reference section.

Figure C.2 (a) shows the schematic of a NOT gate circuit developed using Logic Works. The NOT gate is simply selected from a library of gates available. Its input is labeled A and connected to a CLOCK and the output is labeled B. The timing diagram in (b) shows the simulation output of this circuit for several clock cycles. Here, the NOT gate is modeled to offer a zero delay. In (c) the results of simulation are shown again, except that this time, the NOT gate characteristics are changed to include a nonzero delay.

Figure C.3 (a) shows the schematic of an AND-OR circuit developed using Logic Works. All the gates are simply selected from a library of gates available. The inputs are labeled A and B. A toggle switch is connected to each input and the output OUT is connected to a buffer.

The input combinations can be changed by toggling the switches between 1 and 0 and corresponding output can be observed, as part of the simulation. The timing diagram in (b) shows the simulation output of this circuit for several input combinations. Note that all the intermediate signals can be labeled and monitored during the simulation.

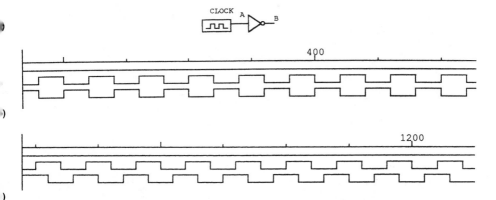

Figure C.2 Logic works simulation of a NOT gate. (a) Schematic; (b) simulation result with zero delay; (c) simulation results with nonzero delay.

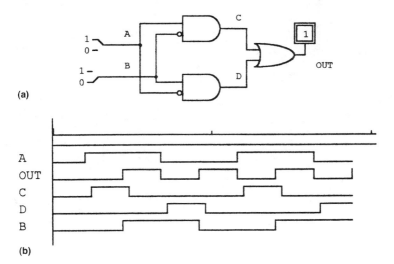

Figure C.3 Simulation of an AND-OR circuit. (a) Schematic; (b) simulation results.

References

Capilano Computing Systems Ltd., *Logic Works 3*, Menlo Park, CA: Addison Wesley, 1996.

Computer, NY: IEEE Computer Society, December 1974 (Special issue on HDLs).

Computer, NY: IEEE Computer Society, June 1977 (Special issue on HDL applications).

Maliniak, L. "A Beginner's Guide to VHDL," *Electronic Design*, October 14, 1994.

Perry, D. L., *VHDL*, New York: McGraw Hill, 1993.

Shiva, S. G. "Computer Hardware Description Languages—A Tutorial," *Proceedings of IEEE*, December 1979, pp. 1605–1615.

Shiva, S. G. "Automatic Hardware Synthesis," *Proceedings of IEEE*, January 1983.

Appendix D

CAD Tools for Designing with PLDs and FPGAs

Several computer-aided design (CAD) tools that aid in designing digital systems using PLDs and FPGAs are available. Here we will examine the capabilities of some of these tools. This Appendix is a brief introduction to the capabilities of these tools and is not intended to be a user's manual for any of them. Just as the introduction of newer PLDs make the existing PLDs obsolete (at a fairly fast rate), CAD tools also become obsolete. Refer to manuals from PLD device manufacturers and CAD tool designers for the latest details. The next Section provides the details of Warp2+, a CAD tool for designing with PLDs and FPGAs. Section D.2 illustrates the use of CAD tools in designing with PLAs and Section D.3 provides similar details for designing with PALs.

D.1 *Warp2+**

Warp2+ is a VHDL (VHSIC Hardware Description Language) compiler for designing with Cypress PLDs. *Warp2+* accepts VHDL input, synthesizes and optimizes the entered design, and outputs a JEDEC map for the desired

---

*This section is extracted from *Programmable Logic Data Book*, Cypress Semiconductor, Note CY 3120/CY3125, pp. 5-2 through 5-5 (1996).

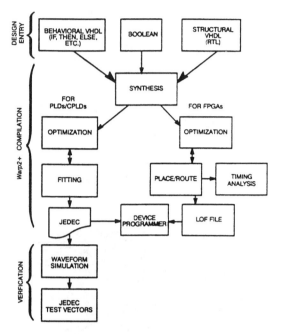

Figure D.1   *Warp2+* design flow (Courtesy of Cypress Semiconductors).

PLD or CPLD (complex PLD), or outputs an LOF file for the desired FPGA (see Figure D.1). For simulation, *Warp2+* provides the graphical waveform simulator called NOVA.

D.2 VHDL

VHDL is a non-proprietary language that is a standard for behavioral design entry and simulation. It is mandated for use by the Department of Defense and is supported by every major vendor of CAE (computer-aided electronic design) tools. VHDL allows designers to learn a single language that is useful for all facets of the design process. VHDL offers designers the ability to describe designs at many different levels. At the highest level, designs can be entered as a description of their behavior. This behavioral description is not tied to any specific target device. As a result, simulation can be done very early in the design to verify correct functionality, which significantly speeds the design process.

 Warp2+ VHDL syntax also includes support for intermediate level entry modes such as state table and Boolean functions. At the lowest level, designs can be described using gate-level RTL (register transfer language) descriptions. *Warp2+* gives the designer the flexibility to intermix all of these entry modes.

In addition, VHDL allows the hierarchical design mode in which new entities are built in terms of other existing entities. This can be done in either "top-down" (designing the highest levels of the system and its interfaces first, then progressing to greater and greater detail) or "bottom-up" (designing elementary building blocks of the system then combining these to build larger and larger parts) methodologies with equal ease. Because VHDL is an IEEE standard, multiple vendors offer tools for design entry, simulation at both high and low levels, and synthesis of designs to different silicon targets. The use of device independent behavioral design entry gives the freedom to retract designs to different devices. The wide availability of VHDL tools provides complete vendor independence as well. Designers can begin their project using *Warp2+* for Cypress PLDs and FPGAs and convert to high volume gate arrays using the same VHDL behavioral description with industry-standard synthesis tools.

While design portability and device independence are significant benefits, VHDL has other advantages. The VHDL language allows users to define their own functions. User-defined functions allow users to extend the capabilities of the language and build reusable libraries of tested routines. As a result the user can produce complex designs faster than with ordinary "flat" languages. VHDL also provides control over the timing of events or processes. VHDL has constructs that identify processes as either sequential, concurrent, or a combination of both. This is essential when describing the interaction of complex state machines.

VHDL is a rich programming language. Its flexibility reflects the nature of modern digital systems and allows designers to create accurate models of digital designs. Because of its depth and completeness, it is easier to describe a complex hardware system accurately in VHDL than in any other hardware description language. In addition, models created in VHDL can readily be transported to other CAE environments. *Warp2+* supports a rich subset of VHDL including loops, to generate statements, full hierarchical designs with packages, as well as synthesis for enumerated types and integers.

D.3 Designing with Warp2+

Warp2+ descriptions specify:

1. The behavior or structure of a design and
2. The mapping of signals in a design to the pins of a PLD/CPLD/ FPGA (optional).

The part of a *Warp2+* description that specifies the behavior or structure of the design is called an entity/architecture pair. Entity/architecture pairs can be divided into two parts: an entity declaration, which declares the design's interface signals and a design architecture, which describes the design's behavior or structure.

If the entity/architecture pair is kept in a separate file, that file is usually referred to as the design entity file. The entity portion of a design entity file is a declaration of what the design presents to the outside world (the interface). For each external signal, the entity declaration specifies a signal name, a direction, and a data type. In addition, the entity declaration specifies a name by which the entity can be referenced in the design architecture.

Code segments from four sample design entity files are given below. The top portion of each example features the entity declaration.

The architecture portion of a design entity file specifies the function of the design. As shown in Figure D.1, multiple design-entry methods are supported in *Warp2+*. A behaviorial description in VHDL often includes well known constructs such as IF . . . THEN, ELSE, and CASE statements. Here is a code segment from a simple state machine design (soda vending machine) that uses behavioral VHDL to implement the design:

```
ENTITY drink Is
    PORT (nickel, dime, quarter, clock: in bit;
    returnDime, returnNickel, GiveDrink: outbit):
END drink;


ARCHITECTURE fsm OF drink IS
TYPE drinkState IS (zero,five,ten,fifteen,twenty,twentyfive,owedime);
SIGNAL drinkstatus;drinkState;


BEGIN
PROCESS BEGIN
    WAIT UNTIL clock = '1';
    giveDrink <= '0';
    returnDime <='0';
    returnNickel <= '0';
CASE drinkStatus IS
WHEN zero =>
IF (nickel = '1') THEN
    drinkStatus <= drinkStatus'SUCC
    (drinkStatus);
    —goto Five
```

```
ELSEIF (dime = '1') THEN
  drinkStatus <= Ten;
ELSEIF (quarter = '1') THEN
  drinkStatus <= TwentyFive;
ENDIF;
WHEN Five =>
IF (nickel = '1') THEN
  drinkStatus <= Ten;
ELSEIF (dime = '1') THEN
  drinkStatus <= Fifteen;
ELSEIF (quarter = '1') THEN
  giveDrink <= '1';
  drinkStatus <= drinkStatus'PRED
  (drinkStatus);
  —goto Zero
ENDIF;

WHEN oweDime =>
  returnDime <= '1';
  drinkStatus <= zero;
  when others =>
  —This ELSE makes sure that the state
  —machine resets itself if
  —it somehow gets into an undefined state.
END CASE;
END PROCESS;

END FSM;
```

VHDL is a strongly typed language. It comes with several redefined operators, such as + and /= (add, not-equal-to). VHDL offers the capability of defining multiple meanings for operators (such as +), which results in simplification of the code written. For example, the following code segment shows that ''count = count+1'' can be written such that count is a bit vector, and 1 is an integer.

```
ENTITY sequence IS
  port (clk: in bit;
  s: inout bit);
  end sequence:

ARCHITECTURE fsm OF sequence IS
SIGNAL count: INTEGER RANGE 0 TO 7:
```

```
BEGIN
PROCESS BEGIN
WAIT UNTIL clk = '1';
CASE count IS
WHEN 0 / 1 / 2 / 3 =>
   s <= '1';
   count <= count + 1;
WHEN 4 =>
   s <= '0';
   count <= count + 1;
WHEN 5 =>
   s <= '1''
   count <= '0';
WHEN others =>
   s <= '0';
   count <= '0';
END CASE;
END PROCESS;

END FSM;
```

In this example, the + operator is overloaded to accept both integer and bit arguments. *Warp2+* supports overloading of operators.

A major advantage of VHDL is the ability to implement functions. The support of functions allows designs to be reused by simply specifying a function and passing the appropriate parameters. *Warp2+* features some built-in functions such as ttf (truth-table-function). The ttf function is particularly useful for state machine or look-up table designs. The following code describes a seven-segment display decoder implemented with the ttf function:

```
ENTITY seg7 IS
   PORT(
   inputs: IN BIT_BECTOR (0 to 3)
   outputs: OUT BIT_BECTOR (0 to 6)
   );
END SEG7;

ARCHITECTURE mixed OF seg7 IS
CONSTANT truthTable:
   x01_table (0 to 11, 0 to 10):=<
   —input & output
"0000" & "0111111",
```

```
"0001" & "0000110",
"0010" & "1011011",
"0011" & "1001111",
"0100" & "1100110",
"0101" & "1101101",
"0110" & "1111101",
"0111" & "0000111",
"1000" & "1111111",
"1001" & "1101111",
"101x" & "1111100",—creates E pattern
"111x" & "1111100",
);

BEGIN
    outputs <= ttf (truthTable,inputs);
END mixed;
```

A third design-entry method available to *Warp2+* users is Boolean equations. The following code describes how a one-bit half adder can be implemented in *Warp2+* with Boolean equations:

```
—entity declaration
ENTITY half_adder IS
    PORT (x, y, : IN BIT;
    sum, carry : OUT BIT);
END half_adder;
—architecture body
ARCHITECTURE behave OF half_adder IS
BEGIN
    sum <= x XOR y;
    carry <= x AND y;
END behave;
```

While all of the design methodologies described thus far are high-level entry methods, structural VHDL provides a method for designing at a very low level. In structural descriptions (all called RTL), the designer simply lists the components that make up the design and specifies how the components are wired together. Figure D.3 displays the schematic of a simple 3-bit shift register and the following code shows how this design can be described in *Warp2+* using structural VHDL.

```
ENTITY shifter3 IS port (
    clk : IN BIT;
    X : IN BIT;
```

```
    q0 : OUT BIT;
    q1 : OUT BIT;
    q2 : OUT BIT;
END shifter3;


ARCHITECTURE struct OF shifter3 IS
SIGNAL q0_temp, q1_temp, q2_temp : BIT;
BEGIN
    d1 : DFF PORT MAP (x,clk,q0_temp);
    d2 : DFF PORT MAP (q0_temp,clk,q1,_temp);
    d3 : DFF PORT MAP (q1_temp,clk,q2_temp);
    q0 <=q0_temp;
    q1 <=q1_temp;
    q2 <=q2_temp;
END struct;
```

All of the design-entry methods described can be mixed as desired. The ability to combine both high- and low-level entry methods in a single file is unique to VHDL. The flexibility and power of VHDL allows users of *Warp2+* to describe designs using whatever method is appropriate for their particular design.

Once the VHDL description of the design is complete, it is compiled using *Warp2+*. Although implementation is with a single command, compilation is actually a multistep process as shown in Figure D.1. The first part of the compilation process is the same for all devices. The input VHDL description is synthesized to a logical representation of the design. .

The second step of compilation is an interactive process of optimizing the design and fitting the logic into the targeted device. Logic optimization of *Warp2+* is accomplished using Espresso algorithms. The optimized design is automatically fed to the *Warp2+* filter if the user is targeting a PLD or CPLD. This filter supports automatic selection of D or T flip-flops. After the optimization and fitting step is complete, *Warp2+* creates a JEDEC file for the specified PLD or CPLD.

If the target device is an FPGA, *Warp2+* outputs a QDIF netlist file after optimization that is read into the *Warp2+* place and route software, SpDE. SpDE determines the placement of logic in the FPGA and routing of the interconnect that maximizes the speed of the operation and minimizes the area utilization of the design. After the place and route is complete, the design timing can be checked by the SpDE's path analyzer and a LOF file is output for programming.

Warp2+ includes Cypress's NOVA Simulator. NOVA features a graphical waveform simulator that can be used to simulate PLD/CPLD designs generated in *Warp2+*. The NOVA simulator provides functional simulation for PLDs/CPLDs and features interactive waveform editing and viewing. The simulator also provides the ability to probe internal nodes, automatically generate clocks and pulses, and generate JEDEC test vectors from simulator wave-forms. FPGA static timing analysis is available with that tool flow. (Higher level simulation support is available with *Warp3* [CY3130].)

The result of *Warp2+* compilation is a JEDEC or LOF file that implements the input design in the targeted device. Using this file, Cypress devices can be programmed on Cypress's Impulse3 programmer or on any qualified third-party programmer.

D.4 Designing with PLAs

The PLA-based design process is illustrated first by designing a binary-to-seven-segment decoder (from Application Note 10 of the Signetics programmable logic devices manual, 1986) shown in Figure D.2. The four-bit data on *W*, *X*, *Y*, and *Z* lines form the input to the PLA, and the seven-segment outputs are produced by the PLA, as shown in Figure D.3. The Signetics PLS153, an 18 × 42 × 10 PLA is used in the design. Here, a common

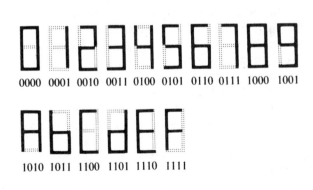

Z Y X W	a b c d e f g
0 0 0 0	1 1 1 1 1 1 0
0 0 0 1	0 1 1 0 0 0 0
0 0 1 0	1 1 0 1 1 0 1
0 0 1 1	1 1 1 1 0 0 1
0 1 0 0	0 1 1 0 0 1 1
0 1 0 1	1 0 1 1 0 1 1
0 1 1 0	1 0 1 1 1 1 1
0 1 1 1	1 1 1 0 0 0 0
1 0 0 0	1 1 1 1 1 1 1
1 0 0 1	1 1 1 1 0 1 1
1 0 1 0	1 1 1 0 1 1 1
1 0 1 1	0 0 1 1 1 1 1
1 1 0 0	1 0 0 1 1 1 0
1 1 0 1	0 1 1 1 1 0 1
1 1 1 0	1 0 0 1 1 1 1
1 1 1 1	1 0 0 0 1 1 1

Figure D.2 BCD-to-seven-segment decoder truth table (Courtesy of Signetics Corporation).

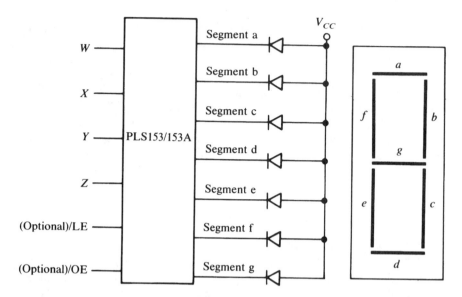

Figure D.3   Segment decoder driving a common anode LED display (Courtesy of Signetics Corporation).

anode display device is used. This device requires a low signal to activate its segments, and hence the PLA should produce active-low output signals.

Figure D.4 shows the configuration of the PLS153. It consists of 42 AND gates and 10 OR gates with fusible link connections for programming I/O polarity and direction. The 8 input lines (I) and the 10 bidirectional I/O lines (B) are connected to the AND matrix. These, along with the 10 direction control gates (D), yield flexible gate configurations ranging from 18 inputs to 10 outputs. On-chip buffers allow the connection of either true (I, B) or complement (I', B') inputs to all AND gates whose outputs can be optionally linked to the OR gates. A set of Exclusive-OR gates connected to the outputs of OR gates allows programming the output polarities to active-low or active-high.

Figure D.5 shows the PLA programming table derived from the truth table in Figure D.2. I_0–I_3 are used for the four-bit input, and B_1–B_7 are used for the seven outputs that are programmed as active-low, as shown in the polarity section of the table. All 16 AND terms are used in this implementation.

Note that all inputs become low when the input corresponds to an eight. Hence, if the implementation uses logic-0s rather than logic-1s, the AND term corresponding to eight is not required, thus saving an AND con-

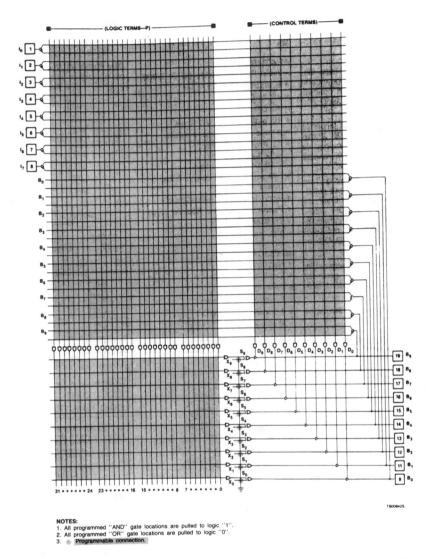

NOTES:
1. All programmed "AND" gate locations are pulled to logic "1".
2. All programmed "OR" gate locations are pulled to logic "0".
3. ● Programmable connection.

Figure D.4　Signetics PLS153A (Courtesy of Signetics Corporation).

82S153/153A PROGRAMMING TABLE   DRAWING ID: B TO 7 SEG DECODER   REV. —
DESIGNER: D.K. WONG   DATE: 12/15/84

	AND				OR POLARITY H H L L L L L L L H			COMMENTS	
	I	B(I)		B(O)		INPUT	OUTPUT
	7 6 5 4 3 2 1 0	9 8 7 6 5 4 3 2 1 0		9 8 7 6 5 4 3 2 1 0			
0	– – – – L L L L	– – – – – – – – – –		· · A A A A A A · ·		0	0
1	– – – – L L L H	– – – – – – – – – –		· · · A A · · · · ·		1	1
2	– – – – L L H L	– – – – – – – – – –		· · A A · A A · A ·		2	2
3	– – – – L L H H	– – – – – – – – – –		· · A A A A · · A ·		3	3
4	– – – – L H L L	– – – – – – – – – –		· · · A A · · A A ·		4	4
5	– – – – L H L H	– – – – – – – – – –		· · A · A A · A A ·		5	5
6	– – – – L H H L	– – – – – – – – – –		· · A · A A A A A ·		6	6
7	– – – – L H H H	– – – – – – – – – –		· · A A A A · · · ·		7	7
8	– – – – H L L L	– – – – – – – – – –		· · A A A A A A A ·		8	8
9	– – – – H L L H	– – – – – – – – – –		· · A A A A · A A ·		9	9
10	– – – – H L H L	– – – – – – – – – –		· · A A A · A A A ·		10	A
11	– – – – H L H H	– – – – – – – – – –		· · · · A A A A A ·		11	b
12	– – – – H H L L	– – – – – – – – – –		· · A · · A A A A ·		12	C
13	– – – – H H L H	– – – – – – – – – –		· · · A A A A · A ·		13	d
14	– – – – H H H L	– – – – – – – – – –		· · A · · A A A A ·		14	E
15	– – – – H H H H	– – – – – – – – – –		· · A · · · A A A ·		15	F
16							
17							
18							
19							
20							
21							
22							
23							
24							
25							
26							
27							
28							
29							
30							
31							
D9							
D8							
D7	L – – – – – – – –	– – – – – –		ENABLE B7 FOR OUTPUT (SEG A)			
D6	L – – – – – – – –	– – – – – –		" B6 " " " B			
D5	L – – – – – – – –	– – – – – –		" B5 " " " C			
D4	L – – – – – – – –	– – – – – –		" B4 " " " D			
D3	L – – – – – – – –	– – – – – –		" B3 " " " E			
D2	L – – – – – – – –	– – – – – –		" B2 " " " F			
D1	L – – – – – – – –	– – – – – –		" B1 " " " G			
D0							
N A M E	IDE	N.Y.X.W		A B C D E F G			

Figure D.5 H/L programming table for binary-to-seven-segment decoder (Courtesy of Signetics Corporation).

nection. The PLA programming table of Figure D.6 shows this implementation. Note that the output polarities are now changed to high and the OR-array is now the complement of that in Figure D.5. This table is now used to program the PLA.

The Signetics CAD system automatically produces the table of Figure D.6 from the truth table input. We will first introduce the components of the CAD system, followed by the results obtained for this design.

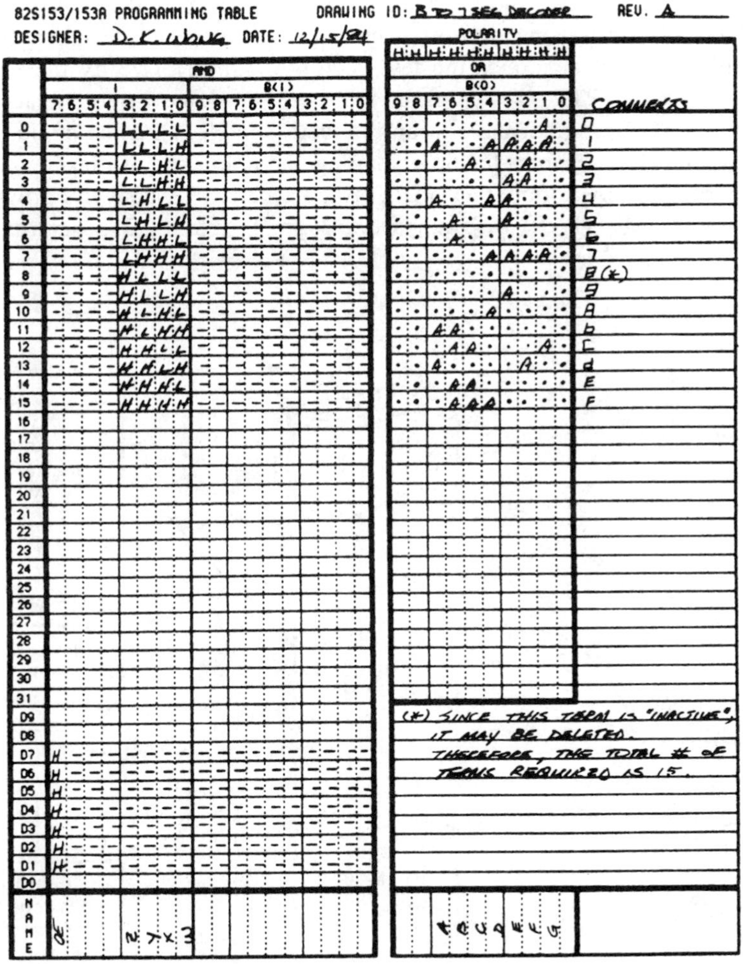

82S153/153A PROGRAMMING TABLE DRAWING ID: B TO 7SEG DECODER REV. A

DESIGNER: D. K. Wong DATE: 12/15/84

Term	AND I 7:6:5:4	AND I 3:2:1:0	AND B(I) 9:8	AND B(I) 7:6:5:4	AND B(I) 3:2:1:0	OR B(O) 9 (H)	8 (H)	7 (H)	6 (H)	5 (H)	4 (H)	3 (H)	2 (H)	1 (H)	0 (H)	COMMENTS
0	- - - -	L L L L	- -	- - - -	- - - -	·	·	·	·	·	·	·	·	A	·	0
1	- - - -	L L L H	- -	- - - -	- - - -	·	·	A	·	·	A	A	A	A	·	1
2	- - - -	L L H L	- -	- - - -	- - - -	·	·	·	·	A	·	·	A	·	·	2
3	- - - -	L L H H	- -	- - - -	- - - -	·	·	·	·	·	·	A	A	·	·	3
4	- - - -	L H L L	- -	- - - -	- - - -	·	·	A	·	·	A	A	·	·	·	4
5	- - - -	L H L H	- -	- - - -	- - - -	·	·	·	A	·	·	A	·	·	·	5
6	- - - -	L H H L	- -	- - - -	- - - -	·	·	·	A	·	·	·	·	·	·	6
7	- - - -	L H H H	- -	- - - -	- - - -	·	·	·	·	·	A	A	A	A	·	7
8	- - - -	H L L L	- -	- - - -	- - - -	·	·	·	·	·	·	·	·	·	·	8 (*)
9	- - - -	H L L H	- -	- - - -	- - - -	·	·	·	·	·	·	A	·	·	·	9
10	- - - -	H L H L	- -	- - - -	- - - -	·	·	·	·	·	A	·	·	·	·	A
11	- - - -	H L H H	- -	- - - -	- - - -	·	·	A	A	·	·	·	·	·	·	b
12	- - - -	H H L L	- -	- - - -	- - - -	·	·	·	A	A	·	·	·	A	·	C
13	- - - -	H H L H	- -	- - - -	- - - -	·	·	A	·	·	·	·	A	·	·	d
14	- - - -	H H H L	- -	- - - -	- - - -	·	·	·	A	A	·	·	·	·	·	E
15	- - - -	H H H H	- -	- - - -	- - - -	·	·	·	A	A	A	·	·	·	·	F
16–31																
D9																(*) SINCE THIS TERM IS "INACTIVE",
D8																IT MAY BE DELETED.
D7	H - - -	- - - -	- -	- - - -	- - - -											THEREFORE, THE TOTAL # OF
D6	H - - -	- - - -	- -	- - - -	- - - -											TERMS REQUIRED IS 15.
D5	H - - -	- - - -	- -	- - - -	- - - -											
D4	H - - -	- - - -	- -	- - - -	- - - -											
D3	H - - -	- - - -	- -	- - - -	- - - -											
D2	H - - -	- - - -	- -	- - - -	- - - -											
D1	H - - -	- - - -	- -	- - - -	- - - -											
D0																
NAME	E	W Y X Z						a	b	c	d	e	f	g		

Figure D.6 Minimized progamming table (Courtesy of Signetics Corporation).

D.5 AMAZE System

AMAZE (Automated Map and Zap Equation) is a software package designed to aid in the use of Signetics programmable logic devices (PLDs) in circuit design. Figure D.7 shows the components of the software system. The components of this system communicate through the PLD standard fuse file, which contains details on all the PLD devices available. The designer enters the design information using the BLAST (Boolean logic and state transfer) module. BLAST is an interactive module that checks the design information and automatically compiles the PLD program table. SIM, the PLD simulator, has two modes of operation. In the manual mode, it simulates the PLD table using the test vectors provided by the designer. In the automatic mode, it generates the test vectors required to verify the design and simulates it. PTE, the program table editor, is used to generate and modify the PLD program table in an interactive mode. DPI, the device programmer interface, enables the downloading and uploading of files in various formats required by commercial PLD programmers. PTP converts programmable array logic (PAL) fuse maps or programs into PLD programs and

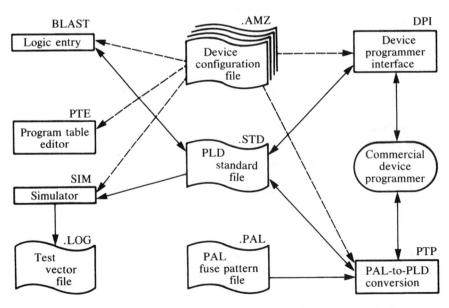

Figure D.7   AMAZE software system (Courtesy of Signetics Corporation).

utilizes the PAL fuse pattern file. The AMAZE software package can be obtained from Signetics in either an IBM personal computer or a DEC VAX version.

Figure D.8 shows the design flow using the AMAZE package. Once the designer completes the conceptual design and selects an appropriate PLA from the library of available PLAs, BLAST is used to assign pins of the PLA to the input and output variables in the design. Then the Boolean equations (corresponding to the combinational logic portion) and the state equations (corresponding to the sequential logic portion) are entered using BLAST. BLAST can also derive the equation information from the logic schematic capture programs available from various vendors. The fuse map is then assembled by BLAST. The design is now verified through simulation using SIM. The test vectors (i.e., the input values) can be generated automatically by SIM or provided by the designer. The fuse map and test vectors are then downloaded to the device programmer.

We will now show the utility of this package, through the binary-to-seven-segment decoder design example.

Figure D.9 shows the assignment of I/O functions to the pins of the PLS153 in an interactive session with BLAST. Once the device number is known, BLAST displays the pin configuration and function of each pin. For those pins that are user definable, the designer can then allocate a label and a function (e.g., *I*, *O*, *B*).

The logic entry function of BLAST is then invoked. Figure D.10 shows this phase. The ''@'' character is the prompt by BLAST. It first asks for the device type. Once the device type is provided, bookkeeping information such as the one shown (@DRAWING through @DESCRIPTION) can be input to BLAST. This serves merely as a documentation of the design. All the product terms required in the design are entered in response to @COMMON PRODUCT TERM. Each product term has a label followed by the AND (∗) of the true or complement (/) of variables specified during pin specification. Then the equations corresponding to each output are entered in response to @LOGIC EQUATION. An output enable input (*OE*) is used. Outputs are enabled when /*OE* goes low, as described by the @I/O DIRECTION section. This completes the entry of design information. The BLAST then produces the fuse map shown in Figure D.11. Figure D.12 shows the output of the simulator.

Signetics is now a subsidiary of Phillips Semiconductors and the AMAZE design package is obsolete as of this writing. Nevertheless the above description illustrates the capabilities of a typical CAD tool for PLA-based design.

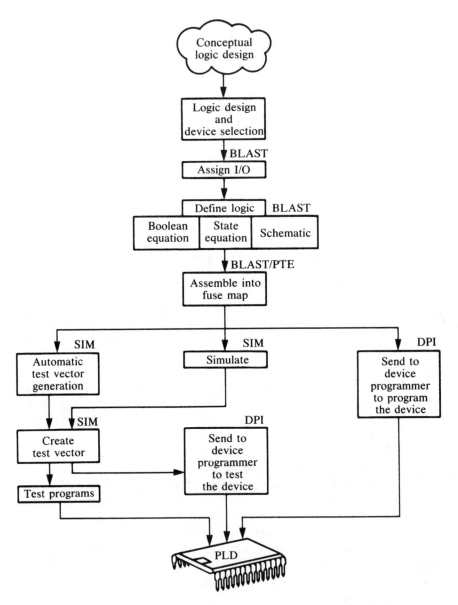

Figure D.8   Design flow (Courtesy of Signetics Corporation).

```
******************* P I N   L I S T *******************
     LABEL       ** FNC **PIN --------- PIN** FNC **    LABEL
  W              ** I  **  1-!        !-20 ** +5V **VCC
  X              ** I  **  2-!        !-19 ** B   **N/C
  Y              ** I  **  3-!        !-18 ** B   **N/C
  Z              ** I  **  4-!  8     !-17 ** B   **/SEG_A
  N/C            ** I  **  5-!  2     !-16 ** B   **/SEG_B
  N/C            ** I  **  6-!  S     !-15 ** B   **/SEG_C
  N/C            ** I  **  7-!  1     !-14 ** /B  **/SEG_D
  /OE            ** I  **  8-!  5     !-13 ** /B  **/SEG_E
  N/C            ** B  **  9-!  3     !-12 ** /B  **/SEG_F
  GND            ** OV **  10-!       !-11 ** /B  **/SEG_G
                          ---------
```

Figure D.9 Pin list of binary-to-seven segment decoder generated by AMAZE (Courtesy of Signetics Corporation).

D.6 Designing with PALs

This section is based on the PALASM CAD tool description from Monolithic Memories Inc. (MMI). MMI is now a subsidiary of Advanced Micro Devices Inc. (AMD). PALASM4 version 1.5 is now available as a freeware package from AMD.

Figure D.13 shows the components of the PAL software. Using a text editor, the designer first creates a PAL design specification (PDS) file. This specification consists of the assignment of selected PAL device pin numbers to various signals in the circuit being designed and the Boolean equations describing the design and uses the PALASM language. There are two versions of the PALASM language: PALASM-1 and PALASM-2. PDS file can be in either format. PALASM-1 is the earlier version of the two. There are considerable differences in the syntax and capabilities of the two. In addition to the features of PALASM-1, PALASM-2 allows the description of newer asynchronous PLA devices and MegaPALs, which are PALs of very high complexity. If the PDS is in PALASM-1, PDSCNVT is used interactively to convert it into PALASM-2 format.

The PALASM-2 compiler analyzes the PDS file input to it for the correct syntax. If an error is detected, the program attempts to indicate where in the input description the error has occurred. If no errors are detected, the program generates an intermediate file that contains the input specification in a hierarchically structured form to enable easy processing by the follow-on programs. This program accepts the input descriptions for all PLA devices currently available from MMI.

XPLOT produces the fuse map and JEDEC data for a specified PAL device from the intermediate file output by the PALASM-2 compiler. It checks the input description for consistency among the equations describing the design and also with the PAL device specified. The architectural description of the PAL is obtained by XPLOT from a file containing the profiles of currently available PAL devices.

```
@DEVICE TYPE
82S153
@DRAWING
*********************** BINARY-TO-7 SEGMENT DECODER
@REVISION
*********************** REV. -
@DATE
*********************** OCT 1, 1984
@SYMBOL
*********************** FILE ID: 7deco
@COMPANY
*********************** SIGNETICS
@NAME
*********************** DAVID K. WONG
@DESCRIPTION
********************************************************************
*    This circuit converts a 4-bit binary code ( HEX ) into        *
*    a 7-segment display. The display is a common anode 7-segment LED. *
*    The output of the 82S153 goes LOW for each segment that is ON. *
********************************************************************
```

```
                                    TRUTH TABLE
                a            Z Y X W      a  b  c  d  e  f  g
            -------          -------------------------------------
    f :           : b        0 0 0 0      0  0  0  0  0  0  1
    :      g      :          0 0 0 1      1  0  0  1  1  1  1
            -------          0 0 1 0      0  0  1  0  0  1  0
    e :           : c        0 0 1 1      0  0  0  0  1  1  0
    :             :          0 1 0 0      1  0  0  1  1  0  0
            -------          0 1 0 1      0  1  0  0  1  0  0
                d            0 1 1 0      0  1  0  0  0  0  0
                             0 1 1 1      0  0  0  1  1  1  1
                             1 0 0 0      0  0  0  0  0  0  0
                             1 0 0 1      0  0  0  0  1  0  0
                             1 0 1 0      0  0  0  1  0  0  0
                             1 0 1 1      1  1  0  0  0  0  0
                             1 1 0 0      0  1  1  0  0  0  1
                             1 1 0 1      1  0  0  0  0  1  0
                             1 1 1 0      0  1  1  0  0  0  0
                             1 1 1 1      0  1  1  1  0  0  0
```

```
@COMMON PRODUCT TERM
ZER = /z * /y * /x * /w  ;
ONE = /z * /y * /x *  w  ;
TWO = /z * /y *  x * /w  ;
THR = /z * /y *  x *  w  ;
FOU = /z *  y * /x * /w  ;
FIV = /z *  y * /x *  w  ;
SIX = /z *  y *  x * /w  ;
SEV = /z *  y *  x *  w  ;
EIG =  z * /y * /x * /w  ;
NIN =  z * /y * /x *  w  ;
AAA =  z * /y *  x * /w  ;
BBB =  z * /y *  x *  w  ;
CCC =  z *  y * /x * /w  ;
DDD =  z *  y * /x *  w  ;
EEE =  z *  y *  x * /w  ;
FFF =  z *  y *  x *  w  ;

@LOGIC EQUATION
/SEG_A =   ONE + FOU + BBB + DDD                    ;
/SEG_B =   FIV + SIX + BBB + CCC + EEE + FFF        ;
/SEG_C =   TWO + CCC + EEE + FFF                    ;
/SEG_D =   ONE + FOU + SEV + AAA + FFF              ;
/SEG_E =   ONE + THR + FOU + FIV + SEV + NIN        ;
/SEG_F =   ONE + TWO + THR + SEV + DDD              ;
/SEG_G =   ZER + ONE + SEV + CCC                    ;

@I/O DIRECTION 150/1, 152/3, 154-9
''
********************************************************************
*              OUTPUTS ARE ENABLED WHEN /OE GOES LOW.             *
*              THEREFORE, D1...D7 = /(/OE)  =  OE    ;             *
********************************************************************
''
D1 = OE  ;
D2 = OE  ;
D3 = OE  ;
D4 = OE  ;
D5 = OE  ;
D6 = OE  ;
D7 = OE  ;

''
********************************************************************
*                    END OF LOGIC EQUATIONS                       *
********************************************************************
```

Figure D.10 Entry of logic equations (Courtesy of Signetics Corporation).

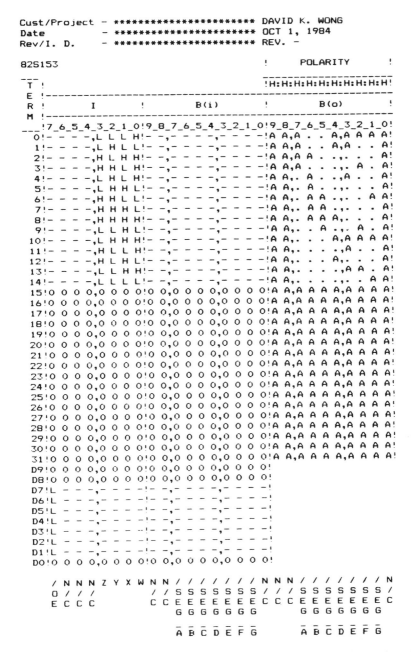

Figure D.11 Program table generated by AMAZE (Courtesy of Signetics Corporation).

```
82S153   A:7deco153.STD
" 4-bit binary to 7-segment decoder simulation
"
"  INPUTS  <=B(I/O)=>   TRACE TERMS
" 76543210 9876543210
"
  00000000 ..LLLLLLH. ;
  00000001 ..HLLHHHH. ;
  00000010 ..LLHLLHL. ;
  00000011 ..LLLLHHL. ;
  00000100 ..HLLHHLL. ;
  00000101 ..LHLLHLL. ;
  00000110 ..LHLLLLL. ;
  00000111 ..LLLHHHH. ;
  00001000 ..LLLLLLL. ;
  00001001 ..LLLLHLL. ;
  00001010 ..LLLHLLL. ;
  00001011 ..HHLLLLL. ;
  00001100 ..LHHLLLH. ;
  00001101 ..HLLLLHL. ;
  00001110 ..LHHLLLL. ;
  00001111 ..LHHHLLL. ;
  10001111 .......... ;
  10000000 .......... ;
"
" X------- ----------    I/O CONTROL LINES
"          IIBBBBBBBI    DESIGNATED I/O USAGE
"          IIBBBBBBBI    ACTUAL I/O USAGE
"
" PIN LIST...
" 08 07 06 05 04 03 02 01 19 18 17 16 15 14 13 12 11 09 ;
```

Figure D.12   Simulator output (Courtesy of Signetics Corporation).

SIM is a functional simulator for the design described in PALASM-2. It calculates the outputs based on the inputs specified through the simulator command set, preprocessed by PALASM-2. The output of SIM is a history file that traces the values of every pin through the simulation sequence. A trace file, which is a subset of the history file that traces only the pins specified by the designer, can also be obtained. SIM can also be used to generate test vectors required by the PLA programming device to test the fuse pattern after the PAL has been programmed.

Hard-Array Logic (HAL) devices are mask-programmable PALs. ZHALs are zero-standby-power CMOS HAL devices. The ZHAL program is used to check whether the design specification fits into the ZHAL device architecture.

In addition to the above software modules, other support software modules, such as MENU, VTRACE, and PC2, are available. MENU is an interactive program that provides a pop-up menu at each phase of the design, thereby simplifying user interaction with the PALASM-2 software. VTRACE produces timing diagrams from the output of SIM. PC2 enables communication between the host machine and the PLD programmers.

The following design examples taken from the MMI PLA handbook (1986) illustrate the utility of the PALASM-2 software in the design of PLA-based digital systems.

Figure D.14 shows the design of a *comparator* circuit that compares two eight-bit data strings, *A* and *B*, for equality.

Software

Following is a summary of all currently available programs.

1. PDSCNVT PALASM 1 to PALASM 2 syntax conversion
2. PALASM 2 PALASM 2 syntax parser
3. XPLOT PALASM 2 fuse map and JEDEC output
4. SIM PALASM 2 simulator
5. ZHAL ZHAL device fit

Supplementary Software

1. MENU Simplified PALASM 2 user interface
2. PC2 Programmer interface program
3. VTRACE Graph display simulator trace output

.Files

1.		.PDS	User PALASM 2 PLD design description
2. PALASM 2		.TRE	PLD intermediate design description
3.		.PDF	PLD architecture description data
4.		.XPT	Contains PLD fuse map data
5.		.JED	Contains PLD fuse JEDEC data
6.		.HST	Contains full simulation history data
7.		.TRF	Contains user simulation trace data
8.		.JDC	Contains both PLD fuse JEDEC data and JEDEC test vectors

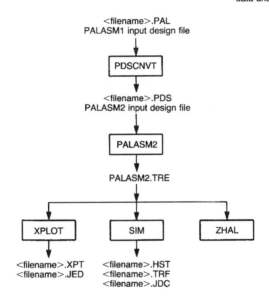

Figure D.13 Components of PALASM software (Courtesy of Monolithic Memories Inc.).

The first portion of the design description (i.e., the PDS file) shown in the figure is for documentation purposes only. The statements starting with a ";" are comments. The keyword CHIP specifies a name (OctalCompare) for the design, and the PAL device (PAL16C1) used in the design. Then the signals in the design are assigned to the pins of the PAL. Note the correspondence of the listing of signal names to pin numbers shown in the logic symbol of the PAL.

Now the equations governing the operation of the circuit being designed are listed. In these equations, *NE* is the output signal, indicating that

PAL Device Design Specification

```
Title     Octal_Comparator
Pattern   OctComp.pds
Revision  A
Author    Mehrnaz Hada
Company   Monolithic Memories Inc., Santa Clara,CA
Date      1/29/85

;The octal comparator establishes when two 8-bit data
;strings (A7-A0) and (B7-B0) are equivalent (EQ=H) or
;equivalent (NE=H).
```

PDS

```
CHIP OctalComparato PAL16C1

A7 A0 B0 A1 B1 A2 B2 A3 B3 GND
A4 B4 A5 B5 EQ NE A6 B6 B7 VCC

EQUATIONS

NE =  A0*/B0  +  /A0* B0              ;A0 :+: B0
   +  A1*/B1  +  /A1* B1              ;A1 :+: B1
   +  A2*/B2  +  /A2* B2              ;A2 :+: B2
   +  A3*/B3  +  /A3* B3              ;A3 :+: B3
   +  A4*/B4  +  /A4* B4              ;A4 :+: B4
   +  A5*/B5  +  /A5* B5              ;A5 :+: B5
   +  A6*/B6  +  /A6* B6              ;A6 :+: B6
   +  A7*/B7  +  /A7* B7              ;A7 :+: B7


SIMULATION

TRACE_ON A7 A6 A5 A4 A3 A2 A1 A0 NE
         B7 B6 B5 B4 B3 B2 B1 B0
```

Simulation commands

```
SETF  A7 /A6 /A5 /A4 /A3 /A2 /A1 /A0     ;A7=H, B7=L
      /B7 /B6 /B5 /B4 /B3 /B2 /B1 /B0
SETF /A7 A6                              ;A6=H, B6=L
SETF /A6 A5                              ;A5=H, B5=L
SETF /A5 A4                              ;A4=H, B4=L
SETF /A4 A3                              ;A3=H, B3=L
SETF /A3 A2                              ;A2=H, B2=L
SETF /A2 A1                              ;A1=H, B1=L
SETF /A1 A0                              ;A0=H, B0=L
SETF /A7 /A6 /A5 /A4 /A3 /A2 /A1 /A0     ;A7=L, B7=H
      B7
SETF /B7 B6                              ;A6=L, B6=H
SETF /B6 B5                              ;A5=L, B5=L
SETF /B5 B4                              ;A4=L, B4=H
SETF /B4 B3                              ;A3=L, B3=H
SETF /B3 B2                              ;A2=L, B2=H
SETF /B2 B1                              ;A1=L, B1=H
SETF /B1 B0                              ;A0=L, B0=H
SETF /B0                                 ;Test all L's
SETF A7 A6 A5 A4 A3 A2 A1 A0             ;Test all H's
     B7 B6 B5 B4 B3 B2 B1 B0
SETF /A7 A6 /A5 A4 /A3 A2 /A1 A0         ;Test even ones
     /B7 B6 /B5 B4 /B3 B2 /B1 B0
SETF A7 /A6 A5 /A4 A3 /A2 A1 /A0         ;Test odd ones
     B7 /B6 B5 /B4 B3 /B2 B1 /B0


;Function Table for PALASM1

;A7 A6 A5 A4 A3 A2 A1 A0 B7 B6 B5 B4 B3 B2 B1 B0 NE EQ
```

Function description

```
; Input A    Input B    Outputs
; 76543210   76543210   NE EQ    Comments
;------------------------------------------------------
; HLLLLLLL   LLLLLLLL    H  L    A7=H, B7=L
; LHLLLLLL   LLLLLLLL    H  L    A6=H, B6=L
; LLHLLLLL   LLLLLLLL    H  L    A5=H, B5=L
; LLLHLLLL   LLLLLLLL    H  L    A4=H, A5=L
; LLLLHLLL   LLLLLLLL    H  L    A3=H, B3=L
; LLLLLHLL   LLLLLLLL    H  L    A2=H, B2=L
; LLLLLLHL   LLLLLLLL    H  L    A1=H, B1=L
; LLLLLLLH   LLLLLLLL    H  L    A0=H, B0=L
; LLLLLLLL   HLLLLLLL    H  L    A7=L, B7=H
; LLLLLLLL   LHLLLLLL    H  L    A6=L, B6=H
; LLLLLLLL   LLHLLLLL    H  L    A5=L, B5=H
; LLLLLLLL   LLLHLLLL    H  L    A4=L, B4=H
; LLLLLLLL   LLLLHLLL    H  L    A3=L, B3=H
; LLLLLLLL   LLLLLHLL    H  L    A2=L, B2=H
; LLLLLLLL   LLLLLLHL    H  L    A1=L, B1=H
; LLLLLLLL   LLLLLLLH    H  L    A0=L, B0=H
; LLLLLLLL   LLLLLLLL    L  H    Test all L's
; HHHHHHHH   HHHHHHHH    L  H    Test all H's
; HLHLHLHL   HLHLHLHL    L  H    Test even checkerboard
; LHLHLHLH   LHLHLHLH    L  H    Test odd  checkerboard
;------------------------------------------------------
```

Figure D.14 Design Example D.1: octal comparator (Courtesy of Monolithic Memories Inc.).

Simulation Results

```
Page :  1
          gggggggggg gggggggggg
     A7   HLLLLLLLLL LLLLLLLHLH
     A6   LHLLLLLLLL LLLLLLLHHL
     A5   LLHLLLLLLL LLLLLLLHLH
     A4   LLLHLLLLLL LLLLLLLHLH
     A3   LLLLHLLLLL LLLLLLLHLH
     A2   LLLLLHLLLL LLLLLLLHHL
     A1   LLLLLLHLLL LLLLLLLHLH
     A0   LLLLLLLHLL LLLLLLLHLH
     NE   HHHHHHHHHH HHHHHHLLLL
     B7   LLLLLLLLHL LLLLLLLHLH
     B6   LLLLLLLLLH LLLLLLLHHL
     B5   LLLLLLLLLL HLLLLLLHLH
     B4   LLLLLLLLLL LHLLLLLHHL
     B3   LLLLLLLLLL LLHLLLLHLH
     B2   LLLLLLLLLL LLLHLLLHHL
     B1   LLLLLLLLLL LLLLHLLHLH
     B0   LLLLLLLLLL LLLLLHLHHL
```

Logic Symbol

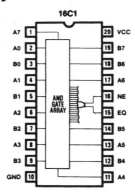

Figure D.14   *(Continued)*

A is not equal to *B*. Only *NE* is described, since *EQ* is implemented by complementing *NE*. The following operators are used:

    =   Assignment
    *   AND
    +   OR
    /   NOT
    :+:   EXCLUSIVE-OR

Again, the inline comments start with a semicolon.

PAL Device Design Specification

```
TITLE      16-BIT ADDRESSABLE REGISTER
PATTERN    ADREG16.PDS
REVISION   A
AUTHOR     John Birkner
COMPANY    Monolithic Memories Inc. Santa Clara, CA
DATE       2/11/85

;      The 16-bit addressable register loads one of 16 registers
;      selected by ADDR[0..3] with data input, DATA.


CHIP ADREG16 PAL32R16
Q0 Q1 Q2 Q3 /E1 NC NC A0 A1 VCC A2 A3 DATA NC /PRLD2 CLK2
Q4 Q5 Q6 Q7 Q8 Q9 Q10 Q11 /E2 NC NC NC NC GND NC NC NC NC /PRLD1 CLK1
Q12 Q13 Q14 Q15
```

EQUATIONS

```
Q0    :=  A0              *Q0              ;hold
      +        A1         *Q0              ;hold
      +             A2    *Q0              ;hold
      +                   A3*Q0            ;hold
      + /A0*/A1*/A2*/A3*DATA               ;load

Q1    := /A0             *Q1              ;hold
      +        A1         *Q1              ;hold
      +             A2    *Q1              ;hold
      +                   A3*Q1            ;hold
      +  A0*/A1*/A2*/A3*DATA               ;load

Q2    :=  A0              *Q2              ;hold
      +       /A1         *Q2              ;hold
      +             A2    *Q2              ;hold
      +                   A3*Q2            ;hold
      + /A0* A1*/A2*/A3*DATA               ;load

Q3    := /A0             *Q3              ;hold
      +       /A1         *Q3              ;hold
      +             A2    *Q3              ;hold
      +                   A3*Q3            ;hold
      +  A0* A1*/A2*/A3*DATA               ;load

Q4    :=  A0              *Q4              ;hold
      +        A1         *Q4              ;hold
      +       /A2         *Q4              ;hold
      +                   A3*Q4            ;hold
      + /A0*/A1* A2*/A3*DATA               ;load

Q5    := /A0             *Q5              ;hold
      +        A1         *Q5              ;hold
      +       /A2         *Q5              ;hold
      +                   A3*Q5            ;hold
      +  A0*/A1* A2*/A3*DATA               ;load

Q6    :=  A0              *Q6              ;hold
      +       /A1         *Q6              ;hold
      +       /A2         *Q6              ;hold
      +                   A3*Q6            ;hold
      + /A0* A1* A2*/A3*DATA               ;load

Q7    := /A0             *Q7              ;hold
      +       /A1         *Q7              ;hold
      +       /A2         *Q7              ;hold
      +                   A3*Q7            ;hold
      +  A0* A1* A2*/A3*DATA               ;load

Q8    :=  A0              *Q8              ;hold
      +        A1         *Q8              ;hold
      +             A2    *Q8              ;hold
      +                  /A3*Q8            ;hold
      + /A0*/A1*/A2* A3*DATA               ;load

Q9    := /A0             *Q9              ;hold
      +        A1         *Q9              ;hold
      +             A2    *Q9              ;hold
      +                  /A3*Q9            ;hold
      +  A0*/A1*/A2* A3*DATA               ;load

Q10   :=  A0              *Q10             ;hold
      +       /A1         *Q10             ;hold
      +             A2    *Q10             ;hold
      +                  /A3*Q10           ;hold
      + /A0* A1*/A2* A3*DATA               ;load

Q11   := /A0             *Q11             ;hold
      +       /A1         *Q11             ;hold
      +             A2    *Q11             ;hold
      +                  /A3*Q11           ;hold
      +  A0* A1*/A2* A3*DATA               ;load

Q12   :=  A0              *Q12             ;hold
      +        A1         *Q12             ;hold
      +       /A2         *Q12             ;hold
      +                  /A3*Q12           ;hold
      + /A0*/A1* A2* A3*DATA               ;load

Q13   := /A0             *Q13             ;hold
      +        A1         *Q13             ;hold
      +       /A2         *Q13             ;hold
      +                  /A3*Q13           ;hold
      +  A0*/A1* A2* A3*DATA               ;load

Q14   :=  A0              *Q14             ;hold
      +       /A1         *Q14             ;hold
      +       /A2         *Q14             ;hold
```

Figure D.15 Design Example D.2: addressable register (Courtesy of Monolithic Memories Inc.).

```
            +              /A3*Q14        ;hold
            +   /A0* A1* A2* A3*DATA       ;load

Q15    := /A0              *Q15           ;hold
            +   /A1        *Q15           ;hold
            +       /A2    *Q15           ;hold
            +              /A3*Q15        ;hold
            +   A0* A1* A2* A3*DATA       ;load

SIMULATION

TRACE_ON Q0 Q1 Q2 Q3 Q4 Q5 Q6 Q7 Q8 Q9 Q10 Q11 Q12 Q13 Q14 Q15
         A0 A1 A2 A3 DATA

SETF E1 E2 /DATA /PRLD1 /PRLD2

          SETF /A0 /A1 /A2 /A3
          CLOCKF CLK1 CLK2

          SETF  A0 /A1 /A2 /A3
          CLOCKF CLK1 CLK2

          SETF /A0  A1 /A2 /A3
          CLOCKF CLK1 CLK2

          SETF  A0  A1 /A2 /A3
          CLOCKF CLK1 CLK2

          SETF /A0 /A1  A2 /A3
          CLOCKF CLK1 CLK2

          SETF  A0 /A1  A2 /A3
          CLOCKF CLK1 CLK2

          SETF /A0  A1  A2 /A3
          CLOCKF CLK1 CLK2

          SETF  A0  A1  A2 /A3
          CLOCKF CLK1 CLK2

          SETF /A0 /A1 /A2  A3
          CLOCKF CLK1 CLK2

          SETF  A0 /A1 /A2  A3
          CLOCKF CLK1 CLK2

          SETF /A0  A1 /A2  A3
          CLOCKF CLK1 CLK2

          SETF  A0  A1 /A2  A3
          CLOCKF CLK1 CLK2

          SETF /A0 /A1  A2  A3
          CLOCKF CLK1 CLK2

          SETF  A0 /A1  A2  A3
          CLOCKF CLK1 CLK2

          SETF /A0  A1  A2  A3
          CLOCKF CLK1 CLK2

          SETF  A0  A1  A2  A3
          CLOCKF CLK1 CLK2

          SETF DATA

          SETF /A0 /A1 /A2 /A3
          CLOCKF CLK1 CLK2
```

Simulation Results

```
Page :  1
      g g cgcgcg  cgcgcgcgcg  cgcgcgcgcg  cgcgcgc
Q0    XXXXLLLLLL  LLLLLLLLLL  LLLLLLLLLL  LLLLLLH
Q1    XXXXXXLLLL  LLLLLLLLLL  LLLLLLLLLL  LLLLLLL
Q2    XXXXXXXXLL  LLLLLLLLLL  LLLLLLLLLL  LLLLLLL
Q3    XXXXXXXXXX  LLLLLLLLLL  LLLLLLLLLL  LLLLLLL
Q4    XXXXXXXXXX  XXLLLLLLLL  LLLLLLLLLL  LLLLLLL
Q5    XXXXXXXXXX  XXXXLLLLLL  LLLLLLLLLL  LLLLLLL
Q6    XXXXXXXXXX  XXXXXXLLLL  LLLLLLLLLL  LLLLLLL
Q7    XXXXXXXXXX  XXXXXXXXLL  LLLLLLLLLL  LLLLLLL
Q8    XXXXXXXXXX  XXXXXXXXXX  LLLLLLLLLL  LLLLLLL
Q9    XXXXXXXXXX  XXXXXXXXXX  XXLLLLLLLL  LLLLLLL
Q10   XXXXXXXXXX  XXXXXXXXXX  XXXXLLLLLL  LLLLLLL
Q11   XXXXXXXXXX  XXXXXXXXXX  XXXXXXLLLL  LLLLLLL
Q12   XXXXXXXXXX  XXXXXXXXXX  XXXXXXXXLL  LLLLLLL
Q13   XXXXXXXXXX  XXXXXXXXXX  XXXXXXXXXX  LLLLLLL
Q14   XXXXXXXXXX  XXXXXXXXXX  XXXXXXXXXX  XXLLLLL
Q15   XXXXXXXXXX  XXXXXXXXXX  XXXXXXXXXX  XXXXLLL
A0    XXLLLLHHLH  HLLHHLLHHL  LHHLLHHLLH  HLLHHLL
A1    XXLLLLLLHH  HLLLLHHHHL  LLLHHHHLLL  LHHHHLL
A2    XXLLLLLLLL  LHHHHHHHHL  LLLLLLLHHH  HHHHHLL
A3    XXLLLLLLLL  LLLLLLLLLH  HHHHHHHHHH  HHHHHLL
DATA  LLLLLLLLLL  LLLLLLLLLL  LLLLLLLLLL  LLLLLHH
```

Figure D.15   (*Continued*)

XPLOT Output

```
PALASM XPLOT, V2.06 - BETA RELEASE
(C) - COPYRIGHT MONLITHIC MEMORIES INC., 1984

Title    : 16-BIT Addressable Register
Pattern  : ADREG16.PDS
Revision : A
Author   : John Birkner
Company  : Monolithic Memories Inc
Date     : 2/11/85


PAL32R16
ADREG16


            111111 11112222 22222233 33333333 44444444 44555555 55556
   01234567 89012345 67890123 45678901 23-56789 01234567 89012345 67890

 0 -------- -------- -------- -------- -------- -------- --------  -X---
 1 -------- -------- -------- -------- -------- -------- --------  -----
 2 X------- -------- -------- -------- -------- -------- --------  -----
 3 ----X--- -------- -------- -------- -------- -------- --------  -----
 4 -X---X-- X------- -------- -------- -------- -------- --------  X---X
 5 XXXXXXXX XXXXXXXX XXXXXXXX XXXXXXXX XXXXXXXX XXXXXXXX XXXXXXXX  XXXXX
 6 XXXXXXXX XXXXXXXX XXXXXXXX XXXXXXXX XXXXXXXX XXXXXXXX XXXXXXXX  XXXXX
 7 XXXXXXXX XXXXXXXX XXXXXXXX XXXXXXXX XXXXXXXX XXXXXXXX XXXXXXXX  XXXXX
 8 XXXXXXXX XXXXXXXX XXXXXXXX XXXXXXXX XXXXXXXX XXXXXXXX XXXXXXXX  XXXXX
 9 XXXXXXXX XXXXXXXX XXXXXXXX XXXXXXXX XXXXXXXX XXXXXXXX XXXXXXXX  XXXXX
10 XXXXXXXX XXXXXXXX XXXXXXXX XXXXXXXX XXXXXXXX XXXXXXXX XXXXXXXX  XXXXX
11 -X---X-- X------- -------- -------- -------- -------- --------  -X--X
12 ----X--- -------- -------- -------- -------- -------- --------  --X--
13 X------- -------- -------- -------- -------- -------- --------  --X--
14 -------- -------- -------- -------- -------- -------- --------  --X--
15 -------- -------- -------- -------- -------- -------- --------  X-X--

16 -------- -------- -------- -------- -------- -------- ------X-  -X---
17 -------- -------- -------- -------- -------- -------- ------X-  ----X
18 X------- -------- -------- -------- -------- -------- ------X-  -----
19 ----X--- -------- -------- -------- -------- -------- ------X-  -----
20 -X---X-- X------- -------- -------- -------- -------- --------  X----
21 XXXXXXXX XXXXXXXX XXXXXXXX XXXXXXXX XXXXXXXX XXXXXXXX XXXXXXXX  XXXXX
22 XXXXXXXX XXXXXXXX XXXXXXXX XXXXXXXX XXXXXXXX XXXXXXXX XXXXXXXX  XXXXX
23 XXXXXXXX XXXXXXXX XXXXXXXX XXXXXXXX XXXXXXXX XXXXXXXX XXXXXXXX  XXXXX
24 XXXXXXXX XXXXXXXX XXXXXXXX XXXXXXXX XXXXXXXX XXXXXXXX XXXXXXXX  XXXXX
25 XXXXXXXX XXXXXXXX XXXXXXXX XXXXXXXX XXXXXXXX XXXXXXXX XXXXXXXX  XXXXX
26 XXXXXXXX XXXXXXXX XXXXXXXX XXXXXXXX XXXXXXXX XXXXXXXX XXXXXXXX  XXXXX
27 -X---X-- X------- -------- -------- -------- -------- --------  -X---
28 ----X--- -------- -------- -------- -------- -------- --X-----  -----
29 X------- -------- -------- -------- -------- -------- --X-----  -----
30 -------- -------- -------- -------- -------- -------- --X-----  ----X
31 -------- -------- -------- -------- -------- -------- --X-----  X----

32 -------- -------- -------- -------- -------- ------X- --------  -X---
33 -------- -------- -------- -------- -------- ------X- --------  ----X
34 -X------ -------- -------- -------- -------- ------X- --------  -----
35 -----X-- -------- -------- -------- -------- ------X- --------  -----
36 X---X--- X------- -------- -------- -------- -------- --------  X---X
37 XXXXXXXX XXXXXXXX XXXXXXXX XXXXXXXX XXXXXXXX XXXXXXXX XXXXXXXX  XXXXX
38 XXXXXXXX XXXXXXXX XXXXXXXX XXXXXXXX XXXXXXXX XXXXXXXX XXXXXXXX  XXXXX
39 XXXXXXXX XXXXXXXX XXXXXXXX XXXXXXXX XXXXXXXX XXXXXXXX XXXXXXXX  XXXXX
40 XXXXXXXX XXXXXXXX XXXXXXXX XXXXXXXX XXXXXXXX XXXXXXXX XXXXXXXX  XXXXX
41 XXXXXXXX XXXXXXXX XXXXXXXX XXXXXXXX XXXXXXXX XXXXXXXX XXXXXXXX  XXXXX
42 XXXXXXXX XXXXXXXX XXXXXXXX XXXXXXXX XXXXXXXX XXXXXXXX XXXXXXXX  XXXXX
43 X---X--- X------- -------- -------- -------- -------- --------  -X--X
44 -----X-- -------- -------- -------- -------- --X----- --------  -----
45 -X------ -------- -------- -------- -------- --X----- --------  -----
46 -------- -------- -------- -------- -------- --X----- --------  -----
47 -------- -------- -------- -------- -------- --X----- --------  X----

48 -------- -------- -------- -------- -------- ------X- --------  -X---
49 -------- -------- -------- -------- -------- ------X- --------  ----X
50 -X------ -------- -------- -------- -------- ------X- --------  -----
51 -----X-- -------- -------- -------- -------- ------X- --------  -----
52 X---X--- X------- -------- -------- -------- -------- --------  -----
53 XXXXXXXX XXXXXXXX XXXXXXXX XXXXX´XX XXXXXXXX XXXXXXXX XXXXXXXX  XXXXX
54 XXXXXXXX XXXXXXXX XXXXXXXX XXXX.XXX XXXXXXXX XXXXXXXX XXXXXXXX  XXXXX
55 XXXXXXXX XXXXXXXX XXXXXXXX XXXXXXXX XXXXXXXX XXXXXXXX XXXXXXXX  XXXXX
56 XXXXXXXX XXXXXXXX XXXXXXXX XXXXXXXX XXXXXXXX XXXXXXXX XXXXXXXX  XXXXX
57 XXXXXXXX XXXXXXXX XXXXXXXX XXXXXXXX XXXXXXXX XXXXXXXX XXXXXXXX  XXXXX
58 XXXXXXXX XXXXXXXX XXXXXXXX XXXXXXXX XXXXXXXX XXXXXXXX XXXXXXXX  XXXXX
59 X---X--- X------- -------- -------- -------- -------- --------  -X---
60 -----X-- -------- -------- -------- --X----- -------- --------  -----
61 -X------ -------- -------- -------- --X----- -------- --------  -----
62 -------- -------- -------- -------- --X----- -------- --------  ----X
63 -------- -------- -------- -------- --X----- -------- --------  X----

64 -------- -------- -------- ------X- -------- -------- --------  -X---
65 -------- -------- -------- ------X- -------- -------- --------  ----X
66 -------- -------- -------- ------X- -------- -------- --------  -----
67 ----X--- -------- -------- ------X- -------- -------- --------  -----
68 -X--X--- X------- -------- -------- -------- -------- --------  X---X
69 XXXXXXXX XXXXXXXX XXXXXXXX XXXXXXXX XXXXXXXX XXXXXXXX XXXXXXXX  XXXXX
70 XXXXXXXX XXXXXXXX XXXXXXXX XXXXXXXX XXXXXXXX XXXXXXXX XXXXXXXX  XXXXX
71 XXXXXXXX XXXXXXXX XXXXXXXX XXXXXXXX XXXXXXXX XXXXXXXX XXXXXXXX  XXXXX
72 XXXXXXXX XXXXXXXX XXXXXXXX XXXXXXXX XXXXXXXX XXXXXXXX XXXXXXXX  XXXXX
73 XXXXXXXX XXXXXXXX XXXXXXXX XXXXXXXX XXXXXXXX XXXXXXXX XXXXXXXX  XXXXX
74 XXXXXXXX XXXXXXXX XXXXXXXX XXXXXXXX XXXXXXXX XXXXXXXX XXXXXXXX  XXXXX
75 -X--X--- X------- -------- -------- -------- -------- --------  -X--X
76 -----X-- -------- -------- --X----- -------- -------- --------  -----
77 X------- -------- -------- --X----- -------- -------- --------  -----
78 -------- -------- -------- --X----- -------- -------- --------  -----
79 -------- -------- -------- --X----- -------- -------- --------  X----

80 -------- -------- ------X- -------- -------- -------- --------  -X---
81 -------- -------- ------X- -------- -------- -------- --------  ----X
82 X------- -------- ------X- -------- -------- -------- --------  -----
83 -------- -------- ------X- -------- -------- -------- --------  -----
84 -X--X--- X------- -------- -------- -------- -------- --------  X----
85 XXXXXXXX XXXXXXXX XXXXXXXX XXXXXXXX XXXXXXXX XXXXXXXX XXXXXXXX  XXXXX
86 XXXXXXXX XXXXXXXX XXXXXXXX XXXXXXXX XXXXXXXX XXXXXXXX XXXXXXXX  XXXXX
87 XXXXXXXX XXXXXXXX XXXXXXXX XXXXXXXX XXXXXXXX XXXXXXXX XXXXXXXX  XXXXX
88 XXXXXXXX XXXXXXXX XXXXXXXX XXXXXXXX XXXXXXXX XXXXXXXX XXXXXXXX  XXXXX
89 XXXXXXXX XXXXXXXX XXXXXXXX XXXXXXXX XXXXXXXX XXXXXXXX XXXXXXXX  XXXXX
90 XXXXXXXX XXXXXXXX XXXXXXXX XXXXXXXX XXXXXXXX XXXXXXXX XXXXXXXX  XXXXX
91 -X--X--- X------- -------- -------- -------- -------- --------  -X---
92 -----X-- -------- --X----- -------- -------- -------- --------  -----
93 X------- -------- --X----- -------- -------- -------- --------  -----
94 -------- -------- --X----- -------- -------- -------- --------  ----X
95 -------- -------- --X----- -------- -------- -------- --------  X----
```

Figure D.15   (*Continued*)

```
 96  --------  ------X-  --------  --------  --------  --------  --------  -X---
 97  --------  ------X-  --------  --------  --------  --------  --------  -----
 98  -X------  ------X-  --------  --------  --------  --------  --------  -----
 99  ----X---  ------X-  --------  --------  --------  --------  --------  -----
100  X----X--  X-------  --------  --------  --------  --------  --------  X---X
101  XXXXXXXX  XXXXXXXX  XXXXXXXX  XXXXXXXX  XXXXXXXX  XXXXXXXX  XXXXXXXX  XXXXX
102  XXXXXXXX  XXXXXXXX  XXXXXXXX  XXXXXXXX  XXXXXXXX  XXXXXXXX  XXXXXXXX  XXXXX
103  XXXXXXXX  XXXXXXXX  XXXXXXXX  XXXXXXXX  XXXXXXXX  XXXXXXXX  XXXXXXXX  XXXXX
104  XXXXXXXX  XXXXXXXX  XXXXXXXX  XXXXXXXX  XXXXXXXX  XXXXXXXX  XXXXXXXX  XXXXX
105  XXXXXXXX  XXXXXXXX  XXXXXXXX  XXXXXXXX  XXXXXXXX  XXXXXXXX  XXXXXXXX  XXXXX
106  XXXXXXXX  XXXXXXXX  XXXXXXXX  XXXXXXXX  XXXXXXXX  XXXXXXXX  XXXXXXXX  XXXXX
107  X----X--  X-------  --------  --------  --------  --------  --------  -X--X
108  ----X---  --X-----  --------  --------  --------  --------  --------  -----
109  -X------  --X-----  --------  --------  --------  --------  --------  -----
110  --------  --X-----  --------  --------  --------  --------  --------  -----
111  --------  --X-----  --------  --------  --------  --------  --------  X---

112  ------X-  --------  --------  --------  --------  --------  --------  -X--
113  ------X-  --------  --------  --------  --------  --------  --------  ----X
114  -X---X-   --------  --------  --------  --------  --------  --------  ----
115  ----X-X-  --------  --------  --------  --------  --------  --------  ----
116  X----X--  X-------  --------  --------  --------  --------  --------  X---
117  XXXXXXXX  XXXXXXXX  XXXXXXXX  XXXXXXXX  XXXXXXXX  XXXXXXXX  XXXXXXXX  XXXXX
118  XXXXXXXX  XXXXXXXX  XXXXXXXX  XXXXXXXX  XXXXXXXX  XXXXXXXX  XXXXXXXX  XXXXX
119  XXXXXXXX  XXXXXXXX  XXXXXXXX  XXXXXXXX  XXXXXXXX  XXXXXXXX  XXXXXXXX  XXXXX
120  XXXXXXXX  XXXXXXXX  XXXXXXXX  XXXXXXXX  XXXXXXXX  XXXXXXXX  XXXXXXXX  XXXXX
121  XXXXXXXX  XXXXXXXX  XXXXXXXX  XXXXXXXX  XXXXXXXX  XXXXXXXX  XXXXXXXX  XXXXX
122  XXXXXXXX  XXXXXXXX  XXXXXXXX  XXXXXXXX  XXXXXXXX  XXXXXXXX  XXXXXXXX  XXXXX
123  X----X--  X-------  --------  --------  --------  --------  --------  -X--
124  --X-X---  --------  --------  --------  --------  --------  --------  -----
125  -XX-----  --------  --------  --------  --------  --------  --------  ----
126  --X-----  --------  --------  --------  --------  --------  --------  ----
127  --X-----  --------  --------  --------  --------  --------  --------  X---
```

```
        OUTPUT PINS:     111222223334
                         1234789012347890
        POLARITY FUSE:   ----------------


        OUTPUT  BANK:    4-40    17-24
           FLUSH FUSE:     X       X

    TOTAL FUSES BLOWN:  5008
```

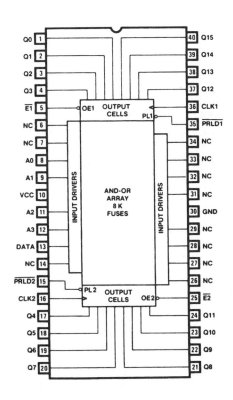

Figure D.15 (*Continued*)

State Machine Design Example

Figure 1 illustrates a simple traffic intersection consisting of two one-way streets, direction 1 and direction 2. Each direction has a signal consisting of red, yellow, and green lamps which are activated with appropriately named active high signals. Also each direction has a sensor which provides an active high signal indicating the presence of an oncoming vehicle. Our controller is to manage this intersection with the sensors as inputs and the lamps as outputs, as shown in Figure 2.

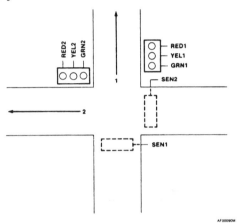

Figure 1. Traffic Intersection

Figure 2 also includes the system clock and an initialize (or reset) signal, which drives the controller to a predefined initial state. This raises two important issues in designing sequential logic with PAL devices. First, all circuit implementations of sequential logic with PAL devices are totally synchronous. This implies that all state variables (flip-flops) change at the same time, precisely after the rising edge of the clock. Second, PAL sequential logic designs should include a means for initialization to implement test programs and ensure reliable circuit operation. The specifics of the controller operations are detailed with a state diagram shown in Figure 3.

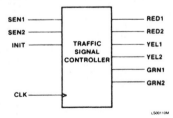

Figure 2

Figure D.16   Design Example D.3: traffic light controller (Courtesy of Monolithic Memories Inc.).

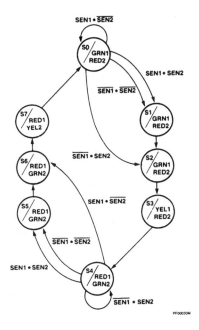

Figure 3. State Diagram — Traffic Signal Controller

Each circle in Figure 3 represents a stable state, i.e. an output configuration lasting at least one clock cycle. Inside the circles is the name of the state (S0 – S7) and the outputs associated with that state. For the sake of simplicity in the state diagram, the transitions involving INIT are omitted; INIT simply drives the circuit to S0 from any state, regardless of other inputs.

Since RED1 = /RED2, RED1 is implemented with one flip-flop and RED2 with an external inverter.

Figure D.16 *(Continued)*

PAL Device Design Specification

```
TITLE        TRAFFIC SIGNAL CONTROLLER
PATTERN      TRAFFIC1.PDS
REVISION     A
AUTHOR       KELVIN CHOW
COMPANY      MONOLITHIC MEMORIES INC., SANTA CLARA
DATE         2/28/85

CHIP TRAFFIC PAL16RP8

CLK SEN1 SEN2 INIT NC NC NC NC NC GND
/OE Q2 Q1 Q0 R1 Y1 G1 Y2 G2 VCC

STRING I1 ' /SEN1*/SEN2*/INIT '
STRING I2 ' /SEN1*SEN2*/INIT  '
STRING I3 ' SEN1*/SEN2*/INIT  '
STRING I4 ' SEN1*SEN2*/INIT   '
STRING I5 ' INIT              '

STATE

S0     = BIN[4](R1,Y1,G1,Y2,G2)
S1     = BIN[4](R1,Y1,G1,Y2,G2)
S2     = BIN[4](R1,Y1,G1,Y2,G2)
S3     = BIN[8](R1,Y1,G1,Y2,G2)
S4     = BIN[17](R1,Y1,G1,Y2,G2)
S5     = BIN[17](R1,Y1,G1,Y2,G2)
S6     = BIN[17](R1,Y1,G1,Y2,G2)
S7     = BIN[18](R1,Y1,G1,Y2,G2)

EQUATIONS

S0     = I1*S1 + I2*S2 + I3*S0 + I4*S1 + I5*S0
S1     = I1*S2 + I2*S2 + I3*S2 + I4*S2 + I5*S0
S2     = I1*S3 + I2*S3 + I3*S3 + I4*S3 + I5*S0
S3     = I1*S4 + I2*S4 + I3*S4 + I4*S4 + I5*S0
S4     = I1*S5 + I2*S4 + I3*S6 + I4*S5 + I5*S0
S5     = I1*S6 + I2*S6 + I3*S6 + I4*S6 + I5*S0
S6     = I1*S7 + I2*S7 + I3*S7 + I4*S7 + I5*S0
S7     = I1*S0 + I2*S0 + I3*S0 + I4*S0 + I5*S0

SIMULATION

TRACE_ON CLK INIT SEN1 SEN2 R1 Y1 G1 Y2 G2

SETF OE INIT
CLOCKF
CLOCKF
CHECK /R1 /Y1 G1 /Y2 /G2

SETF /INIT /SEN1 /SEN2
CLOCKF

SETF SEN1 /SEN2
CLOCKF
CHECK /R1 /Y1 G1 /Y2 /G2

SETF /SEN1 SEN2
CLOCKF
CHECK /R1 /Y1 G1 /Y2 /G2

SETF SEN1 SEN2
CLOCKF
CHECK /R1 Y1 /G1 /Y2 /G2

SETF /SEN1 /SEN2
CLOCKF
CHECK R1 /Y1 /G1 /Y2 G2

SETF /SEN1 SEN2
CLOCKF
CHECK R1 G2

CLOCKF
CHECK R1 G2

CLOCKF
CHECK R1 /Y1 /G1 Y2 /G2

CLOCKF
CHECK /R1 /Y1 G1 /Y2 /G2

CLOCKF
CLOCKF
CLOCKF
CLOCKF

; This simulation was done using the alpha release version
; of Palasm2 software.
```

Simulation Results

```
        g  c  cg c  g  cg cg cg    cg c  c  c    c  c  c
CLK   XXHLHHLLHL LHLLHLLHLL HLLHLHLHLH LHLHLHHL
INIT  HHHHHHHLLL LLLLLLLLLL LLLLLLLLLL LLLLLLLL
SEN1  XXXXXXXLLL HHHLLLLHHL LLLLLLLLLL LLLLLLLL
SEN2  XXXXXXXLLL LLLHHHHHHL LLHHHHHHHH HHHHHHHH
R1    XXXXXLLLLL LLLLLLLLLL LHHHHHHHHH LLLLLLHH
Y1    XXXXXLLLLL LLLLLLLLHH HLLLLLLLLL LLLLHHLL
G1    XXXXXHHHHH HHHHHHHHLL LLLLLLLLLL HHHHLLLL
Y2    XXXXXLLLLL LLLLLLLLLL LLLLLLLLHH LLLLLLLL
G2    XXXXXLLLLL LLLLLLLLLL LHHHHHHHHL LLLLLLHH
```

Logic Symbol

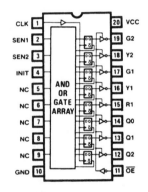

Figure D.16   (*Continued*)

The simulation commands list the signals to be traced, followed by various input conditions.

The set of comment lines following the simulation commands describe the simulation inputs and the function table for the comparator, for documentation purposes.

The results of simulation shown indicate the correct operation of the circuit. Now the fuse plot can be obtained by calling XPLOT.

Figure D.15 shows the design of a 16-bit addressable register using PAL32R16. The PDS file, simulation commands, simulation results, description of the circuit function, and the fuse plot generated by the PALASM-2 software are shown in the figure. In the fuse plot shown, an ''X'' indicates a fuse not blown, while a ''$-$'' indicates a fuse blown.

Figure D.16 shows the design of a sequential circuit (traffic light controller) using the PAL16RP8, a PAL that also contains eight D flip-flops. The state diagram for the traffic light controller is shown in the figure. In order to input the details of the state diagram, the input conditions corresponding to each transition are first defined as STRINGs (I1 through I5), as shown in the PDS file. Each state is defined by listing the outputs provided by the state machine while in that state. For instance, BIN[4] corresponds to 00100, meaning that R1 = 0, Y1 = 0, G1 = 1, Y2 = 0, and G2 = 0 while in state S0. Then the equations define the state transitions.

The simulation commands and simulation results are also shown in the figure. Note that R2 (R2 = R1$'$) is to be implemented externally to the PLA using an inverter.

References

AMAZE Design Software User Manual, Sunnyvale, Calif.: Signetics, 1986.

PAL Devices Data Book and Design Guide, Sunnyvale, Calif.: Advanced Micro Devices, 1996.

PAL/PLE Device Programmable Logic Array Handbook (5th ed.), Santa Clara, Calif.: Monolithic Memories Inc., 1986.

Perry, D. L. *VHDL*, New York, McGraw-Hill, 1991.

Programmable Logic Data Manual, Sunnyvale, Calif.: Signetics, 1986.

Programmable Logic Data Book, San Jose, Calif.: Cypress Semiconductors, 1996.

Programmable Logic Devices Data Handbook, Sunnyvale, Calif.: Signetics/Phillips Semiconductors, 1992.

Shiva, S. G. ''Automatic Hardware Synthesis,'' *Proceedings of IEEE*, 71 (January 1983), 76–86.

Technical Staff of Monolithic Memories, Inc., *Designing with Programmable Array Logic*, New York, McGraw-Hill, 1981.

Index